Anthony W. Thomas, Wolfram Weise

The Structure of the Nucleon

Anthony W. Thomas, Wolfram Weise

The Structure of the Nucleon

WILEY-VCH

Berlin · Weinheim · New York · Chichester · Brisbane · Singapore · Toronto

Authors:
Prof. Dr. Anthony W. Thomas, CSSM, University of Adelaide, Australia
e-mail: athomas@physics.adelaide.edu.au
Prof. Dr. Wolfram Weise, Department of Physics, Technical University of Munich, Germany
e-mail: weise@physik.tu-muenchen.de

Cover illustration copyright by Derek Leinweber and Will Detmold, CSSM, Adelaide, Australia

Illustration of an external electroweak probe scattering from the internal structure of the nucleon. The valence quarks are pictured as red, blue and green and the quark anti-quark fluctuation represents the non-perturbative sea required by spontaneous chiral symmetry breaking. These valence and sea quarks are pictured in a background of (cooled) gluonic topological charge which was generated by the CSSM lattice group using Adelaide University's CM5. We are grateful to Derek Leinweber and Will Detmold for generating this artistic impression of the structure of the nucleon.

1st edition

Library of Congress Card No: applied for

British Library Cataloguing-in-Publication Data: A catalogue record for this book is available from the British Library.

Die Deutsche Bibliothek – CIP Cataloguing-in-Publication-Data
A catalogue record for this publication is available from Die Deutsche Bibliothek

ISBN 3-527-40297-7

Printed on non-acid paper.

Composition: Uwe Krieg, D-10437 Berlin
Printing: Betzdruck GmbH, D-63291 Darmstadt
Bookbinding: Wilhelm Osswald & Co., D-67433 Neustadt (Weinstraße)

Printed in the Federal Republic of Germany.

WILEY-VCH Verlag Berlin GmbH
Bühringstrasse 10
D-13086 Berlin

To Joan and Sabine

Preface

Hadronic physics, the study of strongly interacting systems, is one of the most fascinating and challenging areas of modern science. Although we know both the players, the quarks and gluons, and the rules, local gauge invariance under $SU(3)_{\mathrm{color}}$, we are still a long way from a satisfactory understanding. To a large extent this is a consequence of the extreme non-linearity of the problem. The forces between individual quarks as they are separated are huge; roughly 10 tonnes, no matter how far apart they move. These are forces characteristic of trucks, acting between objects less than 10^{-18} m in size!

In striving to understand atomic nuclei, as well as the behavior of nuclear matter under extremes of density and temperature, from pulsars to relativistic heavy ion collisions, the structure of the nucleon itself plays a fundamental role. One cannot hope to really understand how hadron properties change, or how the hadrons themselves might dissolve into quark matter or a quark-gluon plasma, without a deep understanding of the structure of the nucleon.

The need for the present volume became apparent to us some eight years ago. There were excellent texts on quantum field theory, including the venerable Bjorken and Drell [BD 64], Itzykson and Zuber [IZ 80] and more recently Peskin and Schroeder [PS 95], as well as more specialized texts on gauge field theory [AH 82, LP 85, BL 86] and QCD itself [Mut 98]. There were books and reviews of various nucleon models [Bha 88], as well as reviews of the multitude of tools available to probe hadronic structure. The book of Roberts [Rob 90] gave a fine introduction to deep inelastic scattering and there were numerous other specialist reviews. However, none of these was suitable for use as a comprehensive overview of *all* facets of nucleon structure.

A complete understanding of hadron structure, in general, and nucleon structure, in particular, will only come through the coherent interplay of experiments, including electroweak probes at all energy scales, together with a variety of theoretical approaches, from modeling to lattice QCD. We need a generation of students excited by the challenge of synthesizing the latest developments in fields which have traditionally been regarded as distinct specialities. This volume is intended for anyone who is excited by this challenge.

Our aim has been to produce a modern introduction to all aspects of the investigation of nucleon structure within the framework of the Standard Model. It is meant to be suitable as either a text or reference book for graduate courses in nuclear and particle physics, as well as a handbook for practitioners.

The structure of the book is simple. We begin with a concise outline of the electroweak sector of the Standard Model. Next we turn to the facts, that is, what is known from measurements with electroweak probes; elastic and exclusive inelastic scattering involving electromagnetic and weak probes in Chapters 2 and 3, and deep inelastic scattering in Chapter 4.

In Chapter 5 we outline the basic ideas and concepts of QCD, including asymptotic freedom and the running coupling. It also introduces vital tools, such as the operator product expansion and the renormalization group. It is not meant to be a text book on QCD, any more than the Guide Michelin is a history text. Yet we believe that it will serve as a helpful guide to students of all ages who have not yet mastered QCD, but may be encouraged and assisted to do so.

Chapter 6 presents the major ideas of non-perturbative QCD, from current algebra and spontaneous symmetry breaking to anomalies. It describes the complexities of the QCD vacuum, the QCD sum rules and finally lattice QCD. The vital role of chiral symmetry in understanding hadron structure and low energy hadronic interactions is described in Chapter 7. The discussion includes an outline of chiral effective field theory, notably chiral perturbation theory, the sigma term and various chiral low energy theorems. Only in Chapter 8 do we turn to attempts to model the structure of the nucleon. While none of the models presented is an exact representation of QCD, all of them embody some aspect of the underlying truth. As the good features of each of them are selected and refined with input from lattice QCD and more sophisticated experiments (and vice versa, because new experiments and lattice investigations can all benefit from the insights gained from models), the next decade should see a tremendous improvement in our depth of understanding.

The years of effort which have gone into the preparation of this volume will have been worthwhile if just a few of those who read it are inspired to contribute to this fascinating problem. It should be apparent that in putting it together we have built upon the efforts of many people. It would be impossible to recall, let alone thank, all those who over the years, through their papers, presentations and conversations, have helped develop our understanding of hadron structure within the framework of QCD. Nevertheless, we would like to acknowledge that debt.

On the other hand, there are dozens of people who have made important contributions to the actual writing of this book and it is a pleasure to thank them specifically. The cover illustration was prepared by Will Detmold and Derek Leinweber at the Special Research Centre for the Subatomic Structure of Matter. The majority of the figures were carefully drafted and computerized by Wenzel Schürmann at the Technical University of Munich and we are very grateful for his skill. Ayse Kizilersu and Stewart Wright also helped with several illustrations. Several researchers helped by communicating their data in one form or another and we would particularly like to thank Jürgen Arends, Reinhard Beck, Mike Hasinoff, Nicole d'Hose, Mark Jones and Charles Perdrisat in this regard.

Finding all the errors that can creep into a work like this is a little like finding the last almond on a tree, impossible.[1] Nevertheless, we are extremely grateful to the many students and colleagues who assisted by reading parts of the text critically and pointed out errors or made helpful suggestions. We mention especially, Iraj Afnan, Francois Bissey, Frederick Bonnet, Ingo Bojak, Will Detmold, Thomas Hemmert, Ayse Kizilersu, Derek Leinweber, Dinghui Lu, Gunther Piller, Andreas Schreiber, Fernando Steffens, Lothar Tiator, Kazuo Tsushima, Tony Williams, Stewart Wright and Ross Young. Special thanks go to Dieter Drechsel, Pierre Gui-

[1] Readers are encouraged to let the authors know of any remaining errors they find by e-mail to *Anthony.Thomas@adelaide.edu.au* or *Wolfram_Weise@Physik.TU-Muenchen.de*. Constructive suggestions for improvements are also welcome. We will post corrections and helpful comments and references on the World-Wide Web at *http://www.physics.adelaide.edu.au/cssm/NucleonStructure*.

chon, Norbert Kaiser, Berthold Schoch and Thomas Walcher, who helped enormously through their suggestions and critical comments.

A work such as this is built upon years of research and we would also like to acknowledge the role of the various granting agencies which have supported that work, especially the Australian Research Council, the Alexander von Humboldt Foundation, the Bundesministerium für Bildung und Forschung (BMBF) and the Deutsche Forschungsgemeinschaft (DFG).

Finally, to our wives, Joan and Sabine, who tolerated numerous absences over evenings, weekends and periods up to a month, as we commuted between Munich and Adelaide piecing the manuscript together, we extend our love and appreciation.

Anthony W. Thomas (Adelaide)
Wolfram Weise (Munich)

November 2000

Contents

1 Introduction

1.1 Prelude

This book deals with the structure of protons and neutrons, the building blocks of the atomic nucleus, summarized under the name "nucleons". Exploring the structure of the nucleon is a vital part of our effort to understand the structure of matter as more than 99.9 % of the mass of any object in our solar system is carried by its nucleons.

Our framework is the Standard Model, the highly successful theory of the Strong and Electro-Weak Interactions. The elementary, spin-$\frac{1}{2}$ particles of this theory are six species, or flavors, of quarks and six leptons. Their basic interactions are mediated by a set of gauge bosons: the gluons (strong interactions), the photons (electromagnetic interactions) and the W and Z bosons (weak interactions). Quarks experience all three fundamental interactions. Amongst the leptons, the electron, muon and tau (e, μ and τ) are subject to electromagnetic and weak interactions, while the neutrinos have only weak interactions.

The quarks fall into two distinct groups according to their masses – the light quarks, up, down and strange (denoted u, d, s), and the heavy quarks, charm, bottom and top (denoted c, b, t). Each quark is characterized by a baryon number, B, of one third, while their electric charges are $+\frac{2}{3}e$ for u, c and t, and $-\frac{1}{3}e$ for d, s and b (where e is the magnitude of the electron charge). To each quark corresponds an antiparticle, or antiquark, with charge and baryon number of equal magnitude but opposite sign. In addition, each of the quarks has three possible "color" charges by which it couples to the gluons. Indeed, the whole basis of the theory of the strong interaction within the standard model, namely Quantum Chromodynamics (QCD), is local gauge invariance under changes of color.

As the lightest baryon, the nucleon is the prototype of a system governed by strong interactions. It has baryon number B = 1 and spin s = $\frac{1}{2}$. Early, naive pictures described the nucleon as a simple system composed of three quarks, such that their individual baryon numbers of $\frac{1}{3}$ added up to unity and the quark spins combined to form the total nucleon spin of $\frac{1}{2}$. The modern view of the nucleon, that has developed over several decades of experimental research and theoretical analysis, is that of a complex system composed of three valence quarks and any number of quark-antiquark pairs and gluons. Precisely how these constituents arrange themselves to build up the nucleon is one of the key questions in nuclear and particle physics.

In the context of strong interactions we must mention the special role played by the pion. It is a tightly bound system of quark-antiquark pairs, whose light mass (an order of magnitude smaller than that of the nucleon), reflects an approximate symmetry of QCD with almost massless u and d quarks: namely chiral symmetry. This symmetry is of great importance in understanding the structure of the nucleon and its dynamics at low energy.

Attempts to go beyond the Standard Model by building grand unified theories (or GUTs) usually predict that the proton should be unstable. This has inspired numerous searches for proton decay, so far without reward. As a result we now have an incredibly precise limit on the lifetime of the proton, which is greater than 10^{32} years! By contrast, the neutron is unstable under the weak interaction and has a lifetime of less than 15 minutes. As we shall see, the consequent absence of a free neutron target poses extra challenges in our attempts to probe the detailed structure of the neutron.

Our most reliable knowledge concerning the internal structure of the nucleon has been achieved using high energy beams of electrons, muons and neutrinos. Their electroweak interactions provide a "clean" probe of the quark currents in the nucleon. A significant portion of this book is devoted to a summary of the knowledge of those currents. As well as being a tribute to the enormous experimental effort that has been put into these measurements by literally thousands of people, this summary is meant to set the stage for the next generation of experimental work with new machines and detectors. We mention particularly the new and upgraded electron accelerators which promise to open new frontiers of energy, intensity, duty-factor and polarization.

For the purpose of discussing nucleon structure, the light u- and d-quarks will be of primary importance. For example, the valence quark content of the proton consists of two u-quarks and one d-quark (uud), while that of the neutron is (udd) – as can be easily seen by just counting electric charges. Any number of light quark-antiquark pairs ($u\bar{u}$, $d\bar{d}$ and $s\bar{s}$) may accompany these valence quarks. The heavy quarks ($Q = c, b$ or t) will usually be of little relevance in this context because their large masses prohibit any significant admixtures of $Q\bar{Q}$ pairs in the nucleon wave function. Our basis for the discussion of the interaction of electroweak probes with the nucleon is therefore the electroweak currents of the u-, d- and s-quarks. We begin with a reminder of the electroweak sector of the Standard Model and then continue with a brief summary of the electromagnetic and weak hadronic currents.

1.2 Electroweak Interactions in the Standard Model

The Standard Model of electromagnetic and weak interactions is based on local gauge symmetry under the group $SU(2) \times U(1)$ [We 67, Sal 69, GIM 70]. After spontaneous symmetry breaking through a Higgs mechanism the interaction part of the Lagrangian is:

$$\begin{aligned} \mathcal{L}_{\text{int}} = -e\, J^{\mu}_{\text{EM}}(x)\, \mathcal{A}_{\mu}(x) \; - \; \frac{g}{2\cos\theta_W}\, J^{\mu}_{\text{NC}}(x)\, \mathcal{Z}_{\mu}(x) \\ - \; \frac{g}{2\sqrt{2}}\, J^{\mu}_{\text{CC}}(x)\, \mathcal{W}^{\dagger}_{\mu}(x) \; + \; \text{Hermitian conjugate.} \end{aligned} \tag{1.1}$$

It involves the electromagnetic current J^{μ}_{EM}, the weak charged current J^{μ}_{CC} and the weak neutral current J^{μ}_{NC}, and their couplings to the photon field $\mathcal{A}_{\mu}$, the charged W-boson fields $\mathcal{W}^{\pm}_{\mu}$ and the neutral Z^0-boson field $\mathcal{Z}_{\mu}$. The electromagnetic and weak couplings are related by the Weinberg angle θ_W:

$$\sin\theta_W \;=\; \frac{e}{g}. \tag{1.2}$$

Its cosine gives the ratio of W- and Z-boson masses:

$$\cos\theta_W = \frac{M_W}{M_Z}. \tag{1.3}$$

With respect to the Fermi coupling constant, G_F, of the weak interaction we have

$$\frac{G_F}{\sqrt{2}} = \frac{g^2}{8M_W^2}. \tag{1.4}$$

At present the best values of these parameters are [PDG 98/00]:

$$\begin{aligned} G_F &= 1.16639 \pm 0.00001 \times 10^{-5}\ \text{GeV}^{-2}, \\ \sin^2\theta_W &= 0.2230 \pm 0.0004, \\ M_W &= 80.42 \pm 0.06\ \text{GeV}; \qquad M_Z = 91.188 \pm 0.002\ \text{GeV}. \end{aligned} \tag{1.5}$$

The electroweak currents in Eq. (1.1) include sums over all quarks and leptons. We now specify the quark currents, starting from the electromagnetic ones.

1.3 Electromagnetic Quark Currents

The u-, d- and s-quarks are the building blocks of a fundamental representation of the unitary flavor group $SU(3)_f$. Together they form a triplet field $\psi(x)$:

$$\psi(x) = \begin{pmatrix} \psi_u(x) \\ \psi_d(x) \\ \psi_s(x) \end{pmatrix} \tag{1.6}$$

where $x \equiv x^\mu = (t, \vec{x})$ represents time and space coordinates. Each of the spin $\frac{1}{2}$ quark fields ψ_u, ψ_d and ψ_s is a four-component Dirac field which annihilates a quark or creates an antiquark of a given flavor. (For the present discussion color plays no role and we shall drop *explicit* color labels. We simply note that all electroweak currents (e.g. Eq. (1.9), below), being color blind, involve a trace over color.)

The hypothesis that the quarks are elementary, pointlike Dirac particles with charges $Q_u = +\frac{2}{3}$ and $Q_d = Q_s = -\frac{1}{3}$ (in units of e) immediately implies that their electromagnetic (EM) current is

$$J^\mu_{\text{EM}}(x) = \bar{\psi}(x) Q \gamma^\mu \psi(x) \tag{1.7}$$

with Dirac matrices γ_μ (μ = 0,1,2,3), $\bar{\psi} = \psi^\dagger\gamma_0$, and our conventions for the metric, Dirac matrices and so on are summarized in Appendix A. The charge matrix

$$Q = \begin{pmatrix} 2/3 & 0 & 0 \\ 0 & -1/3 & 0 \\ 0 & 0 & -1/3 \end{pmatrix} \tag{1.8}$$

takes into account the electric charges of the quarks. More explicitly,

$$J^{\mu}_{\text{EM}} = \frac{2}{3}\bar{\psi}_u\gamma^{\mu}\psi_u - \frac{1}{3}(\bar{\psi}_d\gamma^{\mu}\psi_d + \bar{\psi}_s\gamma^{\mu}\psi_s). \tag{1.9}$$

Note that the u- and d-quarks alone form an isospin $I = 1/2$ doublet, with

$$\vec{t}^{\,2}\begin{pmatrix}\psi_u\\ \psi_d\end{pmatrix} = \frac{3}{4}\begin{pmatrix}\psi_u\\ \psi_d\end{pmatrix}, \qquad t_3\begin{pmatrix}\psi_u\\ \psi_d\end{pmatrix} = \frac{1}{2}\begin{pmatrix}\psi_u\\ -\psi_d\end{pmatrix}, \tag{1.10}$$

where $\vec{t} = \frac{1}{2}(\tau_1, \tau_2, \tau_3)$ are the generators of the isospin $SU(2)$ group, described in detail in Appendix A. Their EM current has an isoscalar and an isovector part. The charge operator for the u- and d-quarks is

$$Q = \frac{1}{2}(B + \tau_3), \tag{1.11}$$

where the isoscalar part involves the quark baryon number, $B = 1/3$, and the isovector part is proportional to τ_3. When the s-quark is included, the isoscalar part is generalized to incorporate its strangeness $S = -1$, and the baryon number is replaced by the hypercharge, $Y = B + S$:

$$Q = \frac{Y}{2} + I_3 \tag{1.12}$$

with

$$Y = \frac{\lambda_8}{\sqrt{3}} = \begin{pmatrix} 1/3 & 0 & 0\\ 0 & 1/3 & 0\\ 0 & 0 & -2/3 \end{pmatrix}, \qquad I_3 = \frac{\lambda_3}{2} = \begin{pmatrix} 1/2 & 0 & 0\\ 0 & -1/2 & 0\\ 0 & 0 & 0 \end{pmatrix}, \tag{1.13}$$

in terms of the flavor $SU(3)$ matrices λ_3 and λ_8 (see Appendix A). The electromagnetic current (1.9) can now be written

$$J^{\mu}_{\text{EM}} = \frac{1}{2}J^{\mu}_Y + V^{\mu}_3 \tag{1.14}$$

with the isospin (isovector) current

$$V^{\mu}_3 = \frac{1}{2}(\bar{\psi}_u\gamma^{\mu}\psi_u - \bar{\psi}_d\gamma^{\mu}\psi_d), \tag{1.15}$$

and the hypercharge current

$$J^{\mu}_Y = \frac{1}{3}(\bar{\psi}_u\gamma^{\mu}\psi_u + \bar{\psi}_d\gamma^{\mu}\psi_d) - \frac{2}{3}\bar{\psi}_s\gamma^{\mu}\psi_s. \tag{1.16}$$

The generalization including the heavy (c, b and t) quarks is straightforward but will not be of primary relevance in this book.

1.4 Weak Quark Currents

For many years only charged current (CC) weak interactions, like neutron beta decay, muon capture or neutrino scattering:

$$\bar{\nu}_\mu + p \rightarrow \mu^+ + n, \tag{1.17}$$

were accessible experimentally. However, even neutral current (NC) interactions can now be studied, either through direct reactions such as

$$\nu_\mu + p \rightarrow \nu_\mu + p, \tag{1.18}$$

or through interference effects such as the measurement of parity violation in electron scattering. As a consequence we shall need all of the weak currents associated with quarks.

Where it is necessary to distinguish between the hadronic and the leptonic weak currents we shall add an appropriate subscript – e.g. $J^\mu_{\text{had,W}}$. This will be dropped where it is obvious from the context which current is intended. The full hadronic, weak current will be conveniently separated into neutral (NC) and charged (CC) current pieces:

$$J^\mu_{\text{had,W}} = J^\mu_{\text{NC}} + J^\mu_{\text{CC}}. \tag{1.19}$$

The NC is the easiest to write as it is diagonal in flavor:

$$\begin{aligned} J^\mu_{\text{NC}} &= \bar{\psi}_u \gamma^\mu \left(\frac{1}{2} - \frac{4}{3} \sin^2 \theta_W - \frac{1}{2}\gamma_5 \right) \psi_u + (u \rightarrow c) + (u \rightarrow t) \\ &+ \bar{\psi}_d \gamma^\mu \left(-\frac{1}{2} + \frac{2}{3} \sin^2 \theta_W + \frac{1}{2}\gamma_5 \right) \psi_d + (d \rightarrow s) + (d \rightarrow b). \end{aligned} \tag{1.20}$$

Whereas the neutral current couplings are identical (within experimental accuracy) for u, c and t as well as for d, s and b, the situation for the charged couplings is considerably more complicated. In particular, the quark mass eigenstates with (weak) isospin $-\frac{1}{2}$ (i.e. d, s and b) are not the natural states to use. Instead one must use the weak eigenstates d', s' and b' obtained through a unitary transformation involving the Cabibbo-Kobayashi-Maskawa (CKM) matrix, V [Ca 63, KM 73]:

$$\begin{pmatrix} d' \\ s' \\ b' \end{pmatrix} = V \begin{pmatrix} d \\ s \\ b \end{pmatrix}, \tag{1.21}$$

where V is expressed in terms of three "Cabibbo angles", $\theta_{1,2,3}$, and a phase angle δ. The matrix V has the form

$$V = \begin{pmatrix} c_1 & -s_1 c_3 & -s_1 s_3 \\ s_1 c_2 & c_1 c_2 c_3 - s_2 s_3 e^{i\delta} & c_1 c_2 s_3 + s_2 c_3 e^{i\delta} \\ s_1 s_2 & c_1 s_2 c_3 + c_2 s_3 e^{i\delta} & c_1 s_2 s_3 - c_2 c_3 e^{i\delta} \end{pmatrix}, \tag{1.22}$$

where for ease of writing we have abbreviated $\cos\theta_1$ as c_1, $\sin\theta_2$ as s_2 and so on. (This is one out of several equivalent parametrizations of the CKM matrix. We choose the original one of [KM 73]).

In terms of these weak eigenstates the CC is given by

$$J^{\mu}_{\mathrm{CC}} = \bar{\psi}_u \gamma^{\mu}(1-\gamma_5)\psi_{d'} + \begin{pmatrix} u \to c \\ d' \to s' \end{pmatrix} + \begin{pmatrix} u \to t \\ d' \to b' \end{pmatrix} + \text{h.c.}, \tag{1.23}$$

where h.c. stands for the Hermitian conjugate. For practical purposes θ_2 and θ_3 are small. Thus for most applications in this book we can ignore all but the usual Cabibbo angle $\theta_C = -\theta_1$, writing

$$\begin{pmatrix} d' \\ s' \end{pmatrix} \cong \begin{pmatrix} \cos\theta_C & \sin\theta_C \\ -\sin\theta_C & \cos\theta_C \end{pmatrix} \begin{pmatrix} d \\ s \end{pmatrix}, \tag{1.24}$$

with $\sin\theta_C = 0.220 \pm 0.002$ [PDG 98/00].

Finally, we shall also need the leptonic weak current, $J^{\mu}_{\mathrm{lep,W}}$, which we summarize here:

$$J^{\mu}_{\mathrm{lep,W}} = J^{\mu}_{\mathrm{lep,\,NC}} + J^{\mu}_{\mathrm{lep,\,CC}}. \tag{1.25}$$

Here the charged current is

$$J^{\mu}_{\mathrm{lep,\,CC}} = \bar{\psi}_{\nu_e}\gamma^{\mu}(1-\gamma_5)\psi_e + \begin{pmatrix} \nu_e \to \nu_\mu \\ e \to \mu \end{pmatrix} + \begin{pmatrix} \nu_e \to \nu_\tau \\ e \to \tau \end{pmatrix} + \text{h.c.}, \tag{1.26}$$

and the neutral current is

$$\begin{aligned} J^{\mu}_{\mathrm{lep,\,NC}} &= \bar{\psi}_e \gamma^{\mu}(g_V - g_A\gamma_5)\psi_e + (e \to \mu) + (e \to \tau) \\ &\quad + \frac{1}{2}\bar{\psi}_{\nu_e}\gamma^{\mu}(1-\gamma_5)\psi_{\nu_e} + (\nu_e \to \nu_\mu) + (\nu_e \to \nu_\tau), \end{aligned} \tag{1.27}$$

with

$$g_A = -\frac{1}{2}; \qquad g_V = 2\sin^2\theta_W - \frac{1}{2}. \tag{1.28}$$

Equipped with these basic tools we are now ready to explore the structure of the nucleon.

2 Electromagnetic Structure of the Nucleon

Historically, the first indications of a composite structure of the nucleon came from measurements of the proton and neutron magnetic moments which revealed strong deviations from the values expected for point-like Dirac particles [Ste 33]. The detailed spatial distribution of the electromagnetic current inside the nucleon remained unknown until elastic electron scattering experiments on the proton and deuteron started in the mid-fifties [MH 56]. Somewhat earlier, the first indications of a non-zero charge distribution of the neutron came from the scattering of thermal neutrons on electrons bound in closed-shell atoms.

Today, the electromagnetic structure parameters of the proton and neutron, the sizes of their charge and current distributions and even their electromagnetic polarizabilities are known to remarkable accuracy. In this chapter we summarize these empirical facts and give first simple interpretations which will later be supplemented by more detailed models. The basic tools to obtain this information are elastic electron scattering and Compton scattering on the proton. For the neutron, with its lifetime of less than 15 minutes, such information comes either from scattering on deuterium targets or from neutron-electron scattering experiments.

Inelastic scattering of electrons and photons on hydrogen and deuterium reveals a rich spectrum of resonances which represent excited states of the nucleon. Quite detailed information exists about the electric and magnetic multipoles and the formfactors for transitions from the nucleon ground state to several such resonances. This kind of information provides important constraints for constructing models of the nucleon.

The present chapter is primarily concerned with the exploration of nucleon structure using photons with moderate resolution (at energies up to several GeV and corresponding momentum transfers). Such experiments probe the properties and composition of the matter that builds up the nucleon, at length scales not much smaller than the nucleon size of about 1 fm, but with high precision. Further steps on the way "inward bound" which involve much higher resolution, down to tiny fractions of 1 fm, will follow in Chapter 4 on deep-inelastic lepton scattering.

2.1 Elastic Electron Scattering

The global properties of the charge and current distributions in the nucleon can be explored by elastic electron scattering. The basic process, in the one-photon approximation, is illustrated in Fig. 2.1. Since the electromagnetic interaction is governed by the small fine structure constant $\alpha = e^2/4\pi \simeq 1/137$, the one-photon approximation is good to an accuracy of about 1 %. From basic Quantum Electrodynamics (QED), this process involves the electron current, the

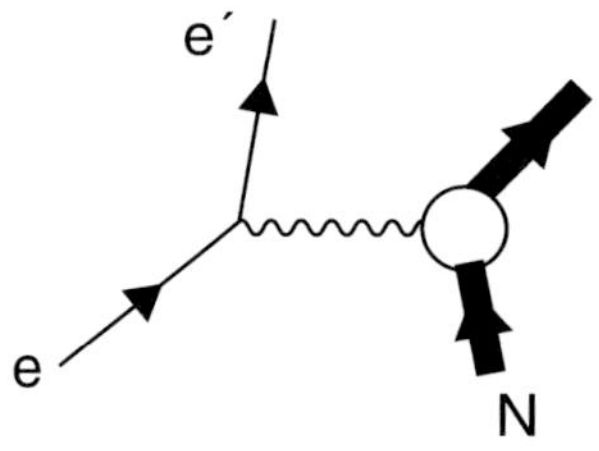

Figure 2.1: Illustration of elastic electron-nucleon scattering through one-photon exchange.

hadronic current J^{μ}_{EM} and the exchanged photon which carries the four-momentum transfer

$$q = p - p' = P' - P, \tag{2.1}$$

where $p \equiv p^{\mu} = (\epsilon, \vec{p}\,)$ and $p' \equiv p'^{\,\mu} = (\epsilon', \vec{p}\,')$ are the four-momenta of the incident and scattered electron, $P \equiv P^{\mu} = (E, \vec{P})$ and $P' \equiv P'^{\mu} = (E', \vec{P}\,')$ are the nucleon four-momenta in the initial and final state. The amplitude corresponding to Fig. 1 is then proportional to $\bar{u}(p')\gamma_{\mu}u(p)\frac{e^2}{q^2}\langle P'|J^{\mu}_{\text{EM}}|P\rangle$, the product of electron and nucleon current matrix elements times the photon propagator.

Elastic scattering means that the nucleon remains in its ground state during the process, so that the energy transfer

$$\nu = \epsilon - \epsilon' = E' - E \tag{2.2}$$

and the three momentum transfer

$$\vec{q} = \vec{p} - \vec{p'} = \vec{P'} - \vec{P} \tag{2.3}$$

are absorbed by the recoiling nucleon as a whole. The squared four-momentum transfer

$$q^2 = \nu^2 - \vec{q}^{\,2} \equiv -Q^2 < 0 \tag{2.4}$$

is a characteristic Lorentz invariant of this process.

Consider first the elastic scattering of a relativistic electron on a spinless, point-like particle of mass M and charge e. Let ϵ be the energy of the incident electron in the laboratory frame, in which the target is at rest. (Here and in the following we neglect the electron mass compared to its energy). Conservation of energy and momentum implies that the energy, ϵ', of the scattered electron is related to its scattering angle θ in the laboratory frame by

$$\epsilon' = \frac{\epsilon}{1 + \frac{2\,\epsilon}{M}\sin^2\frac{\theta}{2}}, \tag{2.5}$$

and the squared four-momentum transfer is

$$Q^2 = 4\,\epsilon\epsilon'\sin^2\frac{\theta}{2}. \tag{2.6}$$

A standard textbook derivation (summarized in Appendix B) then leads to the Mott differential cross section with inclusion of the recoil factor ϵ'/ϵ:

$$\left(\frac{d\sigma}{d\Omega}\right)_{\text{Mott}} = \frac{\alpha^2}{4\,\epsilon^2 \sin^4\frac{\theta}{2}} \cdot \frac{\epsilon'}{\epsilon} \cos^2\frac{\theta}{2}. \tag{2.7}$$

Next we examine the elastic scattering from a point-like spin $\frac{1}{2}$ target with normal (Dirac) magnetic moment. The magnetic scattering from the spin source leads to an additional term which raises the differential cross section at backward angles:

$$\frac{d\sigma}{d\Omega} = \left(\frac{d\sigma}{d\Omega}\right)_{\text{Mott}} \left[1 + \frac{Q^2}{4M^2} \cdot 2\tan^2\frac{\theta}{2}\right]. \tag{2.8}$$

Finally, for a spin 1/2 target with an extended structure and an anomalous magnetic moment one obtains the Rosenbluth cross section [Ros 50]:

$$\frac{d\sigma}{d\Omega} = \left(\frac{d\sigma}{d\Omega}\right)_{\text{Mott}} \left\{F_1^2(Q^2) + \frac{Q^2}{4M^2}\left[F_2^2(Q^2) + 2\big(F_1(Q^2) + F_2(Q^2)\big)^2 \tan^2\frac{\theta}{2}\right]\right\} \tag{2.9}$$

which introduces the Dirac form factor $F_1(Q^2)$ and the Pauli form factor $F_2(Q^2)$. The latter carries the information about the anomalous magnetic moment of the target nucleon. We give precise definitions of these form factors in the next section. It is sometimes more convenient to use the Sachs form factors:

$$G_E(Q^2) = F_1(Q^2) - \frac{Q^2}{4M^2} F_2(Q^2), \tag{2.10}$$

$$G_M(Q^2) = F_1(Q^2) + F_2(Q^2), \tag{2.11}$$

also referred to as electric (G_E) and magnetic (G_M) form factors. In terms of these the Rosenbluth cross section can be written in the form:

$$\frac{d\sigma}{d\Omega} = \left(\frac{d\sigma}{d\Omega}\right)_{\text{Mott}} \left\{\frac{G_E^2(Q^2) + \frac{Q^2}{4M^2} G_M^2(Q^2)}{1 + \frac{Q^2}{4M^2}} + \frac{Q^2}{2M^2} G_M^2(Q^2) \tan^2\frac{\theta}{2}\right\} \tag{2.12}$$

This permits a simple determination of G_E and G_M using a Rosenbluth diagram in which

$$\frac{d\sigma/d\Omega}{(d\sigma/d\Omega)_{\text{Mott}}} = A(Q^2) + B(Q^2)\tan^2\frac{\theta}{2} \tag{2.13}$$

is plotted against $\tan^2\frac{\theta}{2}$.

2.2 Electromagnetic Form Factors of the Nucleon

We come now to a detailed discussion of the electromagnetic form factors of the proton and neutron. When writing nucleon matrix elements of the electromagnetic current, we need some basic Dirac algebra as summarized in Appendix A.

Let $|N_s(P)\rangle$ be the state of a nucleon where $N = p$ refers to a proton, $N = n$ to a neutron, $s = \pm\frac{1}{2}$ denotes the spin projection, and the four-momentum is $P^\mu = (E, \vec{P})$. A free nucleon has the energy $E = \sqrt{\vec{P}^{\,2} + M^2}$, with the proton or neutron mass

$$\begin{aligned} M_p &= 938.27\ \text{MeV/c}^2, \\ M_n &= 939.57\ \text{MeV/c}^2. \end{aligned} \tag{2.14}$$

Free nucleons are described by spinors $u_s(P)$ which satisfy the Dirac equation

$$(\gamma_\mu P^\mu - M)u_s(P) = 0. \tag{2.15}$$

Their explicit form is given in Appendix A.4, and we repeat it here for convenience:

$$u_s(P) = \sqrt{E+M}\begin{pmatrix} \chi_s \\ \frac{\vec{\sigma}\cdot\vec{P}}{E+M}\chi_s \end{pmatrix}, \tag{2.16}$$

where $\vec{\sigma} = (\sigma_x, \sigma_y, \sigma_z)$ and χ_s are the standard Pauli spin matrices and two-component spinors, respectively.

Consider now matrix elements of the electromagnetic current, J_μ^{EM}, between nucleon states. From Lorentz invariance, symmetry under space reflections and charge conservation, the most general form of this matrix element is:

$$\langle N(P')|J^\mu_{\text{EM}}(0)|N(P)\rangle = \bar{u}(P')\left[\gamma^\mu F_1^N(Q^2) + i\sigma^{\mu\nu}\frac{q_\nu}{2M}F_2^N(Q^2)\right]u(P), \tag{2.17}$$

where we have used translational invariance to reduce

$$\langle N(P')|J^\mu_{\text{EM}}(x)|N(P)\rangle = \mathrm{e}^{i(P'-P)\cdot x}\langle N(P')|J^\mu_{\text{EM}}(0)|N(P)\rangle, \tag{2.18}$$

and $\sigma^{\mu\nu}$ is defined in Appendix A.3.

Instead of dealing with protons and neutrons, it is often helpful to make use of their isospin symmetry properties (see Appendix A.1). Isoscalar (S) and isovector (V) combinations of the Dirac and Pauli form factors of the proton and neutron are introduced as follows:

$$F_{1,2}^S = \frac{1}{2}\left[F_{1,2}^p + F_{1,2}^n\right], \tag{2.19}$$

$$F_{1,2}^V = \frac{1}{2}\left[F_{1,2}^p - F_{1,2}^n\right]. \tag{2.20}$$

They will be further discussed in subsections 2.2.5 and 2.2.6.

2.2.1 Low Q^2 Data

Let us now have a first look at the Dirac and Pauli form factors F_1 and F_2 as measured by elastic electron scattering at low momentum transfers ($Q^2 \lesssim 2\ \text{GeV}^2$) (see Figs. 2.2 and 2.3) [BSW 75]. Apart from the very small Dirac form factor of the neutron, these form factors are determined at a respectable level of accuracy. The measured Sachs form factors, $G_E =$

$F_1 - \frac{Q^2}{4M^2}F_2$ and $G_M = F_1 + F_2$, are shown in Figs. 2.4, 2.5 and 2.6. By their pronounced Q^2-dependence, the form factors show clear evidence for the extended charge and current distributions of the nucleons. Their normalizations at $Q^2 = 0$ are given by the nucleon charges and magnetic moments:

$$\begin{aligned} \text{Proton: } G_E^p(0) &= 1, \qquad G_M^p(0) = \mu_p = \quad 2.793; \\ \text{Neutron: } G_E^n(0) &= 0, \qquad G_M^n(0) = \mu_n = -1.913, \end{aligned} \tag{2.21}$$

where $\mu_{p,n}$ are given in units of the nuclear magneton, $\mu_N = e/2M_p$. The isoscalar[1] and isovector combinations

$$G_{E,M}^S = \frac{1}{2}\left[G_{E,M}^p + G_{E,M}^n\right], \tag{2.22}$$

$$G_{E,M}^V = \frac{1}{2}\left[G_{E,M}^p - G_{E,M}^n\right], \tag{2.23}$$

have normalizations

$$\begin{aligned} &G_E^S(0) = G_E^V(0) = \frac{1}{2} \\ &G_M^S(0) \equiv \mu_S = 0.440, \qquad G_M^V(0) \equiv \mu_V = 2.353\ . \end{aligned} \tag{2.24}$$

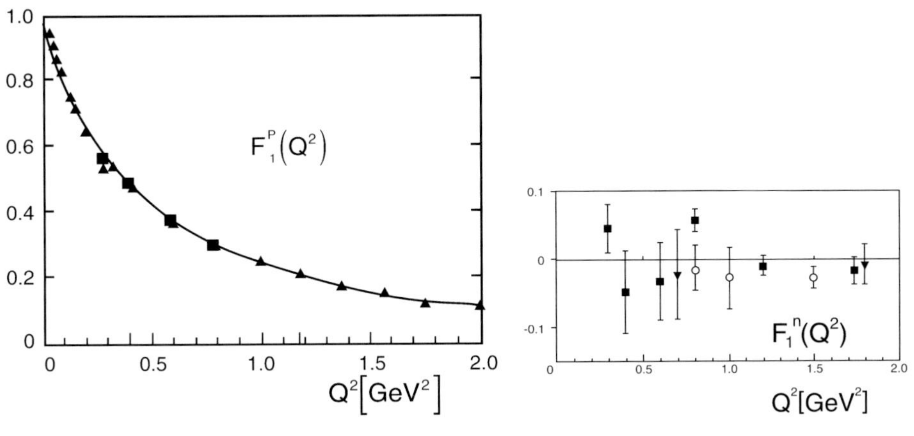

Figure 2.2: Dirac form factors $F_1^{p,n}$ of the proton and the neutron [Arn 86]. The solid curve is a best fit to the data. Note that F_1^n is only marginally different from zero.

The deviation of the neutron electric form factor G_E^n from zero indicates a nontrivial distribution of charges and currents in the neutron. In fact, even though its total charge vanishes, it has a positively charged core and a negative cloud. Earlier deductions of this form factor from deuteron data relied on model dependent assumptions about the deuteron wave function. Better determinations (Fig. 2.5) are possible with experiments that make use of polarization

[1] Note that the isoscalar form factors, $G_{E,M}^S$, should not be confused with the strangeness form factors, $G_{E,M}^s$ (written with a lower case "s") which will be discussed in Chapter 3.

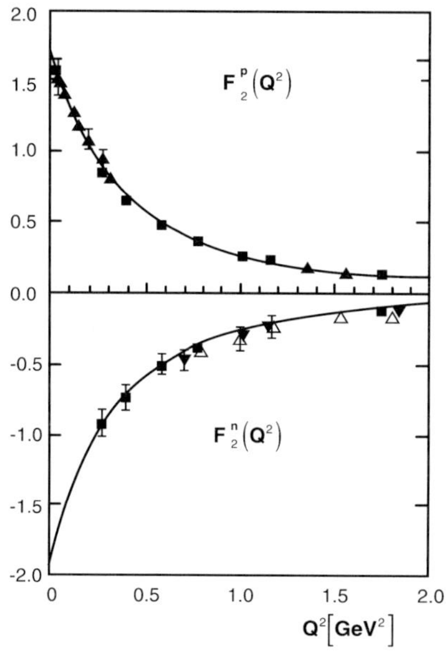

Figure 2.3: Pauli form factors $F_2^{p,n}$ of the proton and the neutron. The solid curves are best fits to the data [Arn 86].

observables. Examples are the scattering of polarized electrons on deuterium, measuring the neutron recoil polarization, and on ^{3}He targets. In such experiments, the detailed angular dependence of the measured asymmetries permits one to map out G_E^n in a way that minimizes the model dependence of the analysis.

Breit Frame

The current matrix elements (2.17) take a particularly simple form in the so-called Breit frame which, for elastic electron scattering, coincides with the electron-nucleon center-of-mass frame. In this reference frame the incoming electron has momentum $\vec{p} = +\vec{q}/2$ and hits a nucleon which has equal but opposite momentum $\vec{P} = -\vec{q}/2$. The scattered electron has momentum $\vec{p}\,' = -\vec{q}/2$, and the nucleon recoils with momentum $\vec{P}\,' = +\vec{q}/2$. The exchanged photon carries momentum $\vec{q}$ but no energy: the four-momentum transfer is $q^\mu = (0, \vec{q}\,)$ so that $Q^2 = \vec{q}\,^2$. A simple exercise leads to the Breit frame matrix elements:

$$\langle N_{s'}(\vec{q}/2)|J^0_{\rm EM}(0)|N_s(-\vec{q}/2)\rangle = 2M\, G_E(\vec{q}\,^2)\delta_{s's}, \tag{2.25}$$

$$\langle N_{s'}(\vec{q}/2)\left|\vec{J}_{\rm EM}(0)\right| N_s(-\vec{q}/2)\rangle = G_M(\vec{q}\,^2)\, \chi^\dagger_{s'}\, i\vec{\sigma}\times\vec{q}\,\chi_s, \tag{2.26}$$

using the normalization given in Appendix A.4. One can now understand the meaning of the terms "electric" and "magnetic" form factors for G_E and G_M. For each Q^2 there is a Breit frame in which the form factors are represented as $G_{E,M}(\vec{q}\,^2)$. Once this information is

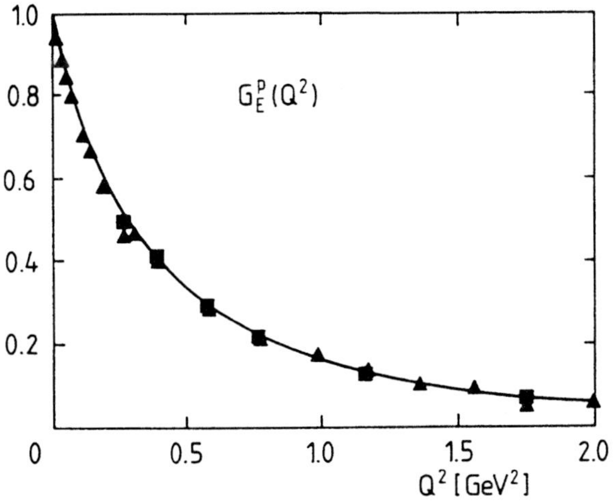

Figure 2.4: Electric form factor of the proton. The solid curve represents a dipole fit, $G_E^p(Q^2) = [1 + Q^2/(0.71\ \mathrm{GeV}^2)]^{-2}$ – from [Arn 86].

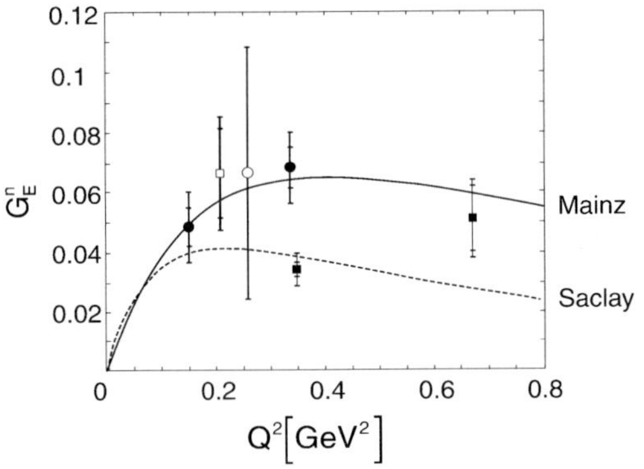

Figure 2.5: Electric form factor of the neutron. Dashed curve: parametrization representing the model dependent Saclay analysis of elastic electron-deuteron scattering using the Paris nucleon-nucleon interaction [Pl 90, Wo 91]. Full dots [He 99, Os 99], open square [Pas 99] and open circle [Ed 94]: deduced from $D(\vec{e}, e'\vec{n})p$ with polarized electron beam, detecting recoil polarization of the neutron. Full squares: deduced from $^3\mathrm{He}(e, e'n)$ experiment using polarized electron beam and polarized helium target [Roh 99, Bec 99]. Solid curve: fit to the Mainz data.

compiled, it can be used as input to calculate the three-dimensional Fourier integral,

$$\rho(\vec{r}) = \int \frac{d^3q}{(2\,\pi)^3}\, e^{-i\vec{q}\cdot\vec{r}} \frac{M}{E(\vec{q}\,)} G_E(\vec{q}^{\,2}), \tag{2.27}$$

which plays a role analogous to the "classical" charge distribution. The corresponding Fourier transform of $G_M(\vec{q}^{\,2})$ would be interpreted as the magnetization density.

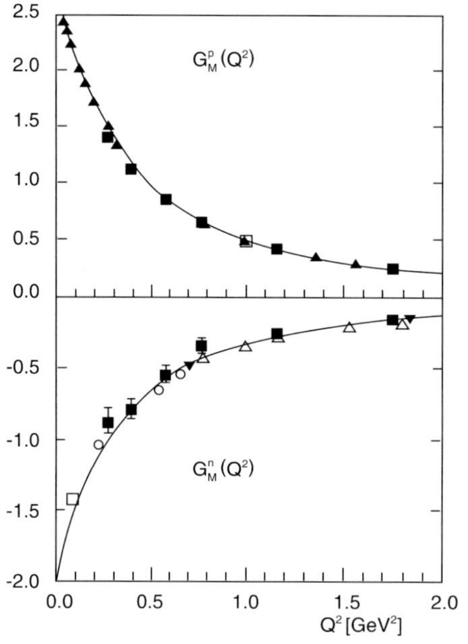

Figure 2.6: Magnetic form factors $G_M^{p,n}$ of the proton and the neutron. The solid curves represent dipole fits $G_M^{p,n}(Q^2) = \mu_{p,n}[1 + Q^2/(0.71\ \mathrm{GeV}^2)]^{-2}$ [Arn 86, Ank 94, Ank 98, Br 95, Wa 89].

Mean Squared Radii

The slopes of the form factors at $Q^2 = 0$ determine mean squared radii:

$$\langle r^2 \rangle = \frac{-6}{G(0)} \frac{dG(Q^2)}{dQ^2}\bigg|_{Q^2=0}, \tag{2.28}$$

for the magnetic radii and the same expression without $1/G(0)$ for the electric form factors. In the Breit frame and for an infinitely heavy, spherical charge distribution this definition coincides with the more familiar relation

$$\langle r^2 \rangle = 4\pi \int_0^\infty dr\, r^4\, \rho(r). \tag{2.29}$$

The empirical proton charge and magnetic radii are [Hö 76, Sim 80, MMD 96]

$$\langle r_E^2 \rangle^{1/2}_{\text{proton}} = (0.86 \pm 0.01)\ \text{fm}, \tag{2.30}$$

$$\langle r_M^2 \rangle^{1/2}_{\text{proton}} = (0.86 \pm 0.06)\ \text{fm}. \tag{2.31}$$

For the neutron magnetic radius one finds:

$$\langle r_M^2 \rangle^{1/2}_{\text{neutron}} = (0.89 \pm 0.07)\ \text{fm}. \tag{2.32}$$

Note that these radii are all equal within experimental errors.

The slope of the neutron electric form factor is determined with high precision from the scattering of slow neutrons on the electron cloud of heavy atoms [Koe 76, Koe 88, Kop 95]. The result is:

$$6 \left.\frac{dG_E^n(Q^2)}{dQ^2}\right|_{Q^2=0} = (0.113 \pm 0.003 \pm 0.004)\ \text{fm}^2, \tag{2.33}$$

with statistical and systematic uncertainties given separately. Note that the sign of this slope is opposite to that of G_E^p, indicating (as mentioned already) a surplus of negative charge in the surface of the neutron which must be compensated by a positive charge in the central region such that the overall charge vanishes.

Because of the smallness of the neutron Dirac form factor F_1^n, most of the slope (2.33) is determined by the Pauli form factor F_2^n. In fact,

$$\left.\frac{dG_E^n(Q^2)}{dQ^2}\right|_{Q^2=0} = \left.\frac{dF_1^n(Q^2)}{dQ^2}\right|_{Q^2=0} - \frac{F_2^n(0)}{4M^2}, \tag{2.34}$$

and the so-called Foldy term, $F_2^n(0)/4M^2 = -0.021\ \text{fm}^2$, alone gives a contribution of $0.127\ \text{fm}^2$ to Eq. (2.33), while the remaining correction from the slope of F_1^n is obviously very small [Fo 58]. There has been – and still is – considerable discussion over the interpretion of this observation. One must be aware of the fact that the distributions of charges and currents are closely linked, and their interpretation changes depending on the frame of reference. In the Breit frame, the only one in which three-dimensional Fourier transforms of the form factors have a simple interpretation, the relativistic effects associated with the motion of the constituents can give rise to a subtle interplay between convection currents and anomalous magnetic structures. For example, in relativistic versions of constituent quark models, contributions from Lorentz boost corrections to the Dirac form factor tend to cancel the Foldy term, leaving room for an interpretation in terms of a neutron charge distribution in which positive central charges are separated from negative surface charges [Is 99].

2.2.2 High Q^2 Data

Elastic electron scattering experiments on proton and deuterium targets have been extended to large squared momentum transfers up to about 30 GeV^2. Resulting form factors extracted from these data are shown in Figs. 2.7 and 2.8.

The global behavior of the proton and neutron magnetic form factors G_M over a wide range of Q^2 is quite well represented by a dipole form

$$G_D(Q^2) = \left(\frac{\Lambda^2}{\Lambda^2 + Q^2}\right)^2 \tag{2.35}$$

with $\Lambda = 0.84$ GeV:

$$\frac{G_M^p(Q^2)}{\mu_p} = \frac{G_M^n(Q^2)}{\mu_n} = G_D(Q^2). \tag{2.36}$$

Deviations from this dipole fit amount to less than 10 % for G_M^p at $Q^2 \lesssim 5$ GeV^2 and just slightly more for G_M^n. The same is true for the proton electric form factor at $Q^2 \lesssim 1$ GeV^2.

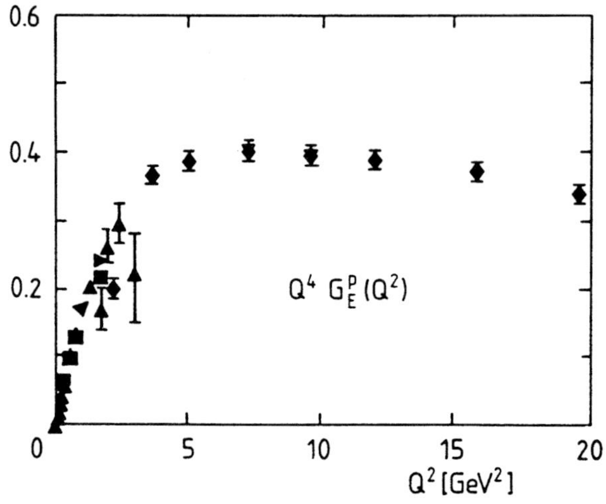

Figure 2.7: Electric form factor of the proton, multiplied by Q^4. See Fig. 2.8 for references.

However, significant differences between G_E^p and G_M^p/μ_p are observed at Q^2 above 1 GeV^2. The accuracy of this observation has been vastly improved with polarization transfer experiments. Consider the process, $\vec{e}p \rightarrow e\vec{p}$, where a longitudinally polarized electron transfers polarization to the recoiling proton. We consider two components of the polarization of the final proton: a longitudinal part, P_l, parallel to the proton momentum and a transverse part, P_t, in the scattering plane spanned by the outgoing momenta. One finds [ACG 81]:

$$P_l = \frac{\epsilon + \epsilon'}{M I_0} \sqrt{\tau(1+\tau)}\, (G_M^p)^2 \tan^2 \frac{\theta}{2}, \tag{2.37}$$

and

$$P_t = -\frac{2}{I_0} \sqrt{\tau(1+\tau)}\, G_E^p G_M^p \tan \frac{\theta}{2}, \tag{2.38}$$

with

$$I_0 = (G_E^p)^2 + \tau \left[1 + 2(1+\tau) \tan^2 \frac{\theta}{2} \right] (G_M^p)^2 \tag{2.39}$$

and $\tau = Q^2/4M^2$. A measurement of P_l and P_t in a polarimeter determines the ratio

$$\frac{G_E^p}{G_M^p} = -\frac{\epsilon + \epsilon'}{2M} \left(\frac{P_t}{P_l} \right) \tan \frac{\theta}{2} \tag{2.40}$$

without the need to know the initial beam polarization and the analyzing power of the polarimeter, so the systematic uncertainties are small. An experiment performed at the Continuous Electron Beam Accelerator Facility (CEBAF) of the Jefferson Laboratory has produced

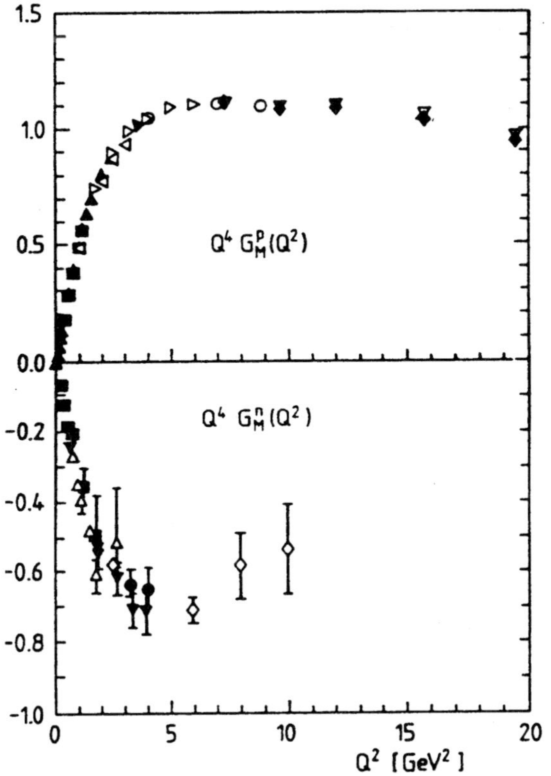

Figure 2.8: Magnetic form factors of the proton and neutron, each multiplied by Q^4. [Arn 86, Fei 81, Lun 93, Roc 82, Sil 93].

the result, Fig. 2.9, which demonstrates that the Q^2 dependence of G_E^p and G_M^p is significantly different. Data of such accuracy place tight constraints on models of the nucleon (see Chapter 8).

At very large Q^2 the asymptotic behavior of the form factor follows qualitatively the expectations of perturbative QCD, as illustrated in Fig. 2.10. To get a first rough idea, consider a picture in which the nucleon is viewed as three quarks interacting through the exchange of gluons. In order to keep the nucleon together in its ground state during the hard electron scattering process, the momentum transfer must be divided among the three quarks in such a way that the nucleon recoils as a whole. This requires minimally the exchange of two gluons between the quarks, each carrying a fraction of Q.

In this picture, the Dirac form factor $F_1(Q^2)$ corresponds to an amplitude in which the short distance (high Q^2) behavior enters through the quark-gluon couplings and the gluon propagators. The quark-gluon couplings have a weak (logarithmic) dependence on Q^2, whereas each of the gluon propagators is proportional to Q^{-2}. Hence one expects a leading asymptotic Q^{-4} behavior which is indeed roughly seen in the data. This feature is referred to as the *dimensional scaling rule*.

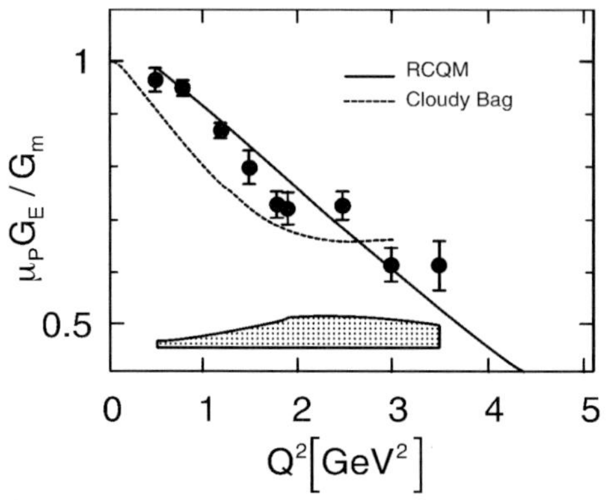

Figure 2.9: The ratio of proton electric and magnetic form factors, $\mu_p G_E^p / G_M^p$, in comparison with selected model calculations (relativistic constituent quark model [Chu 91] and cloudy bag model [Lu 98]). Data: polarization transfer measurement at TJNAF [Jo 00].

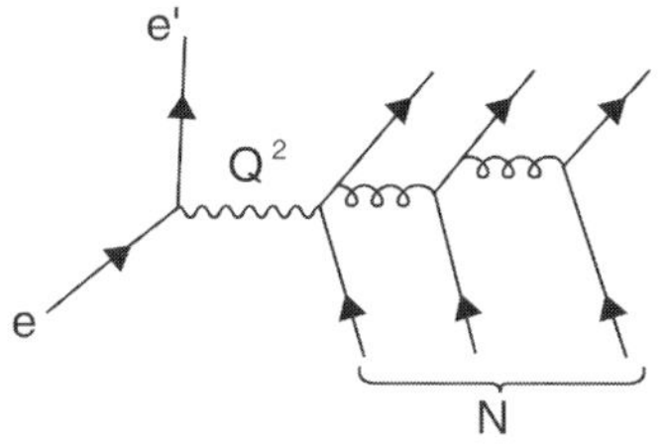

Figure 2.10: Elastic electron scattering from a nucleon in a perturbative QCD picture.

The Pauli form factor F_2 involves helicity flip at the quark level. However, quark helicities are conserved in QCD at high momentum transfers. Contributions from F_2 can therefore only come at next-to-leading order – i.e. at order Q^{-6}. The observed asymptotic behavior of the form factors shown in Figs. 2.7 and 2.8 is compatible with these simple rules.

2.2.3 Light-Front Distributions

Let us give a brief sketch of the basic arguments leading to the dimensional scaling rule, $F_1(Q^2) \sim Q^{-4}$, at asymptotic Q^2. This is conveniently done in the so-called light-front formalism [CZ 84, ER 80, FJ 79, BL 80], a description which is well adapted to hard exclusive reactions such as elastic electron-proton scattering at large Q^2.

Consider a particle with four-momentum $p^\mu = (p^0, p^1, p^2, p^3)$, and introduce light-front momenta,

$$p^+ = p^0 + p^3, \qquad p^- = p^0 - p^3. \tag{2.41}$$

For a nucleon consisting of three valence quarks with light-front momenta $p_i^\pm (i = 1, 2, 3)$, momentum fractions are defined as

$$x_i = \frac{p_i^+}{P^+}, \tag{2.42}$$

where P^+ denotes the corresponding light-front momentum of the whole nucleon.

The essential elements in the description of nucleon form factors at large Q^2 are now the following:

i) For large momentum transfers chosen, for convenience, along the (longitudinal) 3-axis of the original coordinate frame, one can neglect transverse momentum components (those perpendicular to the 3-axis).

ii) For massless point-like spin $\frac{1}{2}$ particles, helicity (the projection of the particle's spin along its direction of motion) is conserved in interactions mediated by vector fields such as photons or gluons. To the extent that quarks can be treated as approximately massless on-shell particles at very high energy and momentum transfers, their helicity is indeed conserved during the hard scattering process. Only the form factor F_1 survives in this limit, whereas F_2 is suppressed.

Under such high-Q^2 conditions, it was shown [LB 79] that nucleon form factors can be written in the following factorized representation:

$$F_1(Q^2) = \int_0^1 dx \int_0^1 dx' \Phi(x)^* T_h \Phi(x'), \tag{2.43}$$

where T_h is the transition operator for the "hard" process, describing the (perturbative) short distance physics. The distribution amplitudes Φ summarize the (non-perturbative) physics of the quark wave functions of the nucleon, expressed in terms of the light-front momentum fractions $x(\equiv x_1,\ x_2,\ x_3)$. The integrals in the convolution $\Phi * T_h * \Phi$ are understood to be taken with $dx = dx_1\, dx_2\, dx_3$, and accordingly $dx' = dx_1'\, dx_2'\, dx_3'$. These integrations are constrained by the condition that the quark momenta add up to the total momentum of the nucleon: $x_1 + x_2 + x_3 = x_1' + x_2' + x_3' = 1$. One of the advantages of working with distributions expressed in light-front coordinates is that Lorentz boosts can be handled easily, and that constraints from momentum conservation are treated properly. This is a prerequisite for calculations of form factors at high Q^2 which involve large recoil momenta of the struck nucleon.

The transition operator T_h in leading order has two gluons exchanged between three quarks, as shown in Fig. 2.10. We denote the exchanged four-momentum carried by the first gluon in the sequence by q_1, that of the second gluon by q_2. Each of the gluon exchange processes introduces the running QCD coupling α_s (see Section 5.4.1), so that T_h becomes:

$$T_h = \frac{\alpha_s(q_1^2)\,\alpha_s(q_2^2)}{q_1^2 q_2^2} f(x, x'). \tag{2.44}$$

When expressed in terms of the light-front momentum fractions, x_i, the particular sequence of Fig. 2.10, taken as an example, is written as

$$T_h \sim \frac{\alpha_s\left((1-x_1)(1-x_1')Q^2\right)\alpha_s(x_3 x_3' Q^2)}{Q^4(1-x_1)^2(1-x_1')^2\, x_3 x_3'}. \tag{2.45}$$

The other possible gluon exchange sequences are found by appropriate rearrangements and permutations of momenta. The origin of the dimensional scaling, $F_1(Q^2) \sim Q^{-4}$, is now evident.

The actual calculation of the formfactor involves the distribution $\Phi(x)$. Starting with three valence quarks, a basic ansatz for its asymptotic form at very large Q^2 is simply

$$\Phi(x_1, x_2, x_3) = 120\, x_1\, x_2\, x_3\, \delta(x_1 + x_2 + x_3 - 1),$$

with the proper normalization factor and a delta function which makes sure that the quark momenta add up to the total nucleon momentum. Radiative QCD corrections by virtual gluon emission and reabsorption (see Chapter 5) introduce a slow logarithmic evolution of Φ with decreasing Q^2. The resulting expression is

$$\Phi(x_1, x_2, x_3, Q^2) = 120\, x_1\, x_2\, x_3\, \delta(x_1 + x_2 + x_3 - 1) \sum_n C_n(Q^2) \mathcal{P}_n(x_i), \quad (2.46)$$

where C_n includes fractional powers of $\alpha_s(Q^2)$ and the $\mathcal{P}_n$ are Appell polynomials ($\mathcal{P}_0 = 1$, $\mathcal{P}_1 = x_1 - x_3$, $\mathcal{P}_2 = 1 - 3x_2$ and so forth). This form of the distribution function has been frequently used in the phenomenological analysis of the nucleon magnetic form factors at large Q^2 shown in Fig. 2.8. The slight decrease of $Q^4 G_M(Q^2)$ at the highest Q^2 may be indicative of radiative corrections in addition to the asymptotic scaling rule. The normalization of this form factor is, however, not fully understood in this picture. The question has been raised whether soft, long-range effects still play an important role in such hard, exclusive processes, even at very large Q^2 [IL 89].

2.2.4 Dispersion Relations

In order to interpret the physics content of the nucleon form factors, it is often useful to represent them in the form of dispersion relations. For this purpose the form factors F_1 and F_2 are considered as functions of the variable $z = q^2$ continued to the complex plane. Instead of restricting the discussion to the region of spacelike momentum transfers, $Q^2 = -q^2 > 0$, physically accessible in electron scattering, we now extend it to positive (timelike) q^2 – even going beyond the real q^2 axis, along which the physical form factors are defined. The key point is that, as a consequence of causality, the complex functions $F_1(z)$ and $F_2(z)$ have certain analytic properties which imply integral relations based on Cauchy's theorem.

In fact, the functions $F_{1,2}$ can have singularities only on the positive real axis for q^2 larger than a characteristic threshold value. Along this cut the form factors develop imaginary parts: the continuum of singularities corresponds to physical states that can be reached in the "crossed channel" $e^+e^- \rightarrow N\bar{N}$ for which $q^2 > 0$ represents the squared total center-of-mass energy. The relevant matrix elements of the electromagnetic current are now of the form $\langle N(P)\, \bar{N}(P')|J^{\mu}_{\mathrm{EM}}|\, 0\rangle$, and the cross section for this process involves the same functions $F_{1,2}$ as introduced previously to describe elastic electron-nucleon scattering, but now taken at positive q^2 with $q = P + P'$. The physical region of electron-positron annihilation with creation of a nucleon-antinucleon pair (or its inverse) is restricted to the range $q^2 \geq 4M^2$. However, the $N\bar{N}$ pair, being a system with baryon number B = 0, couples to a variety of multi-meson states, of which the lightest is two pions. Therefore the cut on the real axis, which determines

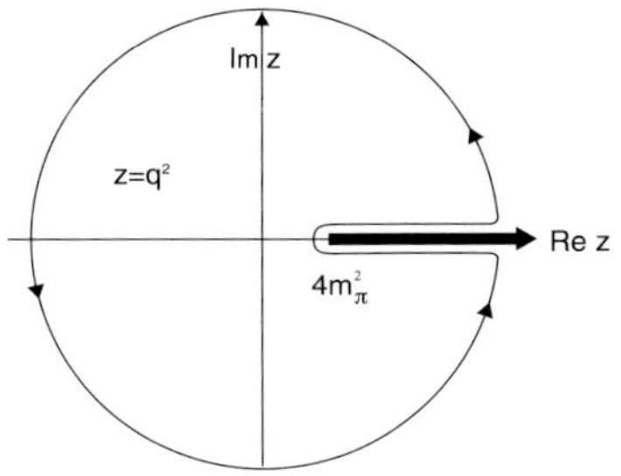

Figure 2.11: Analytic structure of nucleon form factors in the complex $z = q^2$ plane.

the analytic properties of $F_{1,2}$, extends from $q^2 = 4m_\pi^2$ (the threshold for $e^+e^- \to \pi^+\pi^-$) to infinity.

After these preparations, we can now apply Cauchy's integral formula to either of the form factors $F_1(z)$ and $F_2(z)$, closing the contour as shown in Fig. 2.11. In order to make sure that the circle far out in the complex plane gives no contribution to the Cauchy integral, we apply a subtraction at $z = q^2 = 0$. The resulting dispersion relations are:

$$F_{1,2}(q^2) = F_{1,2}(0) + \frac{q^2}{\pi} \int_{t_0}^{\infty} dt \, \frac{\mathrm{Im}\, F_{1,2}(t)}{t(t - q^2 - i\epsilon)}, \tag{2.47}$$

where $t_0 = 4m_\pi^2$, $F_1(0) = 1$ and $F_2(0) = \kappa_p = 1.793$ for the proton. For the neutron we have $F_1(0) = 0$, $F_2(0) = \kappa_n = -1.913$. (It is sometimes assumed that $\mathrm{Im}\, F_1(t)$ decreases rapidly as $t \to \infty$ so that the subtraction in Eq. (2.47) is not needed.)

As mentioned, the imaginary parts of $F_{1,2}$ which enter the dispersion integrals reflect the mass spectrum of physical intermediate states that can be reached when a timelike photon probes the nucleon. The quantities $\frac{1}{\pi}\,\mathrm{Im}\, F(t)$ are often called spectral functions.

Let us look at mean square radii from this point of view. Consider for example the proton charge radius. We have:

$$\langle r_E^2 \rangle_p = \left. \frac{6\, dG_E^p(q^2)}{dq^2} \right|_{q^2=0} = \frac{6}{\pi} \int_{4m_\pi^2}^{\infty} dt \, \frac{\mathrm{Im}\, F_1^p(t)}{t^2} + \frac{3\,\kappa_p}{2M^2}. \tag{2.48}$$

We see that – apart from the small Foldy term $3\,\kappa_p/2M^2 = 0.12\ \mathrm{fm}^2$ – the proton charge radius is governed by the low mass part of the spectrum represented by $\mathrm{Im}\, F_1^p$. If this spectrum is dominated by a hadronic state of mass μ, then this will contribute to the mean square radius a term proportional to $1/\mu^2$.

2.2.5 Vector Meson Dominance

The basic structure of the electromagnetic current (1.9) involves combinations of quark-antiquark pairs which find their direct correspondence in the three lightest vector mesons (the ρ,

ω and ϕ mesons). We list here their masses and some of their decay widths [PDG 98/00]:

$$\begin{aligned} m_\rho &= 769.3 \pm 0.8 \text{ MeV}, \\ \Gamma_{\text{tot}} &= 150.2 \pm 0.8 \text{ MeV}, \\ \Gamma(\rho \to e^+e^-) &= 6.8 \pm 0.3 \text{ keV}; \end{aligned}$$

$$\begin{aligned} m_\omega &= 782.6 \pm 0.1 \text{ MeV}, \\ \Gamma_{\text{tot}} &= 8.44 \pm 0.09 \text{ MeV}, \\ \Gamma(\omega \to e^+e^-) &= 0.60 \pm 0.02 \text{ keV}; \end{aligned}$$

$$\begin{aligned} m_\phi &= 1019.417 \pm 0.014 \text{ MeV}, \\ \Gamma_{\text{tot}} &= 4.46 \pm 0.03 \text{ MeV}, \\ \Gamma(\phi \to e^+e^-) &= 1.30 \pm 0.03 \text{ keV}. \end{aligned}$$

Vector mesons carry spin $J = 1$ and negative parity; they have the same quantum numbers as the photon. Consequently, when a photon probes a nucleon, one should expect vector mesons to play an important role in the spectrum of intermediate hadronic states which contribute to the dispersion integrals in Eqs. (2.47). The quark composition of these vector mesons is:

$$\rho^0 = \frac{1}{\sqrt{2}}(u\bar{u} - d\bar{d}), \tag{2.49}$$

$$\omega = \frac{1}{\sqrt{2}}(u\bar{u} + d\bar{d}), \tag{2.50}$$

$$\phi = s\bar{s}. \tag{2.51}$$

The electromagnetic current can then be rewritten as a sum of terms, each characterized by the quantum numbers of a corresponding vector meson:

$$J_\mu^{\text{EM}} = \frac{1}{\sqrt{2}} J_\mu^{(\rho)} + \frac{1}{3\sqrt{2}} J_\mu^{(\omega)} - \frac{1}{3} J_\mu^{(\phi)}, \tag{2.52}$$

where the $J_\mu^{(V)}$ stand for the following quark currents:

$$J_\mu^{(\omega,\rho)} = \frac{1}{\sqrt{2}}(\bar{\psi}_u \gamma_\mu \psi_u \pm \bar{\psi}_d \gamma_\mu \psi_d), \qquad J_\mu^{(\phi)} = \bar{\psi}_s \gamma_\mu \psi_s. \tag{2.53}$$

When applied to the vacuum, the electromagnetic current operator creates quark-antiquark pairs which have an overlap with the physical vector meson states:

$$\langle\rho| J_\mu^{\text{EM}} |\, 0\rangle = \frac{1}{\sqrt{2}} \langle\rho| J_\mu^{(\rho)} |\, 0\rangle \quad = F_\rho\, \epsilon_\mu, \tag{2.54}$$

$$\langle\omega| J_\mu^{\text{EM}} |\, 0\rangle = \frac{1}{3\sqrt{2}} \langle\omega| J_\mu^{(\omega)} |\, 0\rangle = F_\omega\, \epsilon_\mu, \tag{2.55}$$

$$\langle\phi| J_\mu^{\text{EM}} |\, 0\rangle = -\frac{1}{3} \langle\phi| J_\mu^{(\phi)} |\, 0\rangle \quad = -F_\phi\, \epsilon_\mu, \tag{2.56}$$

where ϵ_μ are the vector meson polarization vectors. These matrix elements enter directly in the decay widths $\Gamma(V \to e^+e^-) = (4\pi\alpha^2/3)F_V^2/m_V^3$. One traditionally writes

$$F_V = \frac{m_V^2}{g_V} \tag{2.57}$$

and finds by comparison with the measured widths:

$$g_\rho = 5.04, \qquad g_\omega = 17.05, \qquad g_\phi = 12.88. \tag{2.58}$$

Note that these constants are defined at the physical vector meson masses, $q^2 = m_V^2$ for each of the $V = \rho, \omega, \phi$. Their ratios roughly reflect the $SU(3)$ symmetry of the quark currents (1.9). Perfect symmetry would imply $g_\rho : g_\omega : g_\phi = 1 : 3 : 3/\sqrt{2}$.

Now, the idea behind the Vector Meson Dominance (VMD) hypothesis [Sak 69, Fey 72] is the following. The matrix element of the electromagnetic current taken between any hadronic states $|A\rangle$ and $|B\rangle$ at a four-momentum transfer, q^2, in the region of the vector meson resonances should be of the form

$$\langle B|J_\mu^{\mathrm{EM}}|A\rangle = \sum_V \frac{m_V^2}{g_V(m_V^2 - q^2 - i\Gamma_V m_V)} \langle B|V_\mu|A\rangle, \tag{2.59}$$

where the presence of the vector mesons is manifest in the characteristic pole structure of the current matrix element and $\langle B|V_\mu|A\rangle$ is the amplitude for the coupling of the vector meson V to the state $|A\rangle$ as it undergoes a transition to $|B\rangle$.

With this in mind, we return to the nucleon electromagnetic form factors beginning with the isoscalar Dirac form factor

$$F_1^S(q^2) = \frac{1}{2}\left[F_1^p(q^2) + F_1^n(q^2)\right] = F_1^S(0) + \frac{q^2}{\pi}\int_{t_0}^{\infty} dt\, \frac{\mathrm{Im}\, F_1^S(t)}{t(t - q^2 - i\epsilon)}, \tag{2.60}$$

where we have written a subtracted dispersion relation with $F_1^S(0) = 1/2$. Actually, the physical right hand cut for this form factor starts at $t_0 = 9m_\pi^2$, because the lightest possible hadronic state with isospin $I = 0$ is a system of three (rather than two) pions (as long as we neglect small isospin breaking effects).

In the isoscalar channel the relevant vector meson is the ω meson. The ϕ meson can also contribute and we include it here for completeness, but its coupling to the nucleon is expected to be much smaller than that of the ω. We treat these vector mesons in the zero-width approximation, given that the total width of the ω meson is only about 8 MeV and the width of the ϕ meson is even smaller. The VMD model then says that the spectrum of hadronic states contributing to $\mathrm{Im}\, F_1^S$ is simply of the form:

$$\mathrm{Im}\, F_1^S(t) = \pi\delta(t - m_\omega^2)\,\frac{m_\omega^2}{g_\omega}\,G_{\omega N} - \pi\delta(t - m_\phi^2)\,\frac{m_\phi^2}{g_\phi}\,G_{\phi N}, \tag{2.61}$$

where $G_{\omega N}$ and $G_{\phi N}$ denote the couplings of the ω and ϕ mesons to the nucleon and we have incorporated the difference in sign between the current matrix elements, Eqs. (2.55, 2.56). Insertion in Eq. (2.60) gives

$$F_1^S(q^2) = \frac{1}{2} + \frac{G_{\omega N}}{g_\omega}\,\frac{q^2}{m_\omega^2 - q^2 - i\epsilon} - \frac{G_{\phi N}}{g_\phi}\,\frac{q^2}{m_\phi^2 - q^2 - i\epsilon}. \tag{2.62}$$

For spacelike $q^2(= -Q^2 < 0)$ one finds that assuming $G_{\phi N} = 0$, the overall behavior of $F_1^S(Q^2)$ is reasonably well reproduced in the intermediate region, $Q^2 \leq 2$ GeV^2 if one chooses $G_{\omega N} \approx 0.6 g_\omega \approx 10$ – a value within the range of ωN coupling constants frequently used in boson exchange potentials of the nucleon-nucleon force. But the mean square radius comes out too small (by about a factor 3/4). A somewhat improved fit is obtained using $G_{\phi N} \approx 0.3 g_\phi \approx 3.9$ together with $G_{\omega N} \approx 14.5$. The detailed dispersion relation analysis of [MMD 96] prefers still larger values for $G_{\omega N}$ and $G_{\phi N}$.

A similar discussion can be carried out for the isoscalar Pauli form factor F_2^S, the one associated with the anomalous tensor part of the EM current. Here the small, isoscalar anomalous magnetic moment translates into a similarly small tensor coupling of the ω meson to the nucleon.

Next, consider the isovector Dirac form factor

$$F_1^V(q^2) = \frac{1}{2}\left[F_1^p(q^2) - F_1^n(q^2)\right] = F_1^V(0) + \frac{q^2}{\pi}\int_{t_0}^{\infty} dt\, \frac{\mathrm{Im}\, F_1^V(t)}{t(t-q^2-i\epsilon)}, \tag{2.63}$$

with $F_1^V(0) = 1/2$ and $t_0 = 4m_\pi^2$. Here we expect the ρ meson to play an important role at moderate q^2. In the zero-width approximation we are led back to

$$\mathrm{Im}\, F_1^V(t) = \pi\delta(t - m_\rho^2)\,\frac{m_\rho^2}{g_\rho}\, G_{\rho N}, \tag{2.64}$$

and

$$F_1^V(q^2) = \frac{1}{2} + \frac{G_{\rho N}}{g_\rho}\,\frac{q^2}{m_\rho^2 - q^2 - i\epsilon}, \tag{2.65}$$

an expression analogous to (2.62). With $G_{\rho N} \approx 0.6 g_\rho \approx 3$ this simplified expression reproduces F_1^V reasonably well for $Q^2 \leq 2$ GeV^2 but gives only about 4/5 of the mean square radius associated with the isovector Dirac form factor. For an improved phenomenology along these lines, see [BRW 86, MKW 87].

Unlike the isoscalar Pauli form factor, the isovector form factor F_2^V is large, as indicated already by the sizeable isovector anomalous magnetic moment $\kappa_V = 1.85$. Correspondingly, the tensor coupling of the interacting isovector two-pion system to the nucleon is far larger than its vector coupling.

2.2.6 Spectral Functions

The idealized vector meson dominance model is obviously not complete. The coupling of the probing virtual photon to the nucleon via vector mesons explains part, but not all of the electromagnetic form factors. From our previous discussion the limitations of VMD became obvious for the isovector form factors. In particular, the large width for the decay $\rho \to \pi^+\pi^-$ plays an important role. An improved description requires a detailed understanding of the dynamics of the $\pi\pi$ system in the isovector channel and its coupling to the nucleon.

In this section we focus primarily on the physics of the spectral functions which enter the dispersion relations for the isovector form factors [Hö 83]. As we draw a more detailed

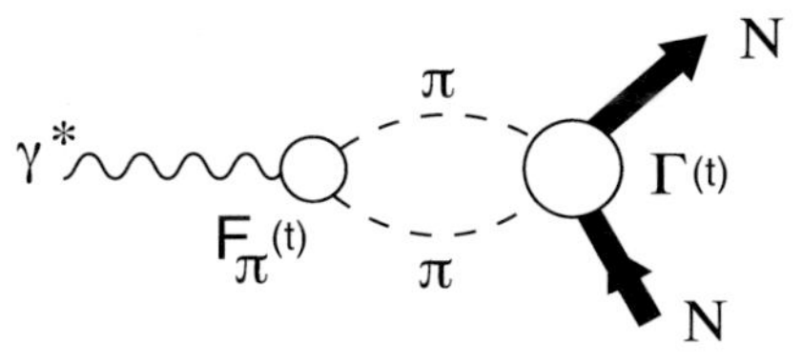

Figure 2.12: Two-pion contribution to the isovector nucleon form factors.

picture of the isovector form factors (see Fig. 2.12), we see that the description must involve two important elements: the photon coupling to the pion as represented by the electromagnetic pion form factor F_π, and the amplitude describing the interaction of the isovector two-pion pair with the nucleon. The Bose symmetry of the $\pi\pi$ system in a state with isospin $I = 1$ implies that its spatial wave function must be odd, so that the minimal relative angular momentum of the pion pair is $l = 1$. Looking at the process in the crossed channel $e^+e^- \to N\bar{N}$, it is then clear that this requires knowledge of the p-wave amplitudes for the process $\pi\pi \leftrightarrow N\bar{N}$. These amplitudes are denoted by $\Gamma_{1,2}$ and tabulated in [Hö 83].

The detailed analysis gives the following relations for the spectral functions, i.e. the imaginary parts, of the isovector Dirac and Pauli form factors:

$$\mathrm{Im}\, F^V_{1,2}(t) = \frac{p_t^3}{\sqrt{t}} F_\pi(t)\, \Gamma^*_{1,2}(t), \tag{2.66}$$

where $p_t = \frac{1}{2}\sqrt{t - 4m_\pi^2}$ is the pion momentum in the crossed (t-)channel. The p_t^3 behavior is characteristic of a p-wave. The pion form factor $F_\pi(t)$ in the timelike region, with $t \geq 4m_\pi^2$, is well determined by $e^+e^- \to \pi^+\pi^-$ measurements. The $\pi\pi \to N\bar{N}$ amplitudes can be derived using methods of analytic continuation and crossing symmetry together with known properties of the $\pi N \to \pi N$ amplitudes.

Eq. (2.66) should be seen as representing the constraints imposed by unitarity. These relations are exactly valid below the threshold for the production of four pions (the $3\,\pi$ channel does not contribute to the I=1 spectral function). In practice they work in the range $4m_\pi^2 \leq t \leq 50m_\pi^2$, covering the region of the ρ resonance.

The result of an analysis for the isovector spectral functions divided by t^2 is shown in Fig. 2.13. The presentation is such that the area under each curve gives $\frac{\pi}{6}\langle r^2\rangle$ for each of the form factors. The pronounced maximum reflecting the ρ meson resonance is clearly visible in these spectral functions, but it is also evident that the $\pi\pi \leftrightarrow N\bar{N}$ dynamics contributes substantially to the non-resonant, low-mass parts of $\mathrm{Im}\, F_1^V$ and $\mathrm{Im}\, G_E^V$. An updated computation of nucleon form factors based on these spectral functions, and incorporating the large-Q^2 behavior as expected from perturbative QCD, has been made for the proton magnetic form factor. This is the one which is best measured up to the highest accessible values of Q^2. The result (see Fig. 2.14) is obviously quite convincing. The detailed analysis also permits a rough delineation between hadronic and quark-gluon descriptions. While the low-mass components of the spectral functions dominate the form factors at $Q \leq 3$ GeV, the asymptotic, perturbative QCD behavior takes over at larger Q. Nevertheless the low-mass spectrum still appears to play a role for the absolute magnitudes of the form factors even at very large Q [MMD 96].

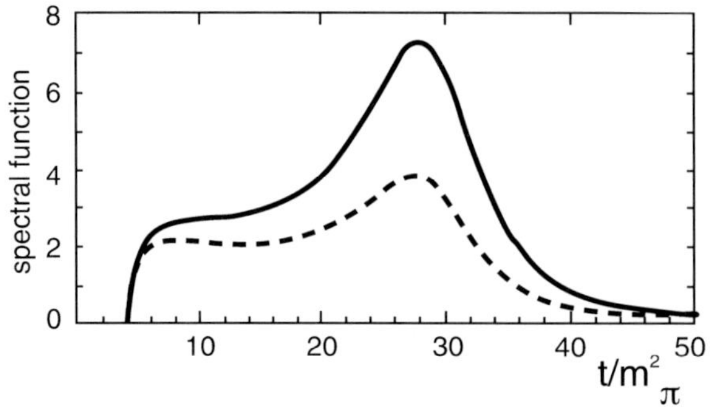

Figure 2.13: Spectral functions of isovector nucleon form factors multiplied by $1/t^2$, in units of $10^{-3} m_\pi^{-4}$. Dashed curve: $\mathrm{Im}\, F_1^V(t)/t^2$; solid curve: $\mathrm{Im}\, G_E^V(t)/t^2$. The curves are based on [Hö 83] as updated by [MMD 96].

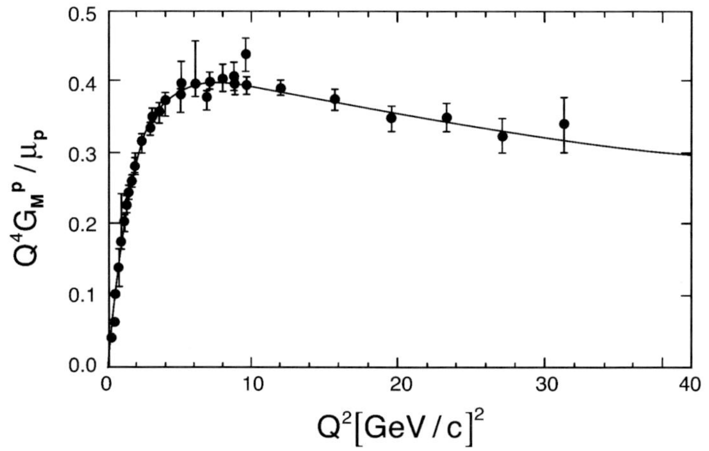

Figure 2.14: Proton magnetic form factor as obtained in a dispersion relation analysis which combines the hadronic low-mass spectrum with an asymptotic quark-gluon description constrained by perturbative QCD [MMD 96]

2.2.7 Form Factors in the Timelike Region

Measurements of proton form factors in the region of positive (i.e. timelike) q^2 have been made using the $e^+e^- \to p\bar{p}$ reaction at electron-positron storage rings, or its inverse using antiproton annihilation on a hydrogen target [Ant 94, Arm 93, Bar 94]. These processes are restricted to squared center-of-mass energies $s = q^2 \geq 4M^2 = 3.52$ GeV^2. The interesting region of intermediate mass vector mesons is therefore not directly accessible.

The magnitudes $|G^p_{E,M}(q^2)|$ of the (complex) form factors can in principle be extracted from the differential cross section for $e^+e^- \rightarrow \bar{p}p$ in the center-of-mass frame:

$$\frac{d\sigma}{d\Omega_{cm}} = \frac{\alpha^2}{4s}\sqrt{1-\frac{4M^2}{s}}\left[|G^p_M(s)|^2\,(1+\cos^2\theta)+\frac{4M^2}{s}\,|G^p_E(s)|^2\sin^2\theta\right]. \quad (2.67)$$

At the nucleon-antinucleon threshold, where only s-wave production contributes, one has $G_M(4M^2) = G_E(4M^2)$, and it is often assumed that this equality still holds above threshold for the magnitudes of the form factors. A typical result for the magnetic form factor is shown in Fig. 2.15 together with a dispersion relation analysis which reproduces both spacelike and timelike parts of the form factor. It illustrates the tendency of the form factor to decrease with increasing $s = q^2$, a behavior that is qualitatively expected, for example, from generalized versions of the VMD model including heavier vector states than just the ρ, ω and ϕ mesons. The Fermilab data [Arm 93] at $s = q^2 > 8$ GeV^2, however, reveal a surprise: the proton magnetic form factor at large timelike q^2 stays at a level about twice that of the spacelike form factor, at the corresponding negative q^2. Such differences should vanish in the asymptotic, perturbative region of QCD. Apparently, for exclusive processes such as the one discussed here, this asymptotic region is not yet reached, even at quite large values of $|q^2|$.

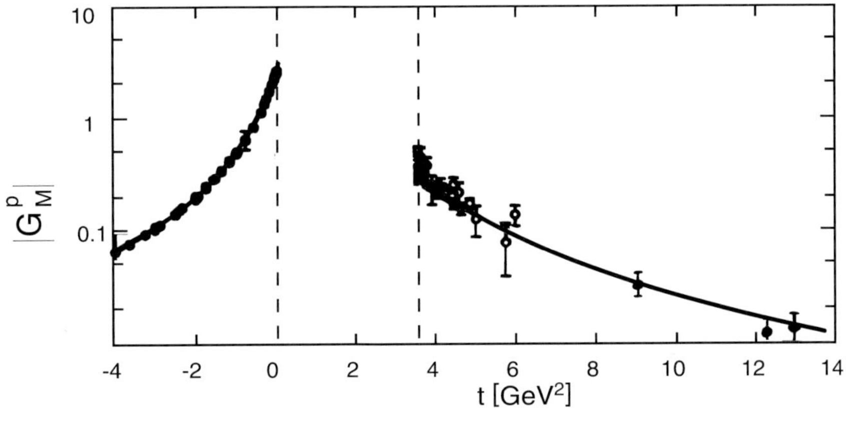

Figure 2.15: Proton magnetic form factor in the spacelike ($t = q^2 < 0$) and in the timelike ($t = q^2 > 0$) region of momentum transfer. The data are from [Ant 94, Arm 93, Bar 94]. The curve is a fit to the data. The region $0 < t < 4M^2$ is not accessible in $e^+e^- \rightarrow \bar{p}p$.

2.3 Nucleon Resonances

The composite nature of the nucleon manifests itself in the fact that its charge and magnetic moment distributions have a finite size, and that the proton and neutron magnetic moments are "anomalous", differing strongly from their expected values for point-like Dirac particles.

A second, complementary manifestation of intrinsic structure is the existence of a spectrum of excited states, the nucleon resonances. A series of such resonances is observed in the mass region $1 - 2$ GeV. Their properties are identified by systematic investigations using electromagnetic and strong probes, primarily with beams of electrons, photons and pions. We concentrate here on the results obtained by inelastic scattering of high energy electrons from a proton target, with energy transfers to the nucleon in the range $\nu < 2$ GeV. Later, in Chapter 4 on deep-inelastic scattering, the substructure of the nucleon will be resolved at still much higher energy and momentum transfer.

2.3.1 Inelastic Electron Scattering

We start with the inclusive inelastic scattering of electrons on nucleons, a process in which only the scattered electron is detected. Unlike elastic scattering for which the energy and momentum transfer are connected by the relation $\nu = Q^2/2M$, we now have two independent variables, ν and Q^2.

In the laboratory frame, the differential cross section for inelastic scattering of high energy electrons from a nucleon is:

$$\frac{d^2\sigma}{d\Omega\, d\epsilon'} = \frac{4\,\alpha^2\epsilon'^2}{Q^4}\left(W_2(\nu,Q^2)\cos^2\frac{\theta}{2} + 2W_1(\nu,Q^2)\sin^2\frac{\theta}{2}\right), \tag{2.68}$$

with

$$Q^2 = 4\,\epsilon\epsilon'\sin^2\frac{\theta}{2}. \tag{2.69}$$

The derivation of this cross section may be found in Appendix B.1. The scattered electron is detected with an energy, ϵ', at a laboratory angle, θ, relative to the incident beam of fixed energy, ϵ. By comparison with the differential cross section (2.8) for scattering on a structureless Dirac particle, we see that the information about nucleon structure is now represented by two response functions (or structure functions) W_1 and W_2. Their variation with the energy transfer ν at fixed scattering angle θ, or fixed Q^2, is of interest here. Scattering at forward angles emphasizes W_2, while isolating W_1 requires backward scattering. A representation of the cross section in the form

$$\frac{d^2\sigma}{d\Omega\, d\epsilon'} = \sigma_0\left(W_2(\nu,Q^2) + 2W_1(\nu,Q^2)\ \tan^2\frac{\theta}{2}\right) \tag{2.70}$$

with

$$\sigma_0(\theta,\epsilon) = \frac{\alpha^2\cos^2\frac{\theta}{2}}{4\,\epsilon^2\sin^4\frac{\theta}{2}}, \tag{2.71}$$

permits a Rosenbluth separation of W_1 and W_2 by plotting the ratio $(d^2\sigma/d\Omega d\epsilon')/\sigma_0$ as a function of $\tan^2\frac{\theta}{2}$.

A typical example of a measured inelastic cross section is shown in Fig. 2.16. The significant feature above the elastic scattering peak is the existence of three pronounced maxima in that cross section. The first peak corresponds to the $\Delta(1232)$, a strong spin-$\frac{3}{2}$ resonance which also dominates the pion-nucleon cross section. The second maximum includes the

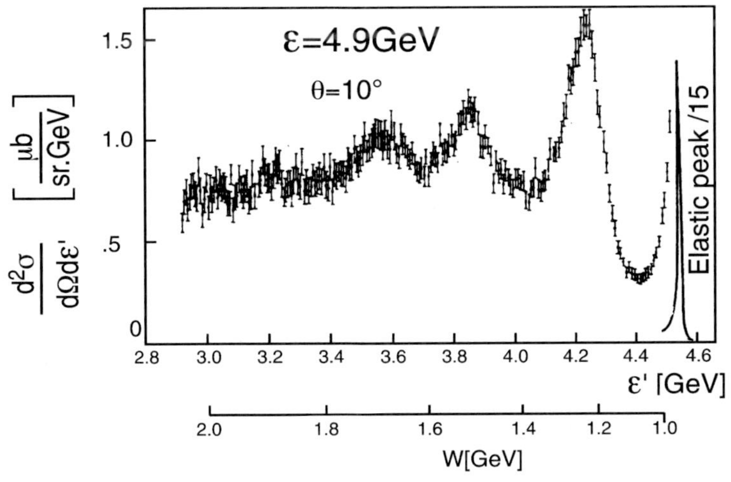

Figure 2.16: Spectrum of scattered electrons in electron-proton scattering at an incident energy $\epsilon = 4.9$ GeV and a laboratory angle of 10 degrees. The differential cross section [Bar 68] is plotted as a function of the outgoing electron energy ϵ' or the invariant mass W with $W^2 = (P+q)^2 = M^2 + 2M\nu - Q^2$.

$N^*(1520)$ and $N^*(1535)$. In the third resonance region, the strongest excitation at low Q^2 is the $N^*(1680)$. The inelastically scattered electron is a source of a virtual photon of squared invariant mass q^2, with energy ν and three-momentum $|\vec{q}| = \sqrt{\nu^2 + Q^2}$. The differential cross section (2.68) can in fact be written equivalently as a cross section for virtual photo-absorption. The virtual photon has both transverse and longitudinal components with respect to the unit vector along $\vec{q}$, its direction of propagation. We denote the corresponding photon-nucleon cross sections by $\sigma_T(\nu, Q^2)$ and $\sigma_L(\nu, Q^2)$. For real photons with $Q^2 = 0$, the longitudinal cross section, σ_L, vanishes and $\sigma_T(\nu, Q^2 = 0)$ becomes the total photo-absorption cross section $\sigma_\gamma(\nu)$. Introducing the parameter, ε, describing the ratio of longitudinal to transverse polarization of the virtual photon:

$$\varepsilon = \left(1 + \frac{2\,\vec{q}^{\,2}}{Q^2}\tan^2\frac{\theta}{2}\right)^{-1}, \tag{2.72}$$

the differential cross section (2.68) can be written:

$$\frac{d^2\sigma}{d\Omega\, d\epsilon'} = \Gamma(\sigma_T + \varepsilon\,\sigma_L), \tag{2.73}$$

with the kinematic factor, Γ, specified in Appendix B.1. (This factor is not without ambiguity for virtual photons. The definition we use coincides with that in [Cl 79].) The transverse and longitudinal photon cross sections are related to the response functions $W_{1,2}$ as follows[2]:

$$\sigma_T = \frac{4\,\pi^2\alpha}{\nu^*} W_1, \tag{2.74}$$

$$\sigma_L = \frac{4\,\pi^2\alpha}{\nu^*} \left(\frac{\vec{q}^{\,2}}{Q^2} W_2 - W_1\right), \tag{2.75}$$

where $\nu^* = \nu - Q^2/2M = (W^2 - M^2)/2M$ is the excitation energy provided by the virtual photon.

Helicity amplitudes

The electromagnetic transition matrix elements from the nucleon ground state to any resonant excited state are conveniently expressed in terms of photon helicity amplitudes. The virtual photon polarization is either transverse or longitudinal. Transverse photons can be described in terms of helicities $\lambda = \pm 1$, corresponding to right- or left-handed circular polarization with respect to the $\vec{q}$-axis. Longitudinal photons have $\lambda = 0$. The respective polarization four-vectors for transverse and longitudinal photons are $\varepsilon^\mu_\pm = (0, \mp 1, -i, 0)/\sqrt{2}$ and $\varepsilon^\mu_0 = (|\vec{q}|, 0, 0, \nu)/Q$. The corresponding components of the current are $J_\pm = \varepsilon^\mu_\pm J_\mu = \pm\frac{1}{\sqrt{2}}(J_x \pm iJ_y)$ and $J_{\lambda=0} = \varepsilon^\mu_0 J_\mu = \frac{Q}{\nu} J_z$. It is common to introduce the following three basic types of transition amplitude connecting a nucleon $N_{\frac{1}{2},m_S}$, of given spin projection m_S, with any nucleon resonance, N^*_{j,m_j}, of spin j and projection m_j:

$$A_{\frac{1}{2}} = \frac{e}{2M}\sqrt{\frac{M}{W^2 - M^2}} \langle N^*_{j,+\frac{1}{2}} |\, J_+ \,| N_{\frac{1}{2},-\frac{1}{2}} \rangle, \tag{2.76}$$

$$A_{\frac{3}{2}} = \frac{e}{2M}\sqrt{\frac{M}{W^2 - M^2}} \langle N^*_{j,+\frac{3}{2}} |\, J_+ \,| N_{\frac{1}{2},+\frac{1}{2}} \rangle, \tag{2.77}$$

$$S_{\frac{1}{2}} = \frac{e}{2M}\sqrt{\frac{M}{W^2 - M^2}} \langle N^*_{j,\frac{1}{2}} |\, J_0 \,| N_{\frac{1}{2},\frac{1}{2}} \rangle, \tag{2.78}$$

with $W^2 = M^2 - Q^2 + 2M\nu$. In terms of these amplitudes, the transverse and longitudinal virtual photon cross sections at resonance can be written

$$\sigma_T(\nu_R, Q^2) = \frac{2M}{\Gamma_R M_R}\left[|A_{\frac{1}{2}}|^2 + |A_{\frac{3}{2}}|^2\right], \tag{2.79}$$

$$\sigma_L(\nu_R, Q^2) = \frac{4M}{\Gamma_R M_R} |S_{\frac{1}{2}}|^2, \tag{2.80}$$

where $\nu_R = (M_R^2 - M^2 + Q^2)/2M$. Here M_R is the invariant mass of the resonance and Γ_R its decay width.

In Section 8.2.7 we present explicit examples in the constituent quark model. When discussing the physics at large momentum transfer, the interpretation of the helicity amplitudes in terms of the underlying transverse and longitudinal photon interactions with the quark constituents of the nucleon is of some interest. Consider a nucleon traveling in the Breit frame

[2] As a sign of caution at this point, we note that different definitions of $W_{1,2}$ appear in the literature.

with momentum $\vec{P} = -\vec{q}/2$. Its helicity, $h = \vec{\sigma} \cdot \vec{P}/|\vec{P}|$, is therefore positive $(h = +1)$ for $m_S = -\frac{1}{2}$ and negative $(h = -1)$ for $m_S = +\frac{1}{2}$. In case of the $A_{\frac{1}{2}}$ amplitude, no helicity flip occurs because the resonant final state recoils with momentum $\vec{P}' = +\vec{q}/2$ and its spin aligned with $\vec{P}'$. Here a photon can be absorbed by a single quark in the nucleon without flipping its spin. The $A_{\frac{3}{2}}$ amplitude, on the other hand, requires a change of helicity, and correspondingly a flip of the quark spin. At large Q^2 the quark mass can be neglected and the photon coupling to the quark vector current conserves helicity for massless Dirac particles. Consequently, at Q^2 of the order of several GeV^2, the helicity changing amplitude $A_{\frac{3}{2}}$ should be suppressed by a power of $1/Q^2$ with respect to the helicity conserving amplitude $A_{\frac{1}{2}}$.

Transition form factors

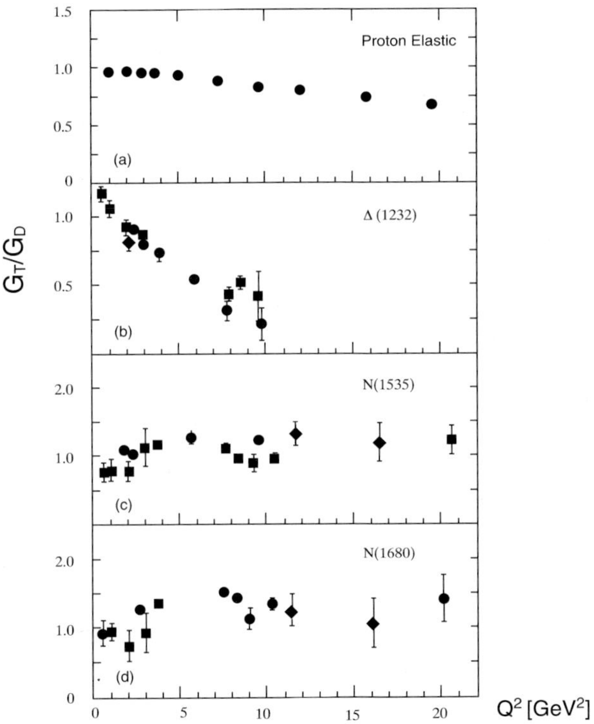

Figure 2.17: Absolute values of transition form factors divided by the dipole form factor, G_D, of Eq. (2.35). The proton elastic form factor is shown for comparison in (a). Figs. (b, c, d) show the form factors for transitions to the $\Delta(1232)$, $N^*(1535)$ and $N^*(1680)$ resonances (from [Sto 93]).

The detailed analysis of the Q^2 dependence of the resonant virtual photon cross sections from the inclusive electron scattering data can be used to extract transition form factors for each of these nucleon resonances. It is usual to introduce transverse transition form factors defined as:

$$|G_T(Q^2)|^2 = \frac{1}{4\pi\alpha} \frac{2M(M_R^2 - M^2)}{Q^2} \left[|A_{\frac{1}{2}}(Q^2)|^2 + |A_{\frac{3}{2}}(Q^2)|^2\right], \tag{2.81}$$

where the kinematic factor is chosen to yield a behavior similar to the elastic proton form factor at low Q^2. Fig. 2.17 shows typical results of such an analysis for transitions to the $\Delta(1232)$, $N^*(1535)$ and $N^*(1680)$ resonances in comparison with the dipole form (2.35) [Sto 93]. While the transition form factors for the higher resonances roughly resemble the global dipole form of the proton charge and magnetic form factors, the $N \to \Delta$ form factor drops faster with increasing Q^2, indicating a larger "size" of the $N \to \Delta$ transition volume than that of the proton itself.

2.3.2 Pion Electroproduction

While inclusive inelastic scattering gives an overview, the detailed investigation of nucleon resonances and their properties requires more exclusive experiments. All resonances seen in Fig. 2.16 decay prominently into a nucleon and one or more pseudoscalar mesons (a pion in almost all cases; an interesting exception is the $N^*(1535)$ which has roughly equal partial decay widths into $N\pi$ and $N\eta$). A detailed separation of the different resonance amplitudes is therefore most efficiently done by studying reactions such as $(e, e'\pi)$ and $(e, e'\eta)$ in which the scattered electron is measured in coincidence with the outgoing meson.

For such processes, the relevant matrix element of the electromagnetic current is now $\langle N_f(P')\pi(p)|J_\mu|N_i(P)\rangle$, where $N_i(P)$ and $N_f(P')$ refer to the initial and final nucleon states, and the four-momentum of the produced pion is denoted by $p = (\omega_\pi, \vec{p}_\pi)$. Current conservation implies that the time and space components of the current are related by $\nu J_0 = \vec{q} \cdot \vec{J}$. The initial and final nucleon can each have two possible spin states, and the photon has one longitudinal and two transverse helicity states, so that there are twelve independent matrix elements to start with. The current matrix element must behave as an axial vector since the pion in the final state has negative intrinsic parity. Further reduction leads to the following general form:

$$\begin{aligned}\langle N_f(P')\pi(p)|J_\mu|N_i(P)\rangle &= i\bar{u}_f(P')\gamma_5\Big[\gamma_\mu\gamma\cdot qB_1 + (P+P')_\mu B_2 + p_\mu B_3 \\ &+ q_\mu B_4 + \gamma_\mu B_5 + (P+P')_\mu\gamma\cdot qB_6 + q_\mu\gamma\cdot qB_7 + p_\mu\gamma\cdot qB_8\Big]u_i(P),\end{aligned} \tag{2.82}$$

where $u_{i,f}$ are the nucleon Dirac spinors and $B_1 \ldots B_8$ are scalar functions of the photon, pion and nucleon momenta. Two of those are redundant because of current conservation, $q \cdot J = 0$, so that six independent amplitudes remain.

Multipoles

Each of the six independent amplitudes, or convenient linear combinations thereof, can be expanded in multipoles, i.e. amplitudes ordered according to angular momentum. This is done in the center-of-mass system of the outgoing pion-nucleon pair. In this frame there are two independent momenta, say the pion momentum $\vec{p}_\pi$ and the momentum $\vec{q}$ of the virtual photon. The only remaining degree of freedom is the spin of the nucleon.

The conventional classification uses the orbital angular momentum, l, of the produced πN (or ηN) system together with its total angular momentum, $j = l \pm \frac{1}{2}$. These assignments have the advantage that they can be directly identified with the quantum numbers of the nucleon resonances which decay into πN. The multipole amplitudes are denoted by E for transverse electric, M for transverse magnetic and L for longitudinal ("Coulomb"). For example, the electric dipole (E1) amplitude involves the final πN in a relative s-state ($l = 0$) and is denoted by E_{0+}. Magnetic dipole (M1) transitions can lead to final πN p-wave ($l = 1$) states with total angular momenta $j = \frac{3}{2}$ or $j = \frac{1}{2}$; the corresponding amplitudes are denoted by M_{1+} or M_{1-}, respectively.

Furthermore, for a πN final state, the nucleon isospin $I = \frac{1}{2}$ combines with the pion isospin $I = 1$ to form total $I = \frac{1}{2}$ or $I = \frac{3}{2}$. Transitions to such final states induced by an isovector photon are marked by an upper index, $E^{(1/2)}$ or $E^{(3/2)}$; transitions induced by isoscalar photons which necessarily lead to $I = \frac{1}{2}$ states are denoted as $E^{(0)}$. Each of the nucleon resonances has a definite isospin. For instance, the spin-$\frac{3}{2}$ isobar of the nucleon, the $\Delta(1232)$, has isospin $I = \frac{3}{2}$. It is reached by a strong magnetic dipole transition, the amplitude of which is denoted by $M_{1+}^{(3/2)}$.

To exemplify the multipole expansion, consider a leading term which arises from the evaluation of the $\gamma_5\vec{\gamma}$ part of the current matrix element. When reduced to an equivalent two-component form, this term is just proportional to the nucleon spin, and its multipole expansion becomes:

$$\begin{aligned}\langle N_f \pi | \vec{J} | N_i \rangle &= \\ &(\chi_f^\dagger \, \vec{\sigma} \, \chi_i) \sum_{l \geq 0} \left[(l M_{l+} + E_{l+}) \, P'_{l+1} + ((l+1) M_{l-} + E_{l-}) \, P'_{l-1} \right] + \cdots \end{aligned} \tag{2.83}$$

plus five more terms involving axial vector combinations of the photon and pion momenta together with the nucleon spin. The detailed derivation of these expansions has been given by Chew, Goldberger, Low and Nambu in [CGLN 57, GW 67]. The multipole amplitudes are functions of the excitation energy ν of the πN system. The derivatives of the Legendre polynomials, $P'_l(\cos\theta_\pi)$, are taken with respect to the cosine of the pion angle, measured in the πN center-of-mass frame, relative to the $\vec{q}$-axis.

A full determination of the six electroproduction amplitudes obviously requires measuring not only the pion momentum, but also a suitable combination of polarization observables, such as the asymmetries of the polarized photon and target proton, and the polarization of the recoil nucleon. Such detailed measurements at the high precision frontier are now becoming possible at new generations of electron accelerators.

Table 2.1 summarizes properties of the lowest nucleon resonances including the multipole amplitudes in which these resonances are seen.

2.3.3 Watson's Theorem

The close connection between the photon-nucleon and pion-nucleon systems implies definite phase relations between the amplitudes in the γN and πN channels. They are manifest in the S-matrix of the coupled $\gamma N \leftrightarrow \pi N$ reactions.

Table 2.1: Properties of nucleon resonances

resonance	M_R[MeV]	Γ[MeV]	J^P	l	multipoles	decay
P_{33}	1232	120	$\frac{3}{2}^+$	1	M_{1+}, E_{1+}, L_{1+}	$N\pi$
P_{11}	1440	350	$\frac{1}{2}^+$	1	M_{1-}, L_{1-}	$N\pi, N\pi\pi$
D_{13}	1520	120	$\frac{3}{2}^-$	2	E_{2-}, M_{2-}, L_{2-}	$N\pi, N\pi\pi$
S_{11}	1535	150	$\frac{1}{2}^-$	0	E_{0+}, L_{0+}	$N\pi, N\eta$
F_{15}	1680	130	$\frac{5}{2}^+$	3	E_{3-}, M_{3-}, L_{3-}	$N\pi, N\pi\pi$

Let the index $\alpha = (ljI)$ stand for a given pion-nucleon partial wave. Below the threshold for two-pion production, the S-matrix for the isolated πN system in each partial wave has the form $\exp(2i\delta_\alpha)$, where the real phase shift δ_α is a function of energy. Now consider the partial wave S-matrix, S_α, for the coupled channels $\pi N \to \pi N$, $\gamma N \to \gamma N$ and $\pi N \leftrightarrow \gamma N$. In the diagonal πN channel the S-matrix element will now have the form $\eta_\alpha \exp(2i\delta_\alpha)$, with $\eta_\alpha < 1$ a real function of energy. Then the unitarity condition, $S^\dagger = S^{-1}$, completely determines the coupled channel S-matrix. This is Watson's theorem [GW 67]. One finds:

$$S_\alpha = \begin{pmatrix} \eta_\alpha & i\sqrt{1-\eta_\alpha^2}\,\mathrm{e}^{i\delta_\alpha} \\ i\sqrt{1-\eta_\alpha^2}\,\mathrm{e}^{i\delta_\alpha} & \eta_\alpha \mathrm{e}^{2i\delta_\alpha} \end{pmatrix}, \tag{2.84}$$

where the elements are grouped such that the Compton scattering channel $\gamma N \to \gamma N$ is in the upper diagonal, the $\pi N \to \pi N$ channel is in the lower diagonal and the two off-diagonal matrix elements represent $\gamma N \to \pi N$ and its inverse. The phase δ_α in Eq. (2.84) is determined only modulo π.

As an example, take the magnetic dipole amplitude for the production of the $\Delta(1232)$ resonance. Watson's theorem implies that the phase of this amplitude is determined by the well-known πN scattering phase shift in the p-wave, $j = I = \frac{3}{2}$ channel:

$$M_{1+}^{(3/2)} = \pm \left| M_{1+}^{(3/2)} \right| \, \exp\left(i\delta_{1+}^{(3/2)} \right). \tag{2.85}$$

The $\Delta(1232)$ can also be excited through an electric quadrupole (E2) transition for which the same theorem holds:

$$E_{1+}^{(3/2)} = \pm \left| E_{1+}^{(3/2)} \right| \, \exp\left(i\delta_{1+}^{(3/2)} \right). \tag{2.86}$$

Hence these M1 and E2 transition amplitudes have a common phase which has been measured with high accuracy in πN elastic scattering. The E2/M1 ratio is then simply given as the ratio of these amplitudes.

2.3.4 Selected Examples

The $\Delta(1232)$ resonance

Let us stay with the spin-$\frac{3}{2}$ isobar of the nucleon yet for a while. For this case the helicity amplitudes are expressed in terms of the relevant M1 and E2 amplitudes as follows:

$$A_{\frac{1}{2}} = -\frac{1}{2}\left(M_{1+}^{(3/2)} + 3E_{1+}^{(3/2)}\right) \tag{2.87}$$

$$A_{\frac{3}{2}} = -\frac{\sqrt{3}}{2}\left(M_{1+}^{(3/2)} - E_{1+}^{(3/2)}\right). \tag{2.88}$$

The factors in these linear combinations are simply related to Clebsch-Gordan coefficients.

At small Q^2 the M1 transition is primarily of the spin-flip type and completely dominates the electric quadrupole transition. The corresponding amplitudes obtained from polarization measurements with real photons are shown in Fig. 2.18 as a function of the photon energy. Watson's theorem requires that the zero of ReM_{1+} must coincide with the zero of ReE_{1+}. The shape of the small (negative) E2 amplitude indicates the presence of a non-resonant background in addition to the Δ resonance contribution.

The detailed behavior of the E2 term is of some interest in its own right since it points to a deformation or asymmetry of the charge density as it undergoes the $N \to \Delta$ transition. At resonance the E2/M1 ratio is found to be $(-2.5 \pm 0.2) \cdot 10^{-2}$ [Be 97]. At large Q^2 one

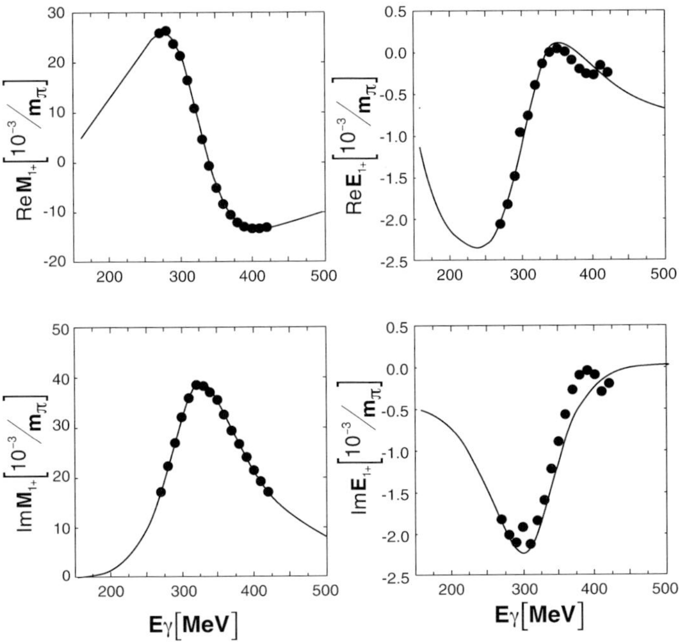

Figure 2.18: The resonant multipole amplitudes $M_{1+}^{(\frac{3}{2})}$ and $E_{1+}^{(\frac{3}{2})}$. Data [Be 97] are compared with a dispersion relation analysis (solid curves) [HDT 96a].

expects, as already mentioned, that only the helicity conserving amplitude $A_{\frac{1}{2}}$ survives. This means that the E2/M1 ratio should approach unity as $Q^2 \to \infty$. Precisely how this turnover takes place, from the dominant $A_{\frac{3}{2}}$ with $E2 \ll M1$ at small Q^2 to $A_{\frac{1}{2}}$ with $E2 \approx M1$ at large Q^2, is a challenging question. Together with the observed rapid decrease of the $N \to \Delta$ transition form factor which, by Watson's theorem, is proportional to $|M1|^2 + 3|E2|^2$, this indicates vastly different scales for spin-flip and non-flip processes inside nucleons. Clarification of these interesting phenomena requires detailed measurements of the asymmetry $(|A_{\frac{1}{2}}|^2 - |A_{\frac{3}{2}}|^2)/(|A_{\frac{1}{2}}|^2 + |A_{\frac{3}{2}}|^2)$ over a broad range of Q^2 [Sto 99, Bu 00].

Higher resonances: $N^*(1520)$ and $N^*(1680)$

In the language of the harmonic oscillator, the $\Delta(1232)$ resonance is primarily a "0 $\hbar\omega$", spin-flip excitation. The $N^*(1520)$ resonance, with $j = \frac{3}{2}$, is of a different kind. It represents a "1 $\hbar\omega$" electric dipole excitation accompanied by a magnetic quadrupole. The $j = \frac{5}{2}$ resonance $N^*(1680)$ is an example of a "2 $\hbar\omega$" excitation induced primarily by an electric quadrupole. Both of these excited states differ from the nucleon ground state through the characteristic orbital motion of their constituents. The helicity amplitudes, $A_{\frac{1}{2}}$ and $A_{\frac{3}{2}}$, for

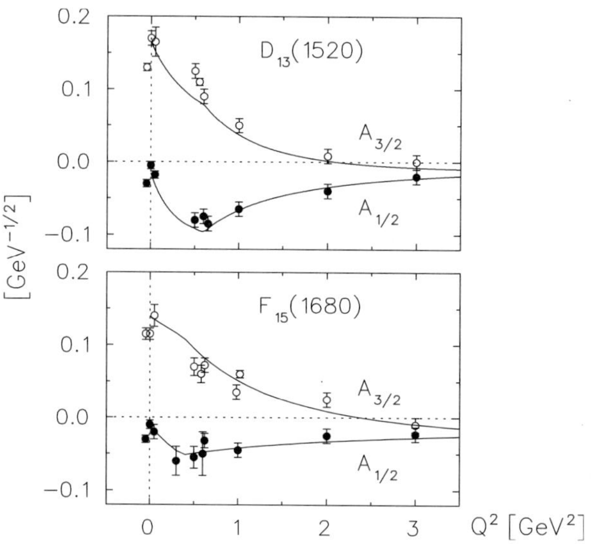

Figure 2.19: Helicity amplitudes as a function of Q^2 for the $D_{13}(1520)$ and $F_{15}(1680)$ resonances [Bu 89]. The curves are the result of a model calculation [Dr 99].

the transitions to these resonances, are shown in Fig. 2.19. At very large Q^2 the helicity non-flip amplitude, $A_{\frac{1}{2}}$, is expected to dominate over the helicity flip amplitude, $A_{\frac{3}{2}}$. At the real photon point, $Q^2 = 0$, the situation is opposite, with $A_{\frac{1}{2}}$ vanishing. These features show up most prominently in the helicity asymmetry, $(|A_{\frac{1}{2}}|^2 - |A_{\frac{3}{2}}|^2)/(|A_{\frac{1}{2}}|^2 + |A_{\frac{3}{2}}|^2)$, which changes from -1 at $Q^2 = 0$ to +1 at $Q^2 \to \infty$, passing through zero typically in the range between 0.5 and 1 GeV^2. This strong Q^2 variation indicates a rapid change in the intrinsic angular momentum structure of the nucleon as it undergoes such transitions. We refer to Section 8.2.7 for a more detailed quark model discussion. These selected examples, taken from a large variety of measurements, should suffice to demonstrate that detailed investigation of the transition amplitudes to nucleon resonances can be expected to impose strong constraints on the construction of nucleon models discussed in Chapter 8.

2.4 Compton Scattering

In this section we summarize information about nucleon structure that can be derived from the scattering and absorption of real photons. At energies below the threshold for pion production, a photon can only scatter elastically from a nucleon. Above pion production threshold the total cross section reveals pronounced resonances that have already been encountered in the analysis of inelastic electron scattering.

Low energy Compton scattering experiments on hydrogen targets are used to determine the proton electric and magnetic polarizabilities. These quantities are a measure of the response to low-frequency, long-wavelength electromagnetic fields. For neutrons, this information is extracted from the scattering of low energy neutrons on atoms.

Unitarity connects the imaginary part of the forward Compton scattering amplitude with the total photon-nucleon cross section. The real part of that amplitude satisfies a dispersion relation, analogous to the Kramers-Kronig relation familiar from atomic physics. Such dispersion relations are the basis of sum rules which provide a global, model independent analysis of the experimental data.

The theme of this section is twofold: first we give an overview of photon scattering and total photon-nucleon cross sections, including a dispersion relation analysis. Then we concentrate on the low-energy properties of the Compton amplitude and discuss the electromagnetic polarizabilities of the nucleon.[3]

2.4.1 Scattering Amplitudes and Cross Sections

Consider a photon of energy ω and momentum $\vec{k}$ in the laboratory frame. Its four-momentum is $k^\mu = (\omega, \vec{k}\,)$, with $k^2 = \omega^2 - \vec{k}^{\,2} = 0$. In our previous discussion of inelastic electron scattering, the lab frame energy of the virtual photon was expressed as $\nu = P \cdot q/M$. Now we have

$$\omega = \frac{P \cdot k}{M} \tag{2.89}$$

in terms of the nucleon four-momentum P and its mass M.

The photon-nucleon interaction is described by matrix elements involving the product, $eJ_\mu \mathcal{A}^\mu$, of the electromagnetic current $J_\mu \equiv J_\mu^{\mathrm{EM}}$ with the photon field,

$$\mathcal{A}^\mu(x) = \varepsilon^\mu \, e^{-ik\cdot x}, \tag{2.90}$$

[3] Throughout this section we will use the fine structure constant $\alpha = e^2/4\pi \approx 1/137$. In order to avoid notational confusion, electric polarizabilities will be denoted by α_E and not by α as sometimes done in the literature.

where ε^μ is the photon polarization vector. For real photons one can choose the time component $\mathcal{A}^0$ to vanish and we have the transversality condition $\vec{\varepsilon} \cdot \vec{k} = 0$. Standard rules,

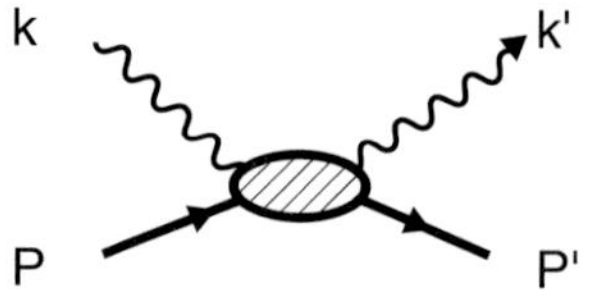

Figure 2.20: Compton scattering on the nucleon.

summarized in Appendix B.3, lead to the T-matrix for Compton scattering on the nucleon (see Fig. 2.20)

$$T(\gamma N \to \gamma N) = 8\pi M\, \mathcal{F} = e^2\, \varepsilon'^{*\mu} \varepsilon^\nu T_{\mu\nu}, \tag{2.91}$$

with the Compton tensor

$$T_{\mu\nu} = i \int d^4x\, e^{ik'\cdot x} \langle N(P')|\mathcal{T}\, J_\mu(x) J_\nu(0)|N(P)\rangle. \tag{2.92}$$

Here k' and ε' refer to the outgoing photon, P and P' are the initial and final nucleon four-momenta, and $\mathcal{T}$ denotes the time ordering of the electromagnetic current operators. The differential cross section for Compton scattering in the laboratory frame is

$$\frac{d\sigma}{d\Omega}(\omega, \theta) = \left(\frac{\omega'}{\omega}\right)^2 |\mathcal{F}|^2. \tag{2.93}$$

Energy and momentum conservation relates the energies of the scattered and incident photon to the laboratory scattering angle by

$$\omega' = \frac{\omega}{1 + \frac{\omega}{M}(1 - \cos\theta)}. \tag{2.94}$$

Scattering from a point-like spin-$\frac{1}{2}$ particle is described by the Klein-Nishina cross section

$$\left(\frac{d\sigma}{d\Omega}\right)_{\text{KN}} = \frac{\alpha^2}{4M^2}\left(\frac{\omega'}{\omega}\right)^2 \left[\frac{\omega'}{\omega} + \frac{\omega}{\omega'} + 4|\vec{\varepsilon}\,'^* \cdot \vec{\varepsilon}\,|^2 - 2\right], \tag{2.95}$$

where the average over the target spin has been taken but the dependence on the photon polarizations is retained. Averaging over photon polarizations leads to

$$\left(\frac{d\sigma}{d\Omega}\right)_{\text{KN}} = \frac{\alpha^2}{2M^2}\left(\frac{\omega'}{\omega}\right)^2 \left[\frac{\omega'}{\omega} + \frac{\omega}{\omega'} - \sin^2\theta\right]. \tag{2.96}$$

If the target has an anomalous magnetic moment κ, further terms up to fourth order in κ are added. The full result for Compton scattering on a point-like nucleon, referred to as the Powell cross section, is summarized in Appendix B.3.

Forward scattering amplitudes

In the case of forward scattering the amplitude (2.91) reduces to

$$\mathcal{F}(\omega, \theta = 0) = \chi_f^\dagger \left[f(\omega)\, \vec{\varepsilon} \cdot \vec{\varepsilon}^{\,\prime *} - i g(\omega)\, \vec{\sigma} \cdot (\vec{\varepsilon} \times \vec{\varepsilon}^{\,\prime *}) \right] \chi_i, \tag{2.97}$$

with nucleon Pauli spinors $\chi_{i,f}$. This introduces the spin-flip amplitude g and the non-flip amplitude f. Crossing symmetry requires that the T-matrix is symmetric under exchange of the in- and outgoing photons ($\omega \to -\omega$ and $\vec{\varepsilon}^{\,\prime *} \leftrightarrow \vec{\varepsilon}$). Therefore the scattering amplitudes have the following crossing properties:

$$f(-\omega) = f(\omega), \qquad g(-\omega) = -g(\omega). \tag{2.98}$$

The amplitudes f and g can in principle be separated by a measurement using circularly polarized photons incident on nucleons polarized parallel or antiparallel with respect to the photon momentum. Consider for example a right-handed photon. Its helicity, $\lambda = +1$, adds with the nucleon spin aligned in parallel to give total angular momentum $\frac{3}{2}$, while in the antiparallel case the spin along the helicity axis is $\frac{1}{2}$. The corresponding amplitudes are

$$f_{\frac{1}{2}} = f + g, \qquad f_{\frac{3}{2}} = f - g. \tag{2.99}$$

Optical theorem

Unitarity relates the imaginary part of $f(\omega)$ to the total photon-nucleon cross section:

$$\sigma_{\text{tot}}(\omega) = \frac{4\pi}{\omega}\, \text{Im}\, f(\omega). \tag{2.100}$$

Similarly, the total absorption cross sections for circularly polarized photons on polarized nucleons are

$$\sigma_{\frac{1}{2},\frac{3}{2}}(\omega) = \frac{4\pi}{\omega}\, \text{Im}\, f_{\frac{1}{2},\frac{3}{2}}(\omega), \tag{2.101}$$

Low energy expansion

At small photon energies the forward scattering amplitudes f and g can be expanded in powers of ω:

$$f(\omega) = -\frac{Z^2 e^2}{4\pi M} + (\alpha_E + \beta_M)\, \omega^2 + \mathcal{O}(\omega^4), \tag{2.102}$$

$$g(\omega) = -\frac{e^2 \kappa^2}{8\pi M^2}\, \omega + \gamma\, \omega^3 + \mathcal{O}(\omega^5), \tag{2.103}$$

with $Z = 1$ for the proton and $Z = 0$ for the neutron. The leading terms are unambiguously determined by low energy theorems [Gel 54, Lo 54] based on gauge invariance. For example, $f(0) = -\alpha/M$ is the Thomson limit which expresses the fact that, at a photon wavelength very large compared to any dimensions of the proton, the scattering amplitude can only involve the fine structure constant and the proton mass. The spin-flip amplitude vanishes at $\omega = 0$. Its leading term of order ω is proportional to κ^2, the squared anomalous magnetic moment. The terms of order ω^2 introduce the electric (α_E) and magnetic (β_M) polarizabilities. To order ω^3 the spin (or vector) forward polarizability γ enters.

Total cross sections

The photon-proton total cross section $\sigma_{\text{tot}}(\omega)$ has been measured systematically from the pion photoproduction threshold up to ω of order 300 GeV, with additional high energy data points from HERA available at $\omega \approx 10^4$ GeV. The data, summarized in Fig. 2.21, show the region of nucleon resonances ($\omega \leq 2$ GeV) followed by a continuum with a nearly constant cross section of about 125 μb which extends further on to a smoothly (logarithmically) rising asymptotic cross section. As already mentioned, the nucleon resonances excited by photoabsorption

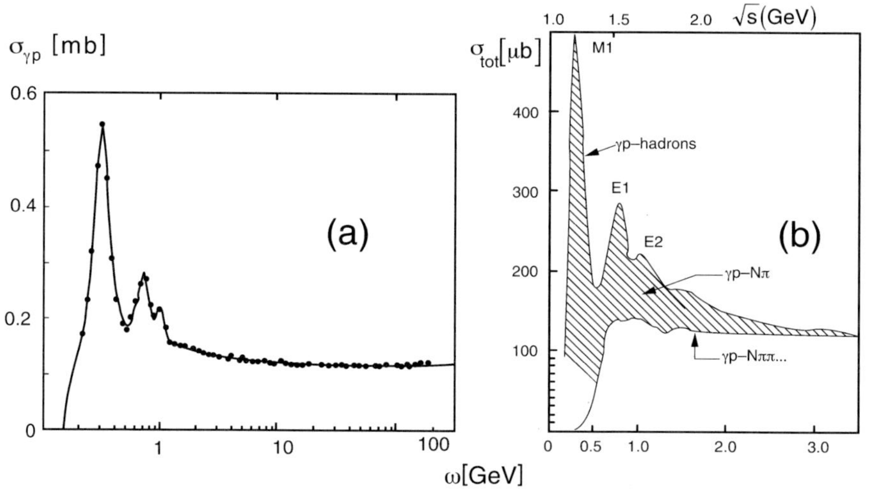

Figure 2.21: Total photon-proton cross section as a function of photon energy. The curve in (a) is a fit to the data which is used in a dispersion relation analysis [AFW 88]. The low energy cross section and its decomposition is shown in (b) [Dr 92]

decay predominantly into πN and $\pi\pi N$ channels. At energies above 2 GeV the contributions from final states with two or more pions give most of the cross section.

Total cross sections for neutrons are not directly measurable. They are extracted from photoabsorption measurements on deuterium targets, using the known proton data as input. The total γp and γd cross sections averaged over nucleon spin and photon polarization are determined quite accurately. Measurements of the cross sections $\sigma_{\frac{1}{2}}$ and $\sigma_{\frac{3}{2}}$ using polarized photons on a polarized proton target are in progress. An analysis of the difference $\sigma_{\frac{3}{2}} - \sigma_{\frac{1}{2}}$ in the resonance region [HDT 96a, HDT 96b] using existing information from pion photoproduction arrives at the result shown in Fig. 2.22 together with data from MAMI [Are 98]. The results indicate again the prominent role of the $\Delta(1232)$ resonance in the spin $\frac{3}{2}$ channel, but there is also a substantial contribution of opposite sign from the E_{0+} amplitude at lower energy.

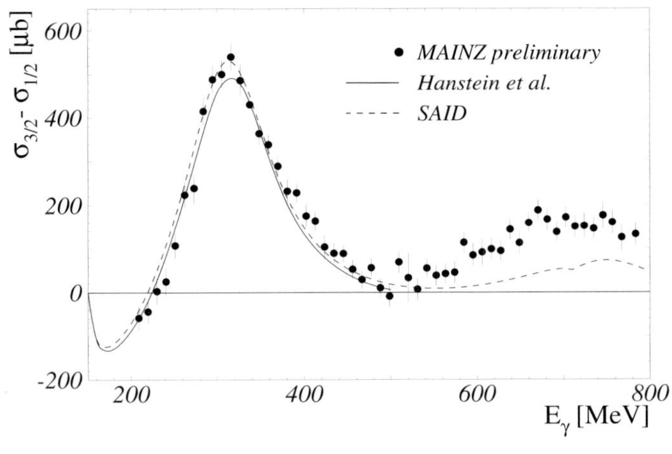

Figure 2.22: Cross section difference $\sigma_{\frac{3}{2}} - \sigma_{\frac{1}{2}}$ in the resonance region. Data from [Are 98, Ahr 00]. The solid curve corresponds to a dispersion relation calculation [HDT 96a, HDT 96b]. The dashed curve is based on a multipole analysis of single pion photoproduction amplitudes.

2.4.2 Dispersion Relations and Sum Rules

Kramers-Kronig relations

The analytic properties of the forward Compton scattering amplitude, together with unitarity and crossing symmetry, can be used to derive dispersion relations for $f(\omega)$ and $g(\omega)$. For the f amplitude this is just the celebrated Kramers-Kronig relation familiar from optics. It connects the real part of f with a dispersion integral over $\mathrm{Im}\, f$. By the optical theorem, the imaginary part can be replaced by the total cross section. One finds:

$$\begin{aligned} \mathrm{Re}\, f(\omega) &= f(0) + \frac{\omega^2}{2\,\pi^2}\, \mathcal{P} \int_{\omega_{\mathrm{th}}}^{\infty} d\omega' \, \frac{\sigma_{\mathrm{tot}}(\omega')}{\omega'^2 - \omega^2}\,, \\ f(0) &= -\frac{Z^2 \alpha}{M}\,, \end{aligned} \tag{2.104}$$

where $\mathcal{P}$ denotes the principal value integral and $\omega_{\mathrm{th}} = m_\pi(1 + m_\pi/2M) \approx 150$ MeV is the threshold energy for pion production. A subtraction has been made at $\omega = 0$ to ensure that the dispersion integral converges even if the total cross section rises logarithmically. The subtraction constant $f(0)$ is determined by a low-energy theorem, the well-known Thomson limit.

The Kramers-Kronig relation (2.104) has been used to extract the real part of the forward Compton amplitude, given the total cross section data. We show the result of such an analysis in Fig. 2.23. The expected strong energy dependence of $\mathrm{Re} f(\omega)$ in the resonance region shows up clearly, with a characteristic change of sign close to the $\Delta(1232)$ resonance. The analyticity of the spin-flip amplitude g together with the crossing relation, Eq. (2.98), can be

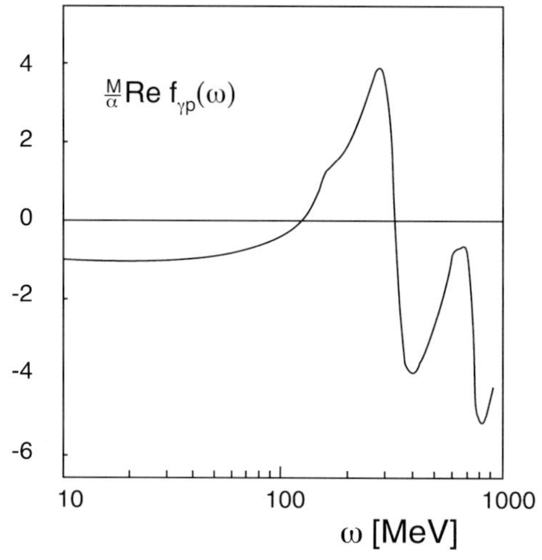

Figure 2.23: Real part of the photon-proton forward scattering amplitude deduced from the dispersion relation (2.104)

used to write down a dispersion relation for this amplitude as well:

$$\begin{aligned} \text{Re } g(\omega) &= \frac{2\,\omega}{\pi}\mathcal{P}\int_{\omega_{\text{th}}}^{\infty} d\omega'\,\frac{\text{Im}\, g(\omega')}{\omega'^2-\omega^2} \\ &= \frac{\omega}{4\,\pi^2}\mathcal{P}\int_{\omega_{\text{th}}}^{\infty}\frac{d\omega'\;\omega'}{\omega'^2-\omega^2}\left[\sigma_{\frac{1}{2}}(\omega')-\sigma_{\frac{3}{2}}(\omega')\right], \end{aligned} \tag{2.105}$$

where the last step follows from the unitarity condition, Eq. (2.101). Here it is assumed that the difference of cross sections for polarized photons on polarized nucleons decreases sufficiently fast at large ω' that the dispersion integral converges even without a subtraction.

Sum rules for electromagnetic polarizabilities

The low-energy expansion of Eq. (2.103), together with the Kramers-Kronig relation, immediately leads to a sum rule for the electric and magnetic polarizabilities of the nucleon [Bal 60]:

$$\alpha_E+\beta_M=\frac{1}{2\,\pi^2}\int_{\omega_{\text{th}}}^{\infty} d\omega\,\frac{\sigma_{\text{tot}}(\omega)}{\omega^2}. \tag{2.106}$$

Their sum is directly connected with the low energy excitation spectrum seen in photoabsorption.

An earlier evaluation using the empirical γp cross section [Da 70] gave:

$$(\alpha_E+\beta_M)_{\text{proton}}=(14.2\pm0.5)\;\cdot 10^{-4}\,\text{fm}^3. \tag{2.107}$$

A more recent analysis based on updated results of photoproduction measurements yields [Bab 98]:

$$(\alpha_E+\beta_M)_{\text{proton}}=(13.69\pm0.14)\;\cdot 10^{-4}\,\text{fm}^3. \tag{2.108}$$

The corresponding neutron value is extracted using the measured γ–deuteron total cross section, with some uncertainties related to subtracting the proton contributions [Da 70]:

$$(\alpha_E + \beta_M)_{\text{neutron}} = (15.8 \pm 0.5) \cdot 10^{-4}\,\text{fm}^3, \tag{2.109}$$

while a smaller value, $(\alpha_E + \beta_M)_{\text{neutron}} = (14.4 \pm 0.7) \cdot 10^{-4}$ fm^3, was found in [Bab 98]. These numbers set important constraints for the discussion of the individual electric and magnetic polarizabilities.

Gerasimov-Drell-Hearn sum rule

Using the low energy expansion, Eq. (2.103), one arrives at the Gerasimov-Drell-Hearn (GDH) sum rule [Ger 66, DH 66]:

$$\frac{\alpha}{M^2}\kappa^2 = \frac{1}{2\pi^2}\int_{\omega_{\text{th}}}^{\infty}\frac{d\omega}{\omega}\left[\sigma_{\frac{3}{2}}(\omega) - \sigma_{\frac{1}{2}}(\omega)\right]. \tag{2.110}$$

This connects the anomalous magnetic moment of the nucleon with the spectrum of its excited states, as seen in polarization measurements. The left hand side of this sum rule has

$$\kappa_p^2 = 3.21 \quad \text{(proton)}, \qquad \kappa_n^2 = 3.66 \quad \text{(neutron)}. \tag{2.111}$$

One would naturally expect that the strong magnetic dipole excitation of the nucleon, the $\Delta(1232)$, plays a major role in this discussion. This is indeed the case. The first, immediate observation is that the GDH sum rule is strongly dominated by isovector contributions. The squared anomalous magnetic moment for either proton or neutron is $\kappa_{p,n}^2 = \kappa_V(\kappa_V \pm 2\,\kappa_S) + \kappa_S^2$ with the isovector and isoscalar anomalous moments, $\kappa_V = 1.85$, $\kappa_S = -0.06$. By far the largest term in the GDH sum rule is therefore the pure isovector part, $\kappa_V^2 = 3.43$. The mixed $\kappa_S\kappa_V$ term is roughly an order of magnitude smaller, and the pure isoscalar term is negligible.

Let us examine how much of the isovector term is actually given by the $\Delta(1232)$ alone. We recall the helicity amplitudes $A_{\frac{1}{2}}$ and $A_{\frac{3}{2}}$ from Section 2.3.4. The cross sections $\sigma_{\frac{1}{2}}$ and $\sigma_{\frac{3}{2}}$ are proportional to the absolute squares of these amplitudes. Taking only the dominant magnetic dipole amplitude for the $N \to \Delta$ transition into account, one finds

$$\sigma_{\frac{3}{2}} - \sigma_{\frac{1}{2}} = 8\pi\frac{|\vec{p}_\pi|}{|\vec{k}_\gamma|}\left|M_{1+}^{\Delta}\right|^2. \tag{2.112}$$

With the empirical M_{1+}^{Δ} amplitude as input, direct evaluation of the sum rule (2.110) gives a contribution of about 3.6 to κ^2 in Eq. (2.110) from the $\Delta(1232)$ alone. Next in importance is the $N^*(1535)$ excitation together with the low-energy electric dipole continuum. Other resonances contribute far less and with varying signs.

We note that the GDH sum rule is also related to the problem of the spin structure of the nucleon as seen in polarized deep-inelastic scattering. A challenging research problem in this context is the question how the integral (2.110) evolves from the real photon point, $Q^2 = 0$, to very large Q^2 for highly virtual photons [Ede 00].

Spin polarizability sum rule

Expanding the dispersion relation (2.105) for $g(\omega)$ to order ω^3 and comparing with Eq. (2.103) one finds the sum rule for the spin polarizability[4]

$$\gamma = \frac{1}{4\pi^2}\int_0^\infty d\omega\,\frac{\sigma_{\frac{1}{2}}(\omega) - \sigma_{\frac{3}{2}}(\omega)}{\omega^3}. \tag{2.113}$$

In comparison with the GDH sum rule, the relevant information is now concentrated much more on the low energy side by the $1/\omega^3$ weight under the integral. It is therefore not only the $\Delta(1232)$ resonance that enters prominently in the spin polarizability, but also the low energy pion photoproduction continuum described by the electric dipole amplitude E_{0+}.

2.4.3 Electromagnetic Polarizabilities of the Nucleon

Proton polarizabilities

The electric and magnetic polarizabilities α_E and α_M can be determined separately by a measurement of the differential cross section for Compton scattering at low energy. To derive this cross section, let us first introduce the amplitude $\mathcal{F}_{\text{point}}$ for Compton scattering from a point nucleon (but with the anomalous magnetic moment taken into account). The differential cross section in the laboratory system,

$$\left(\frac{d\sigma}{d\Omega}\right)_{\text{point}} = \left(\frac{\omega'}{\omega}\right)^2 \left|\mathcal{F}_{\text{point}}\right|^2, \tag{2.114}$$

is given in detail in Appendix B.3. At $\omega = 0$ the point proton amplitude has the usual Thomson scattering limit:

$$\mathcal{F}_{\text{point}}(\omega = 0) = -\frac{\alpha}{M}\vec{\varepsilon}\cdot\vec{\varepsilon}\,'^{*}. \tag{2.115}$$

The intrinsic nucleon structure introduces characteristic deviations from the point-like amplitude which first appear at $\mathcal{O}(\omega^2)$. To this order the amplitude (2.91) becomes:

$$\mathcal{F} = \mathcal{F}_{\text{point}} + \alpha_E\,\omega\omega'\,\vec{\varepsilon}\cdot\vec{\varepsilon}\,'^{*} + \beta_M\,(\vec{\varepsilon}\times\vec{k})\cdot(\vec{\varepsilon}\,'^{*}\times\vec{k}') + \mathcal{O}(\omega^3). \tag{2.116}$$

Note that $\vec{\varepsilon}$ and $\vec{\varepsilon}\,'^{*}$ are the directions of the in- and outgoing electric fields. Likewise, $\vec{\varepsilon}\times\vec{k}$ and $\vec{\varepsilon}\,'^{*}\times\vec{k}'$ are in the directions of the in- and outgoing magnetic fields. The polarizabilities α_E and β_M measure the response of the nucleon to those fields.

The difference between the full Compton cross section and the point cross section to $\mathcal{O}(\omega^2)$ is now readily evaluated. Taking the sum and average over the final and initial photon

[4] We follow the convention frequently used in the literature which gives (somewhat awkwardly) a negative value for γ.

polarizations one finds:

$$\frac{d\sigma}{d\Omega} - \left(\frac{d\sigma}{d\Omega}\right)_{\text{point}} = -\frac{\alpha}{M}\left(\frac{\omega'}{\omega}\right)^2 \omega\omega' \left[\frac{\alpha_E + \beta_M}{2}(1+\cos\theta)^2 + \frac{\alpha_E - \beta_M}{2}(1-\cos\theta)^2\right]. \quad (2.117)$$

We see that forward scattering determines the sum of the polarizabilities while large angle scattering gives their difference.

Several measurements of $d\sigma/d\Omega$ for low energy Compton scattering on the proton have been performed to determine the difference between the electric and magnetic proton polarizabilities [Fed 91, Zi 92, Hal 93, MacG 95]. An example is shown in Fig. 2.24. This analysis gives $\alpha_E - \beta_M \simeq 10^{-3}$ fm^3, with an uncertainty of about 10 percent. If we adopt the sum rule (2.107), this permits one to extract the polarizabilities separately, with the following result:

$$\alpha_E^p = (12.1 \pm 1.3) \cdot 10^{-4}\,\text{fm}^3, \quad (2.118)$$

$$\beta_M^p = (2.1 \mp 1.3) \cdot 10^{-4}\,\text{fm}^3. \quad (2.119)$$

The uncertainties quoted include both statistical and systematic errors. They are anticorrelated because of the constraint from (2.107). The striking feature is the smallness of the magnetic polarizability in comparison with the electric one. At first sight this appears counter-intuitive, given the strong magnetic excitation that connects the nucleon and the $\Delta(1232)$. Clearly, there is interesting nucleon structure information in these polarizabilities.

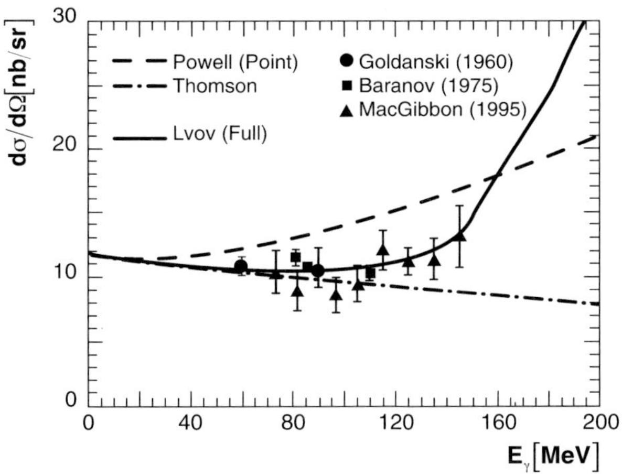

Figure 2.24: Differential cross section for Compton scattering on the proton as a function of photon energy $E_\gamma = \omega$ at $\theta = 90^0$ in the laboratory frame. The data are taken from [MacG 95]. The curves refer to the pure Thomson cross section (dash-dotted), the point (Powell) cross section (dashed) and a full calculation (solid line) including the proton polarizabilities [LPS 97].

An estimate of the proton (forward) spin polarizability γ^p can be made using the data, Fig. 2.22, to determine the integral (2.113) which gives:

$$\gamma^p \simeq -10^{-4}\,\text{fm}^4. \quad (2.120)$$

This number emerges from a subtle compensation between a large negative contribution from the Δ resonance and a sizeable positive contribution from the low energy electric dipole amplitude. Such cancellations are found, for example, in a calculation [Ede 98] which combines Δ excitation and the scattering from the pion cloud of the nucleon (see Chapter 7). The resulting forward spin polarizability is close to the empirical one. An evaluation using dispersion relation techniques [Dr 00] finds $\gamma^p \simeq -0.8 \cdot 10^{-4}\,\mathrm{fm}^4$.

In general, the polarized elastic photon-proton scattering is described by six helicity amplitudes which imply six polarizability parameters [Hem 98]. Three of those are the electric, magnetic and (forward) spin polarizabilities already encountered. The forward spin polarizability is actually a combination, $\gamma \equiv \gamma_0 = \gamma_1 - \gamma_2 - 2\,\gamma_4$, of polarizability parameters related to three of the helicity amplitudes. Additional information about the spin structure of the nucleon can be obtained from polarized Compton scattering at backward angles. It involves the (backward) spin polarizability $\gamma_\pi = \gamma_1 + \gamma_2 + 2\,\gamma_4$ which enters the backward scattering amplitude

$$\mathcal{F}(\theta = \pi) = \mathcal{F}_{\mathrm{Born}} + (\alpha_E - \beta_M)\omega\omega'\,\vec{\varepsilon}\cdot\vec{\varepsilon}\,'^* - i\gamma_\pi\,\omega^3\vec{\sigma}\cdot(\vec{\varepsilon}\times\vec{\varepsilon}\,'^*) + \mathcal{O}(\omega^4). \quad (2.121)$$

An extraction of the backward spin polarizability of the proton, as performed in [Gal 00], gives

$$\gamma_\pi^p = -(36.8 \pm 2.4)\cdot 10^{-4}\,\mathrm{fm}^4, \quad (2.122)$$

more than an order of magnitude larger than the forward spin polarizability. A value $\gamma_\pi^p \simeq -4.4 \cdot 10^{-3}\,\mathrm{fm}^4$ is expected [Hol 98] from the presence of the neutral pion pole in this process, and corrections bring it close to the observed number.

Neutron polarizabilities

Neutron polarizabilities can be indirectly determined from the scattering of very low energy neutrons in the Coulomb field of a heavy nucleus. The Coulomb field polarizes the neutron. Near the surface of a nucleus with charge Ze the interaction potential caused by a polarized neutron is $V = -\alpha_E^n Z^2 e^2/2r^4$. This translates into a corresponding term of the neutron-atom scattering amplitude. A method to extract the neutron electric polarizability, α_E^n, is to perform a precision measurement of the total neutron-nucleus cross section as a function of the incoming neutron momentum p_n. In an expansion of the cross section in powers of p_n the linear term depends only on α_E^n. The detailed analysis for neutron energies ranging between 50 eV and about 50 keV yields [Ros 90, Sch 91]:

$$\alpha_E^n = (12.6 \pm 3.5)\cdot 10^{-4}\,\mathrm{fm}^3. \quad (2.123)$$

Together with the sum rule (2.109) the deduced magnetic polarizability is

$$\beta_M^n = (3.2 \mp 3.5)\cdot 10^{-4}\,\mathrm{fm}^3. \quad (2.124)$$

Within the given errors, the neutron and proton polarizabilities are roughly equal. However, the quoted uncertainties may be considerably too low [Koe 95].

Discussion

We now wish to have a first look at the polarizabilities with the aim of finding a simple order-of-magnitude understanding. We begin with the electric polarizability and discuss the case of a static electric field. The energy shift induced by this field is proportional to the polarizability for which the quantum mechanical expression is:

$$4\,\pi\,\alpha_E = 2 \sum_{I \neq N} \frac{|\langle I|d_z|N\rangle|^2}{E_I - E_N}, \tag{2.125}$$

where $\vec{d} = e\vec{r}$ is the electric dipole operator, $|N\rangle$ is the nucleon ground state, $|I\rangle$ is any excited state that can be reached by an E1 transition and $\Delta E = E_I - E_N$ is the corresponding excitation energy. Nucleon resonance (N^*) states with spin and parity $J^\pi = \frac{1}{2}^-, \frac{3}{2}^-$ fall into this category. A number of such N^* resonances exist in the range of excitation energies $\Delta E \sim (600 - 800)$ MeV. If we use closure in Eq. (2.125) with $\Delta E \simeq 600$ MeV and the empirical proton charge radius, we find:

$$\alpha_E^p \simeq \frac{2}{3}\frac{\alpha}{\Delta E}\langle r_E^2\rangle_p \sim 10^{-3}\,\text{fm}^3, \tag{2.126}$$

which is roughly the correct order of magnitude. A more detailed estimate must also take into account the pion photoproduction continuum observed in the electric dipole cross section at lower energy.

The discussion of the magnetic polarizability is more subtle. At first sight one expects that the magnetic response should be dominated by the strong M1 transition from the nucleon ground state to the $\Delta(1232)$, spin-$\frac{3}{2}$, isobar. This should give a large paramagnetic (spin-flip) susceptibility

$$4\,\pi\,\beta_\Delta = 2\,\frac{|\langle \Delta|m_z|N\rangle|^2}{M_\Delta - M}, \tag{2.127}$$

where $\vec{m}$ is the magnetic dipole operator. Its $N \to \Delta$ transition matrix element can be expressed [EW 88] in terms of the isovector magnetic moment, $\mu_V = (\mu_p - \mu_n)/2 = 2.353$. A straightforward estimate leads to

$$\beta_\Delta \simeq \frac{\alpha}{M_\Delta - M}\left(\frac{\mu_V}{M}\right)^2 \sim 1.2 \cdot 10^{-3}\,\text{fm}^3. \tag{2.128}$$

This large paramagnetic term is in obvious disagreement with the empirical β_M's. There must be a substantial diamagnetic counterpart of opposite sign in order to be compatible with the small observed magnetic polarizabilities. We shall see in Chapter 7 that this is indeed the case when we enter a detailed discussion of the pionic structure of the nucleon and its relation to the approximate chiral symmetry of the strong interaction (see Section 7.3.7).

2.4.4 Virtual Compton Scattering

Virtual Compton scattering (VCS) is a new and potentially powerful probe of nucleon structure. The concept of polarizabilities is now generalized to virtual photons. The prototype process is electron scattering on a proton with an extra photon in the final state: $e + p \to e' + p + \gamma$. In essence this corresponds to the scattering of an initial virtual photon (with four-momentum $q^\mu = (q^0, \vec{q})$ determined by the in- and outgoing electron momenta) on a proton, $\gamma^* + p \to \gamma + p$, leading to a real photon (with four-momentum k^μ) in the final state, as illustrated in Fig. 2.25.

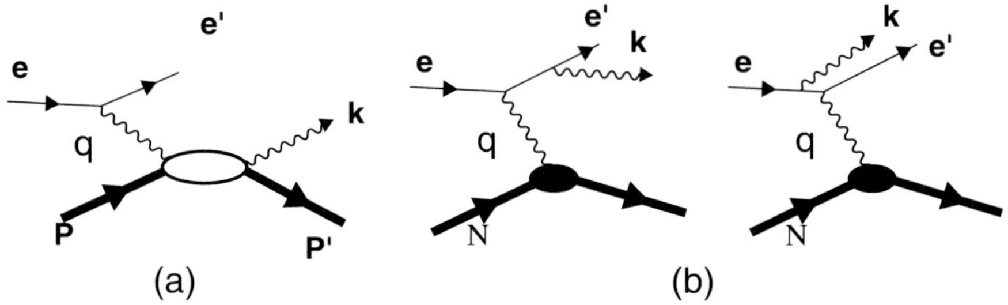

Figure 2.25: Sketch of the $e + p \to e' + p + \gamma$ reaction. Diagram (a) represents the genuine virtual Compton scattering (VCS) process; diagrams (b) show the Bethe-Heitler background reaction.

In practice the VCS subprocess of interest (Fig. 2.25(a)) must be separated from the Bethe-Heitler (BH) background reactions (Fig. 2.25(b)) in which the real photon is radiated from the scattered electron. The BH amplitude can be calculated using the known proton form factors. The difference between the total $e+p \to e' + p + \gamma$ amplitude and the BH amplitude isolates the VCS signal:

$$T_{\text{VCS}} = T_{\text{tot}}(e + p \to e' + p + \gamma) - T_{\text{BH}} \tag{2.129}$$

The VCS amplitude itself has the form of a generalized Compton amplitude:

$$T_{\text{VCS}} = e^3 \varepsilon_\mu'^* \eta_\nu T^{\mu\nu}(k, P'; q, P), \tag{2.130}$$

where now

$$\eta_\nu = \frac{\bar{u}_{e'} \gamma_\nu u_e}{q^2}, \tag{2.131}$$

refers to the virtual photon field produced by the source current of the scattered electron and replaces the polarization vector ε^ν in the standard Compton amplitude (2.91). Separating the vector η^ν into its transverse and longitudinal components relative to the $\vec{q}$ - axis, the VCS amplitude takes the following form:

$$T_{\text{VCS}} = i\left(\vec{\eta}_T \cdot \vec{\mathcal{M}}_T + \frac{q^2}{q_0^2}\eta_L \mathcal{M}_L\right). \tag{2.132}$$

One can then identify twelve structure functions A_i, eight transverse and four longitudinal, associated with the set of tensors that can be formed by combining nucleon spin, photon polarization and momentum vectors:

$$
\begin{aligned}
\vec{\eta}_T \cdot \vec{\mathcal{M}}_T = & \, A_1 \, \vec{\varepsilon}^{\,\prime *} \cdot \vec{\eta}_T + A_2 \, \vec{\varepsilon}^{\,\prime *} \cdot \hat{q} \, \vec{\eta}_T \cdot \hat{k} \\
& + iA_3 \, \vec{\sigma} \cdot \vec{\varepsilon}^{\,\prime *} \times \vec{\eta}_T + iA_4 \, \vec{\sigma} \cdot \hat{k} \times \hat{q} \, \vec{\varepsilon}^{\,\prime *} \cdot \vec{\eta}_T \\
& + iA_5 \, \vec{\sigma} \cdot \vec{\varepsilon}^{\,\prime *} \times \hat{q} \, \vec{\eta}_T \cdot \hat{k} - iA_6 \, \vec{\sigma} \cdot \vec{\varepsilon}^{\,\prime *} \times \hat{k} \, \vec{\eta}_T \cdot \hat{k} \\
& - iA_7 \, \vec{\sigma} \cdot \vec{\eta}_T \times \hat{k} \, \vec{\varepsilon}^{\,\prime *} \cdot \hat{q} - iA_8 \, \vec{\sigma} \cdot \vec{\eta}_T \times \hat{q} \, \vec{\varepsilon}^{\,\prime *} \cdot \hat{q}, \\
\mathcal{M}_L = & \, A_9 \, \vec{\varepsilon}^{\,\prime *} \cdot \hat{q} + iA_{10} \, \vec{\sigma} \cdot \hat{k} \times \hat{q} \, \vec{\varepsilon}^{\,\prime *} \cdot \hat{q} \\
& + iA_{11} \, \vec{\sigma} \cdot \vec{\varepsilon}^{\,\prime *} \times \hat{q} + iA_{12} \, \vec{\sigma} \cdot \vec{\varepsilon}^{\,\prime *} \times \hat{k}.
\end{aligned}
\tag{2.133}
$$

More detailed and complete expressions are given in the review [Gui 98]. Further analysis using crossing symmetry and charge conjugation invariance reveals that actually only six of these twelve structure functions are independent [Dr 97].

The low energy expansion of the VCS amplitude can be expressed in terms of six generalized polarizabilities (GP). Following Guichon et al. [Gui 96] these GP's are denoted by $P(\rho' L', \rho L; S)$ where ρL and $\rho' L'$ refer to the multipolarities of the in- and outgoing photons (e.g. E1, M1 for electric or magnetic dipole etc.), and $S = 0, 1$ is the total spin that can be formed by combining the spins of the initial and final nucleon. The GP's are defined such that, for real photons, they lead back to the polarizabilities we have already encountered. For example, $P(E1, E1; S = 0)$ is proportional to α_E, $P(M1, M1; S = 0)$ to β_M at low energy and in the limit $q^2 = 0$, when the incoming photon becomes real. Other quantities, such as $P(E1, M2; S = 1)$ or $P(M1, E2; S = 1)$, are proportional to certain combinations of vector (or spin) polarizabilities in the limiting case of real photons.

Evidently, a complete knowledge of the GP's amounts to the maximum information one can extract concerning low-energy nucleon structure using electromagnetic probes. In the region below pion production threshold, through the process illustrated in Fig. 2.25(a), one probes the amplitudes for the virtual excited states of the nucleon. Mapping out these polarizabilities as a function of q^2 is of particular interest since detailed model calculations differ in their predictions of this q^2 dependence, whereas they are often indistinguishable at $q^2 = 0$.

Experiments at MAMI (Mainz), BATES (MIT) and at the Jefferson National Laboratory are now aimed at systematically exploring virtual Compton scattering. To illustrate the level of accuracy that is being reached, we show in Fig. 2.26 an example of multi-differential cross section data taken at Mainz [d'Ho 98, Ro 00]. The exclusive $e + p \to e' + p' + \gamma$ reaction is determined by the following independent kinematical variables: the incoming virtual photon four-momentum and its degree of longitudinal polarization is set by the kinematics of the incident and scattered electron; the final state kinematics is then determined by the measurement of $\theta_{\gamma\gamma}$, the angle between the incoming virtual and outgoing real photon, once the azimuthal angle ϕ between the electron and hadron scattering planes is fixed. The measured cross section is analyzed first by using the Bethe-Heitler and nucleon Born term amplitudes only. The difference between the precise data and this analysis points to the polarizability effects for which one is looking.

In practice, such an analysis proceeds as follows. Subtracting the known Bethe-Heitler plus Born cross section from the measured cross section data, one can extract, for unpolarized

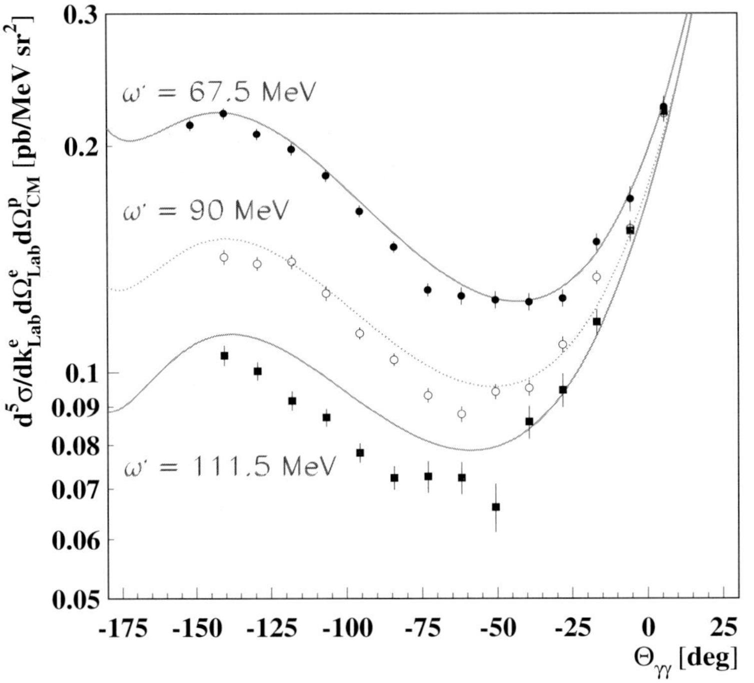

Figure 2.26: Examples of differential $e + p \rightarrow e' + p' + \gamma$ cross sections using incident virtual photons with $Q^2 = 0.33$ GeV^2, $|\vec{q}| = 600$ MeV. The outgoing real photons are detected at different energies, $\omega = |\vec{k}|$, as a function of the angle $\theta_{\gamma\gamma}$ between the in- and outgoing photon directions. The electron and hadron scattering planes are chosen to coincide. The curves are the result of a calculation using the Bethe-Heitler and nucleon Born term amplitudes only [d'Ho 98, Ro 00].

VCS, the difference between the full and BH+Born amplitudes,

$$\begin{aligned} \mathcal{M}_0 - \mathcal{M}_0^{\text{BH+Born}} &= v_{LL}(\theta_{\gamma\gamma}, \phi, \varepsilon) \left(P_{LL}(q^2) - \frac{P_{TT}(q^2)}{\varepsilon} \right) \\ &\quad + v_{LT}(\theta_{\gamma\gamma}, \phi, \varepsilon)\, P_{LT}(q^2). \end{aligned} \tag{2.134}$$

The known functions v_{LL} and v_{LT} depend only on kinematic variables, including the polarization parameter ε. The P_{LL}, P_{TT} and P_{LT} involve known electromagnetic form factors and linear combinations of the unknown generalized polarizabilities. By plotting $(\mathcal{M}_0 - \mathcal{M}_0^{BH+Born})/v_{LT}$ as a function of $\xi = v_{LL}/v_{LT}$, the intercept at $\xi = 0$ gives P_{LT}, while the slope determines $P_{LL} - P_{TT}/\varepsilon$.

To give an impression of what one can expect, we present in Table 2.2 some results from the first measurements at Mainz, together with theoretical results using different approaches: a chiral perturbation theory calculation (see Chapter 7) which emphasizes the role of the pion cloud of the nucleon, and a non-relativistic constituent quark model calculation (see Chap-

Table 2.2: VCS observables involving generalized polarizabilities at $|q^2| = 0.33$ GeV^2. Data from MAMI (Mainz) [d'Ho 99]. Theoretical results: chiral perturbation theory [Hem 97] and non-relativistic quark model [Liu 96, Ro 00].

	$P_{LL}(q^2) - P_{TT}(q^2)/\varepsilon$ [GeV^2]	$P_{LT}(q^2)$ [GeV^2]
Exp.	$23.7 \pm 2.2 \pm 4.9$	$-5.0 \pm 0.8 \pm 2.5$
Chiral Pert. Th.	26.0	-5.3
Quark model	11.1	-3.5

ter 8). Given sufficiently accurate data, detailed VCS measurements do have the potential to discriminate between different models of the nucleon.

3 Weak Probes of Nucleon Structure

With a strength a billion times smaller than the electromagnetic interaction it is not surprising that the weak interaction was not used to investigate nucleon structure until much later than electron scattering. Of course, nuclear β-decay is one of the classical forms of radioactivity and has been widely studied since 1934 [CJ 34]. The first evidence for maximal parity violation in the charged current weak interaction came from β-decay. However the limited momentum transfer in this process means that we essentially determine the appropriate form factor at zero momentum transfer. Unfortunately, when it comes to the study of the nucleon itself, the neutron is unstable and, being uncharged, difficult to handle experimentally. Thus, while neutron decay does give us the best determination of the axial charge of the nucleon, for the weak vector coupling one still needs to rely on carefully chosen nuclear transitions.

The pioneering work of Reines and Cowan [RC 53] on inverse β-decay was eventually followed by the first observations of muon neutrino charged current interactions at Brookhaven National Laboratory (BNL) in 1962 [Dan 62]. The essential scattering processes involved in these early neutrino scattering experiments were

$$\nu_\mu + n \quad \rightarrow \quad \mu^- + p, \tag{3.1}$$

$$\bar{\nu}_\mu + p \quad \rightarrow \quad \mu^+ + n. \tag{3.2}$$

The 12-foot bubble chamber at the Zero Gradient Synchrotron (ZGS) at Argonne National Laboratory (ANL) was crucial to the determination of accurate form factor data for the nucleon because it could be filled with hydrogen or deuterium [Man 74]. In the 1970's the neutral current was also exploited through the challenging (ν, ν') reaction. With the development of new, high quality, intense beams of electrons it has also become feasible to use parity violation as a tool to extract the matrix elements of the weak neutral current.

To conclude this brief introduction we note that some components of the weak matrix elements can only be determined in rather specialized experiments. For example, to obtain direct information on the induced pseudoscalar form factor it has been necessary to use reactions such as $(e, e'\pi^+)$ and radiative muon capture.

3.1 Matrix Elements of the Weak Currents

In Chapter 1 we presented the structure of the quark-level weak currents in some detail. It is clear from Eqs. (1.20) and (1.23) that the nucleon matrix elements of the weak currents can always be expressed as linear combinations of the vector and axial vector matrix elements

$$V_a^\mu(x) = \bar{\psi}(x)\gamma^\mu \frac{\lambda_a}{2}\psi(x), \tag{3.3}$$

$$A_a^\mu(x) = \bar{\psi}(x)\gamma^\mu\gamma_5 \frac{\lambda_a}{2}\psi(x), \tag{3.4}$$

with flavor $SU(3)$ matrices λ_a as given in Appendix A.2. For example, the most widely studied weak matrix element is the charged current matrix element connecting the neutron and the proton

$$\langle p(P')|J_{\mathrm{CC}}^\mu(0)|n(P)\rangle = \cos\theta_C \langle p(P')|V_+^\mu(0) - A_+^\mu(0)|n(P)\rangle, \tag{3.5}$$

where $V_\pm = V_1 \pm iV_2$ and $A_\pm = A_1 \pm iA_2$. Similarly, the matrix element of the weak neutral current can be expressed in terms of $V_{0,3,8}^\mu$ and $A_{0,3,8}^\mu$, where the subscript 0 indicates the flavor singlet operator with the 3×3 unit matrix replacing $\lambda_a/2$ in Eqs. (3.3) and (3.4).

The unity of the weak and electromagnetic interactions of the nucleon is now apparent. We have already studied the vector matrix elements $\langle N(P')|V_3^\mu(0)|N(P)\rangle$ and $\langle N(P')|V_8^\mu(0)|N(P)\rangle$ in Chapter 2. To the extent that charge independence is a good symmetry for the nucleon, one expects that the matrix elements of $V_\pm^\mu$ and V_3^μ are essentially the same – being related by an isospin rotation. The hypothesis that this is exactly so is known as the conserved vector current hypothesis, CVC [GZ 56]. In practice it is usual to assume CVC and to take the vector matrix elements from the electromagnetic form factors. The weak interaction may then be used to study the axial form factors which are not directly accessible in any other way.

3.2 Axial Matrix Elements in Charged Current Neutrino Reactions

On the basis of Lorentz covariance and parity the most general form of the neutron to proton axial matrix element is

$$\langle p(P')|A_+^\mu(0)|n(P)\rangle = \bar{u}_p(P')\left[G_A(Q^2)\gamma^\mu\gamma_5 + i\frac{G_T(Q^2)}{2M}\sigma^{\mu\nu}\gamma_5 q_\nu + \frac{G_P(Q^2)}{2M}\gamma_5 q^\mu\right]u_n(P), \tag{3.6}$$

where the form factors G_A, G_T and G_P are real because of time-reversal invariance. The nucleon mass M is introduced for dimensional convenience. (Note that we have chosen the momentum transfer q to be the momentum transferred from the leptons to the target nucleon, $q \equiv P' - P$ and $Q^2 = -q^2$.) The axial tensor term, G_T, is often omitted because it has the opposite behavior under charge symmetry from the other two terms.

The relevance of charge symmetry is that it turns protons into neutrons (and vice-versa) through a rotation by 180° about the y-axis in isospace (i.e., $\mathcal{C} = e^{i\pi I_2}$). Thus $\mathcal{C}A_+^\mu\mathcal{C}^{-1}$ describes the same transition as $A_+^{\mu\dagger}$. Indeed, if G_T is zero one finds

$$\mathcal{C}A_+^\mu\mathcal{C}^{-1} = -A_+^{\mu\dagger}. \tag{3.7}$$

This simple relationship is destroyed if $G_T \neq 0$ because it transforms with the opposite sign. Such a transformation defines a second-class current and we shall ignore second-class terms from now on. Nevertheless, we emphasize the importance of those experiments which place accurate experimental limits on G_T [Wil 00].

The first of the structure-dependent form factors in Eq. (3.6), $G_A(Q^2)$, is known as the axial form factor and it has been studied in considerable detail. While the second (non-zero) form factor, $G_P(Q^2)$, which is known as the induced pseudoscalar form factor, is much harder to study – being negligible at the momentum transfer of β-decay – it is nevertheless very important.

The concept of chiral symmetry, which will play an important role in the later theoretical chapters (e.g., guiding the construction of many models of nucleon structure) began with the success of the PCAC (Partially Conserved Axial Current) hypothesis of Gell-Mann and Lévy and Nambu [GL 60, Na 60]. In practical terms this means that the matrix elements of $\partial_\mu A_i^\mu$ with $i = 1, 2, 3$ should be proportional to the squared pion mass, m_π^2, and so vanish in the chiral limit ($m_\pi^2 \to 0$ as the quark masses $m_q \to 0$). This is clearly not true for the first term in Eq. (3.6), but if the induced pseudoscalar term has a pion pole (with $Q^2 = -q^2$ and f_π = 92.4 MeV the pion decay constant)

$$G_P(Q^2) \to \frac{4\,M f_\pi g_{\pi NN}(Q^2)}{Q^2 + m_\pi^2}, \tag{3.8}$$

as illustrated in Fig. 3.1, we find[1]

$$q_\mu \langle p(P')|A_+^\mu(0)|n(P)\rangle = 2\bar{u}_p(P') \left[M G_A(Q^2) - \frac{Q^2 f_\pi g_{\pi NN}(Q^2)}{Q^2 + m_\pi^2} \right] \gamma_5 u_n(P). \tag{3.9}$$

Thus the matrix element of the divergence of the axial current vanishes as $m_\pi^2 \to 0$ if $G_A(Q^2)$ and the pion-nucleon form factor $g_{\pi NN}(Q^2)$ are related by

$$M G_A(Q^2) = f_\pi g_{\pi NN}(Q^2). \tag{3.10}$$

At $Q^2 = 0$ this is known as the Goldberger-Treiman relation and it is satisfied at the level of 3 % (with $g_A = G_A(0) = 1.267 \pm 0.004$, $g_{\pi NN} = g_{\pi NN}(0) = 13.2 \pm 0.1$ and $M =$ 0.939 GeV).

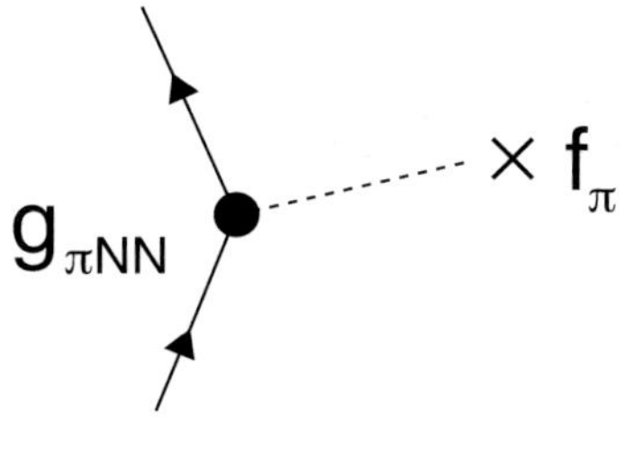

Figure 3.1: Pion pole model for the induced pseudoscalar form factor.

[1] We neglect a small non-pole correction given by Adler and Dothan [AD 66]. See also [BHM 01]

In many applications, where it is not possible to explicitly separate the axial and induced pseudoscalar form factors, it is common to incorporate Eqs. (3.8) and (3.10) into Eq. (3.6), so that

$$\langle p(P')|A_{+}^{\mu}(0)|n(P)\rangle \approx \bar{u}_p(P')\left[G_A(Q^2)\gamma^{\mu}\gamma_5 + \frac{2MG_A(Q^2)}{Q^2+m_{\pi}^2}\gamma_5 q^{\mu}\right]u_n(P). \quad (3.11)$$

Note that the second term is small for any process involving nucleons because $\gamma_5 q^{\mu}$ picks up the lower component of the nucleon spinor and hence a factor of the momentum over twice the nucleon mass. On the other hand, the first term reduces to the nucleon spin operator, $\vec{\sigma}$, in the non-relativistic limit.

3.2.1 Charged Current Neutrino Scattering

Given the hadronic matrix elements of the weak current, it is straightforward to calculate the differential cross section for the processes $n(\nu_{\mu},\mu^{-})p$ and $p(\bar{\nu}_{\mu},\mu^{+})n$. Once again a deuteron target (typically a liquid deuterium bubble chamber) must be used for the former, because of the lack of a free neutron target. However, by using the resolving power of the bubble chamber to tag a low momentum, spectator proton

$$\nu_{\mu} + d \rightarrow \mu^{-} + p + p_{\text{spectator}}, \quad (3.12)$$

one can reduce the experimental uncertainties.

To simplify the final cross section we introduce the following notation. The usual Lorentz invariants s and u are

$$s = (p_{\nu} + P)^2; \qquad u = (p_{\nu} - P')^2, \quad (3.13)$$

where $p_{\nu} = (E, \vec{p}_{\nu})$ is the momentum of the ν (or $\bar{\nu}$) and $P(P')$ the momentum of the initial (final) nucleon. In particular, we need the combination

$$s - u = 4\,ME - Q^2 - m^2, \quad (3.14)$$

with m the mass of the final lepton (which can usually be ignored). It is also convenient to define

$$\tau = \frac{Q^2}{4\,M^2}; \qquad \tilde{G}_A = G_A - \tau\,G_P. \quad (3.15)$$

Then we find that the laboratory differential cross section for an (anti-) neutrino of energy E is [Lle 72]:

$$\frac{d\sigma_{\nu(\bar{\nu})}}{dQ^2} = \frac{M^2 G_F^2 \cos^2\theta_C}{8\,\pi E^2}\left(\frac{m_W^2}{m_W^2+Q^2}\right)^2\left[\mathcal{A} \pm \frac{s-u}{M^2}\mathcal{B} + \frac{(s-u)^2}{M^4}\mathcal{C}\right], \quad (3.16)$$

with

$$\begin{aligned}\mathcal{A} = {} & 4\,\tau\Big[(1+\tau)G_A^2 - 4\,(G_E^V)^2 + 4\,\tau(G_M^V)^2\Big] \\ & + \frac{m^2}{M^2}\left[\tilde{G}_A^2 - 4\,(G_E^V)^2\right] \\ & + \frac{m^4}{M^4}\left[\frac{\tau}{4}\,G_P^2 - (G_M^V)^2 - \frac{1}{4}\,G_A^2 - \frac{1}{2}\,G_P G_A\right],\end{aligned} \quad (3.17)$$

$$\mathcal{B} = 8\,\tau\, G_A\, G_M^V, \tag{3.18}$$

and

$$\mathcal{C} = \frac{1}{4}\, G_A^2 + (F_1^V)^2 + \tau\,(F_2^V)^2. \tag{3.19}$$

Note that the vector form factors are, of course, the isovector combinations defined in Eqs. (2.20) and (2.23). In addition, the last two terms in Eq. (3.17) are usually discarded in the experimental analyses as they are suppressed by powers of $m^2/M^2 \approx 1\,\%$.

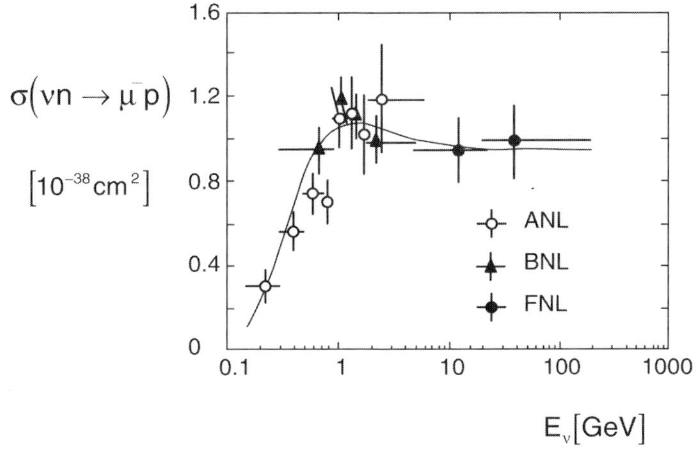

Figure 3.2: Energy dependence of the total cross section for quasi-elastic, charged-current neutrino scattering [Kit 83].

Table 3.1: Dipole mass parameter M_A (in GeV) in the axial form factor extracted from ν_μ quasi-elastic scattering.

Exp.	M_A [GeV]
BNL	1.07 ± 0.06
ZGS	1.00 ± 0.05
Fermilab	$1.05\ ^{+0.12}_{-0.16}$
Average:	1.03 ± 0.04

Figure 3.2 summarizes the total cross section data from the three major groups working on the problem in the 80's at Argonne, Brookhaven and Fermilab [Kit 83]. All of the analyses of the data take the vector form factors from electron scattering data and employ the approximation (3.11) to replace G_P by G_A. As the former contributes only a few percent of the cross section this is a very good approximation. Finally, the data are not sufficiently accurate to

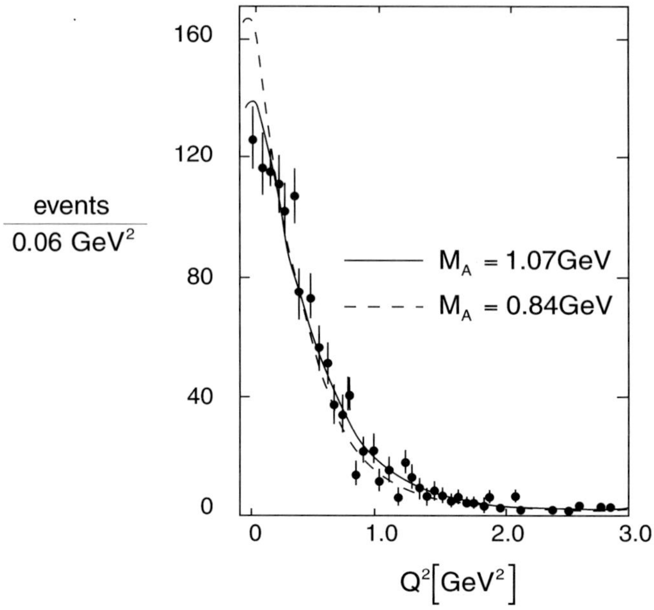

Figure 3.3: Differential cross section for quasi-elastic neutrino scattering on a neutron in the deuteron, showing the sensitivity to M_A [Bak 81].

determine the detailed shape of $G_A(Q^2)$. It is usually parametrized as a dipole form

$$G_A(Q^2) = \frac{G_A(0)}{\left(1 + \frac{Q^2}{M_A^2}\right)^2}, \tag{3.20}$$

by analogy with the vector form factors. Attempts by Miller *et al.* [Mil 82] to fit with either a monopole or tripole form led to fits that were worse by 1.5 standard deviations. With $G_A(0)$ fixed at 1.26 from neutron β-decay (as explained below) the data is then analyzed for M_A with remarkable consistency. Table 3.1 summarizes the values of M_A obtained by the three groups. They are very near to each other, considering the complexity of these experiments, and the combined average determines M_A to within about 3 %. Figure 3.3 shows a typical fit to the quasi-elastic differential cross section data which led to the determination of M_A.

It is possible to relax the CVC constraint on the vector form factors and search simultaneously for the best-fit dipole masses M_A *and* M_V. The most accurate values (all given in GeV) from Baker *et al.* ($M_V = 0.86 \pm 0.07, M_A = 1.04 \pm 0.14$) and Miller *et al.* ($M_V = 0.96 \pm 0.04$, $M_A = 0.80 \pm 0.10$) are consistent with CVC in that M_V measured in electron scattering is 0.84 GeV. There is no evidence for a dependence of M_A on the neutrino energy between 0.5 and 200 GeV. Nor is there any evidence for a second-class current [Ahr 88].

3.2.2 Inelastic Charged Current Interactions

We already know, from the discussion of inelastic electromagnetic interactions, that there is a wealth of information to be had from inelastic scattering. Unfortunately, the difficulty of using the weak interaction as a probe of hadron structure is the low event rates and these make a systematic study of inelastic scattering very hard. The only final state studied in any detail is the production of the $\Delta(1232)$ resonance, e.g.,

$$\nu_\mu p \to \mu^- \Delta^{++} \to \mu^- \pi^+ p. \tag{3.21}$$

In Fig. 3.4 we show the total cross section for neutrino inelastic scattering as a function of the incident neutrino energy, for two cuts on the invariant mass (W) of the final πN pair. The choice of 1.4 GeV is usually used to define Δ production. A complete formalism for

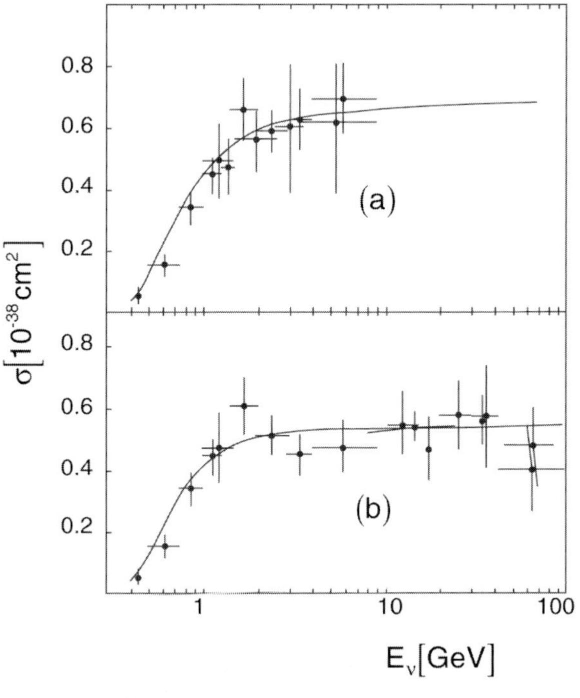

Figure 3.4: The total cross section for inelastic neutrino scattering, $\nu_\mu p \to \mu^- p\pi^+$, for two choices of the invariant mass of the πN pair: a) below 1.6 GeV and b) below 1.4 GeV [Fog 79].

dealing with inelastic processes involving neutrinos, massive leptons and photons, in terms of multipole amplitudes was first set down by Adler [Ad 68]. He also worked through the various constraints imposed by PCAC, including the famous relation between the $Q^2 \to 0$ limit of the inelastic cross section and the off-shell cross section for πN scattering, $\sigma_{\rm off}$ (in which the initial pion has invariant mass $-Q^2$) [Ad 64]

$$\left.\frac{d\sigma^{\nu,\bar\nu}}{dQ^2 dW}\right|_{Q^2\to 0} = \frac{G_F^2 \cos^2\theta_C f_\pi^2 \epsilon'}{2\pi^2 |\vec{k}\,|\epsilon} \sigma_{\rm off}\,(\pi^\pm p \to \pi^\pm p). \tag{3.22}$$

Here ϵ and ϵ' are the initial and final lepton energies (in the laboratory) for the reactions $\nu(\bar{\nu})p \to \mu^{\mp} p \pi^{\pm}$ and $|\vec{k}\,|$ is the 3-momentum transfer in the πN center of mass system.

However, if we restrict our consideration to the production of the $\Delta(1232)$, the connection with models of baryon structure is simplest through the form factors in the Rarita-Schwinger formalism [Lle 72]. We may write the matrix element for Δ production as

$$\langle \mu^- \Delta^{++} | H_W | \nu_\mu p \rangle \;=\; \frac{G_F \cos\theta_C}{\sqrt{2}} j_\mu \, \langle \Delta^{++} | V_+^\mu - A_+^\mu | p \rangle, \tag{3.23}$$

with j_μ the matrix element of the weak leptonic current. Assuming CVC, the vector matrix element can be related directly to the matrix element of the isovector component of the photon (or pion electroproduction) matrix element

$$\langle \Delta^{++} | V_+^\mu | p \rangle \;=\; \sqrt{3} \langle \Delta^+ | V_{\text{EM}}^\mu | p \rangle. \tag{3.24}$$

To be explicit, assuming CVC we have the vector matrix element

$$\langle \Delta^+ | V_{\text{EM}}^\mu | p \rangle j_\mu \;=\bar{u}_\nu(P') \left[\frac{C_3^V}{M} \gamma_\mu + \frac{C_4^V}{M^2} P'_\mu + \frac{C_5^V}{M^2} P_\mu \right] \gamma_5 u(P) F^{\mu\nu} f_{\text{BW}}(s_{\pi N}), \tag{3.25}$$

while the axial matrix element is:

$$\begin{aligned} \langle \Delta^+ | A_3^\mu | p \rangle j_\mu &= \bar{u}_\nu(P') \left[\frac{C_3^A}{M} \gamma_\mu + \frac{C_4^A}{M^2} P'_\mu \right] u(P) F^{\mu\nu} \\ &\quad + \left(C_5^A j^\nu + \frac{C_6^A}{M^2} q^\nu q_\mu j^\mu \right) \bar{u}_\nu(P') u(P) f_{\text{BW}}(s_{\pi N}) \end{aligned} \tag{3.26}$$

Here $u_\nu(P')$ is the Rarita-Schwinger spinor for the final Δ, $F_{\mu\nu} = q_\mu j_\nu - q_\nu j_\mu$ and $f_{\text{BW}}(s_{\pi N})$ is a Breit-Wigner distribution written in terms of the invariant mass of the πN system ($\sqrt{s_{\pi N}}$).

As we have seen in Chapter 2, for the $\Delta(1232)$ the dominant vector transition is a magnetic dipole, M1, in which case C_5^V is zero and $C_4^V(Q^2) = -(M/\sqrt{s_{\pi N}}) C_3^V(Q^2)$. Thus there is only one vector form factor to consider. The electroproduction data in the Δ-region is well fitted with

$$C_3^V(Q^2) = \frac{2.05}{\left(1 + \frac{Q^2}{0.54\ \text{GeV}^2}\right)^2}. \tag{3.27}$$

In the case of the axial vector matrix element, PCAC implies the constraint that its divergence should vanish as $m_\pi^2 \to 0$, and this leads to the conclusion (as in the nucleon case) that the induced pseudoscalar term, C_6^A, must have a pion pole

$$\frac{C_6^A}{M^2} \approx \frac{g_{\Delta N \pi} f_\pi}{\sqrt{6} M (Q^2 + m_\pi^2)}, \tag{3.28}$$

(where $g_{\Delta N\pi}$ is the coupling constant for $\Delta^{++} \to p\pi^+$) and that C_5^A and C_6^A should be related at $Q^2 = m_\pi^2 = 0$ – i.e., $C_5^A \to g_{\Delta N\pi} f_\pi / \sqrt{6} M$, the analogue of the Goldberger-Treiman relation in Eq. (3.10). On the basis of various quark models and dispersion relations

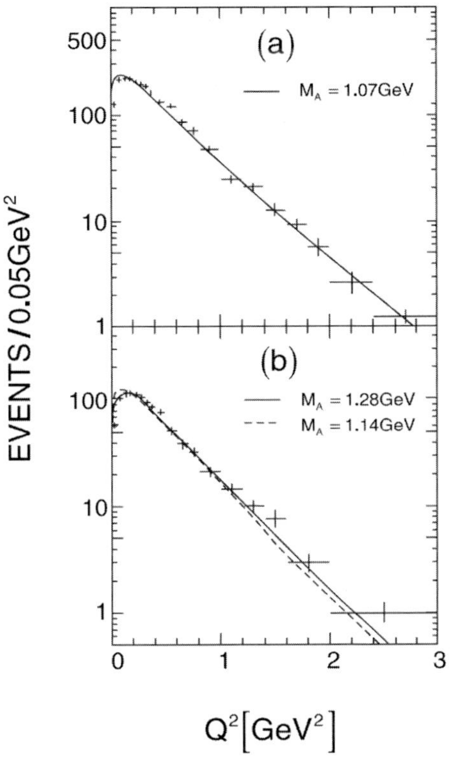

Figure 3.5: The differential cross section for a) proton and b) Δ^{++} production in inelastic neutrino scattering on the deuteron, as a function of Q^2 [Kit 90].

it is generally thought that $C^A_{3,4}$ should be small. By analogy with the matrix element of the axial current of the nucleon, the remaining Q^2-dependence of the dominant axial form factors $C^A_{5,6}$ (apart from the pion pole in Eq. (3.28)) is usually parametrized as a single dipole with mass $M_A^{(\Delta)}$. Figure 3.5 shows the typical quality of the fit obtained in this framework. It is very interesting that the value of $M_A^{(\Delta)}$ obtained from such fits to the most recent data seems to be around 1.3 GeV (specifically, $1.28 + 0.08 \ - 0.10$ GeV [Kit 90] and (1.31 ± 0.12) GeV from BEBC [Jon 89]). This is much harder than the axial vector form factor of the nucleon itself, in marked contrast with the vector matrix element shown in Fig. 2.17, where the NΔ transition form factor was much *softer* than for the nucleon itself. One might expect that this comparison should provide a strong constraint on any model of the structure of the nucleon and the $\Delta(1232)$.

For the student who wishes to study the relationships between the Rarita-Schwinger form factors, the usual helicity amplitudes and the various vector and axial vector multipole amplitudes we refer to the very helpful reviews [SVH 73, Zu 71] as well as the experimental papers themselves.

3.3 Beta-Decay

The classical way to study the weak interaction is, of course, β-decay. There the momentum transfer is so low that only the vector (γ^μ) and axial vector ($\gamma^\mu\gamma_5$) pieces of the weak current contribute and we need the form factors F_1^V and G_A at essentially $Q^2 = 0$. It is traditional to call $2F_1^V(0)$ (which is unity according to CVC and the definition (2.19)) g_V, while $G_A(0)$ is usually written g_A. Beta-decay is then described entirely in terms of these two constants

$$g_V = 2F_1^V(0); \qquad g_A = G_A(0), \tag{3.29}$$

and the CKM matrix, V, given in Eq. (1.22). Any deviation of g_V from unity would signal a break-down of CVC, while a deviation of V from unitarity would indicate new physics beyond the standard model. Finally, as we shall see in Chapter 7, using current algebra and PCAC one can write a dispersion relation for g_A known as the Adler-Weisberger relation. This predicts $g_A = 1.26$.

3.3.1 Experimental Determination of g_A/g_V

Using the zero-momentum transfer limit of the matrix element of the weak current[2]

$$\langle p|V_+^\mu(0) - A_+^\mu(0)|n\rangle = \bar{u}_p(g_V\gamma^\mu - g_A\gamma^\mu\gamma_5)u_n, \tag{3.30}$$

one can calculate the decay rate of the neutron according to standard Feynman rules. For a neutron at rest, with spin in the direction $\vec{s}_n$, and with the final e and $\bar{\nu}$ having velocities $\vec{v}_e$ and $\vec{v}_\nu$ respectively, one finds [Dub 91]:

$$\frac{dR}{dp_e d\Omega_e d\Omega_\nu} = \frac{G_F^2 V_{ud}^2}{(2\,\pi)^5}\,[\alpha + \beta\vec{v}_e\cdot\vec{v}_\nu + \gamma\vec{s}_n\cdot\vec{v}_e + \delta\vec{s}_n\cdot\vec{v}_\nu]\,p_e^2(E_{\text{max}} - E_e)^2. \tag{3.31}$$

Here $E_{\text{max}} = M_n - M_p \simeq 1.3$ MeV is the end-point of the electron spectrum and the coefficients are

$$\begin{aligned} \alpha &= g_V^2 + 3g_A^2, & \beta &= g_V^2 - g_A^2, \\ \gamma &= 2(g_Ag_V - g_A^2), & \delta &= 2(g_Ag_V + g_V^2). \end{aligned} \tag{3.32}$$

Clearly then an accurate measurement of the angular correlation between the outgoing electron and neutrino can yield a value of $|g_A/g_V|$ without any need for neutron polarization. To check the sign requires a spin-dependent measurement. In fact, the most accurate determination of g_A/g_V comes from the determination of the coefficient γ, with the current best value being [PDG 98/00]

$$\frac{g_A}{g_V} = 1.267 \pm 0.004. \tag{3.33}$$

---

[2]Note that we prefer to make the $V - A$ structure of the weak interaction explicit in Eq. (3.30). This has the advantage that g_A is a positive number. On the other hand the Particle Data Group (for example) chooses a plus sign on the right hand side of Eq. (3.30) but makes g_A negative.

3.3.2 Determination of V_{ud}

In Eq. (3.31) we have specifically written $G_F^2 V_{ud}^2 g_V^2$ as the strength of the weak vector coupling, rather than $G_F^2 \cos^2\theta_C$, to emphasize that there are two tests involved here. Firstly, the test that g_V is indeed one is a test of CVC. The extraction of V_{ud} is connected with the possibility of physics beyond the standard model. Clearly by integrating the differential decay rate to obtain the lifetime of the neutron one can, in principle, determine the product $V_{ud}g_V$. In fact, all of the velocity-dependent terms in Eq. (3.31) average to zero when computing the total rate, so that:

$$\tau^{-1} = \frac{G_F^2 V_{ud}^2}{2\pi^3}(g_V^2 + 3g_A^2) m_e^5 f(E_{\text{max}}/m_e), \tag{3.34}$$

with f the standard, dimensionless integral over the momentum of the outgoing electron

$$f = \int_0^{p_{\text{max}}} dpp^2 \left[(p_{\text{max}}^2 + 1)^{\frac{1}{2}} - (p^2+1)^{\frac{1}{2}}\right]^2, \tag{3.35}$$

and where $E_{\text{max}} = (p_{\text{max}}^2 + 1)^{1/2} m_e$. The quantity usually quoted is the product $ft_{1/2}$, known as the ft value. (Note that the lifetime, τ, and the half-life, $t_{1/2}$, are related by $t_{1/2} = (\ln 2)\tau$.)

Measuring the lifetime of a free neutron has proven a tremendous challenge over the years.

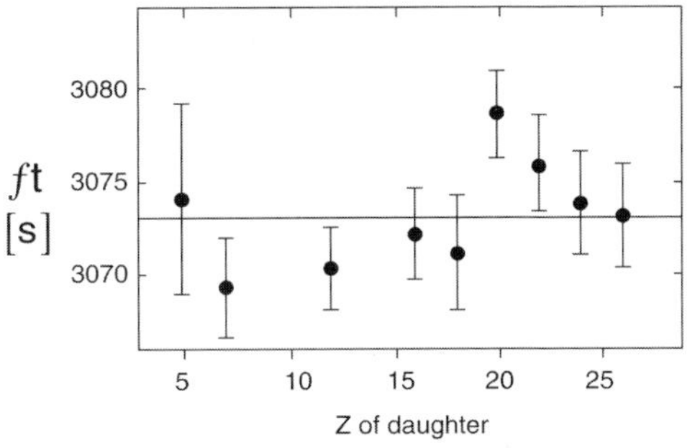

Figure 3.6: The ft values, corrected for radiative and electromagnetic effects, extracted from nuclei undergoing super-allowed Fermi β-decay [To 95].

At present the best value is [PDG 98/00]

$$\tau = (887 \pm 2)\ \text{sec}, \tag{3.36}$$

but there are hopes that new experiments may reduce the error by a further order of magnitude. Until such an improvement can be made the free neutron measurements cannot compete with the most accurate method for determining $V_{ud}g_V$, namely super-allowed Fermi β-decay.

There are many fine reviews of super-allowed Fermi β-decay to which we refer for details [Wil 93, To 95]. However, the concept is easily understood. There are nine accessible, β-unstable nuclei which are members of an isospin-1 multiplet with spin-parity 0^+. Because of the spin and parity there is no contribution from the weak axial current. Assuming that the structure of a bound nucleon is not significantly different from a free one, one can very accurately estimate the nuclear and electromagnetic corrections to these special matrix elements. The ft value (corrected for both nuclear and radiative effects) is then

$$(ft)^{-1} = (\ln 2)^{-1} \frac{G_F^2}{2\pi^3} V_{ud}^2\, g_V^2 m_e^5, \tag{3.37}$$

and one can clearly extract $V_{ud}\, g_V$. Usually g_V is set to one and the measurement is assumed to yield V_{ud}. The consistency of the values extracted from the nine available nuclei is shown in Fig. 3.6. The average values of ft and V_{ud} thus determined are:

$$ft = (3073 \pm 3)\ \text{sec}; \qquad |V_{ud}| = 0.9736 \pm 0.0006. \tag{3.38}$$

There is considerable interest in this value of V_{ud} because, when combined with $|V_{us}| = 0.2205 \pm 0.0018$ and $|V_{ub}| = 0.0032 \pm 0.0009$, one obtains the best test of the unitarity of the CKM matrix:

$$|V_{ud}|^2 + |V_{us}|^2 + |V_{ub}|^2 = 0.9965 \pm 0.0015. \tag{3.39}$$

The deviation from unity is usually viewed as a tantalizing indication of physics beyond the standard model. We note in conclusion that it could also be an indication of a breakdown of CVC for either the nucleon ($g_V \neq 1$) or the bound nucleon ($g_V^{\text{bound}} \neq 1$). Clearly every one of these possibilities is extremely important.

3.4 Muon Capture

Muon capture is a time honoured technique for studying the weak matrix elements of the nucleon. Ordinary muon capture (OMC),

$$\mu^- + p \to \nu_\mu + n, \tag{3.40}$$

leads to a two-body final state with a 5.2 MeV neutron and hence a unique momentum transfer ($Q^2 = 0.88 m_\mu^2$). It has been studied extensively because a low energy muon is much more efficient as a probe than the neutrino, being readily captured into an atomic state when passing through gaseous or liquid hydrogen. Within a picosecond of being captured it has undergone an atomic cascade to form a $\mu^- p$ atom in the 1s-state, with $S = 0$ because this is lower in energy than the $S = 1$ state. (Note that the splitting between these two levels is only $0.18\ eV$, too small for a radiative transition to be effective. However, collision induced transitions totally deplete the $S = 1$ level at densities above 10^{20}/cc.)

In a liquid hydrogen target the $\mu^- p$ atom rapidly forms a $(p\mu p)^+$ molecule. Again there are two possible molecular states, the ortho state with total angular momentum $\frac{1}{2}$ and the para state with total angular momentum $\frac{3}{2}$. Although the latter is lower in energy, the ortho state is formed first and the conversion rate is very low, so the stopped μ^- is found predominantly in the ortho $(p\mu p)^+$ molecule.

Although most stopped muons decay in the usual way, approximately 1 in every 1,000 is captured. For capture from the ortho $(p\mu p)^+$ molecule the rate is

$$\Lambda_{\text{ortho}} = 2\,\gamma_0 \left(\frac{3}{4}\Lambda_s + \frac{1}{4}\Lambda_t \right), \tag{3.41}$$

where $2\gamma_0 = 1.01 \pm 0.01$ is calculated from the $\mu^- p$ overlap in the molecular wave function and $\Lambda_{s,t}$ is the capture rate for a $\mu^- p$ pair with $S = 0$ or 1. With the $V - A$ structure of the standard model Λ_s dominates (i.e., $\Lambda_s \gg \Lambda_t$). In early experiments the signal for muon capture was to detect the slow, outgoing neutron, but the most recent experiments concentrate on the change in the muon lifetime. As $G_A(Q^2 = 0.88m_\mu^2)$ is very well determined from β-decay (and the slope in Q^2 from neutrino quasi-elastic scattering), a measurement of the singlet capture rate in OMC:

$$\Lambda_S = \frac{152 \text{ sec}^{-1}}{g_V^2 + 3\,g_A^2}(\alpha + \beta < \vec{\sigma}_\mu \cdot \vec{\sigma}_p >), \tag{3.42}$$

with the coefficients given by

$$\begin{aligned} \alpha &= \tilde{G}_V^2 + 3\tilde{G}_A^2 + \tilde{G}_P^2 - 2\,\tilde{G}_A\tilde{G}_P, \\ \beta &= 2\,(\tilde{G}_V\tilde{G}_A - \frac{1}{3}\,\tilde{G}_V\tilde{G}_P) - 2\,(\tilde{G}_A^2 - \frac{2}{3}\tilde{G}_A\tilde{G}_P), \end{aligned} \tag{3.43}$$

enables a determination of $\tilde{G}_P \equiv G_P(Q^2 = 0.88m_\mu^2)$. Note that all other quantities with a tilde in eq.(3.43) are evaluated at this momentum transfer too (in particular, $\tilde{G}_V = 2G_E^V(Q^2 = 0.88m_\mu^2)$). Thus one expects $\Lambda_s = 675 \text{ sec}^{-1}$ and $\Lambda_t = 12.5 \text{ sec}^{-1}$. Traditionally the induced pseudoscalar coupling in muon capture is defined in terms of $q^\mu\gamma_5/m_\mu$, rather than $q^\mu\gamma_5/2M$ - as in Eq. (3.6). In view of this and the PCAC hypothesis which led to Eq. (3.11), the result of the analysis is usually called g_P/g_A where

$$\frac{g_P}{g_A} = \frac{m_\mu}{2M}\frac{G_P(0.88m_\mu^2)}{g_A} \approx \frac{2Mm_\mu}{m_\pi^2 + 0.88m_\mu^2}, \tag{3.44}$$

and the last step follows from PCAC, neglecting the small difference between $G_A(0)$ and $G_A(0.88m_\mu^2)$. The current best value for g_P/g_A, obtained from OMC, comes from the Saclay group [Bar 81] who were able to measure the ortho-to-para transition rate, $\lambda_{op} = (4.1 \pm 1.4) \cdot 10^4 \text{ sec}^{-1}$, and correct for it:

$$\left.\frac{g_P}{g_A}\right|_{\text{Saclay}} = 5.6 \pm 2.4. \tag{3.45}$$

(Note that without the correction for the ortho-para conversion they would have obtained $g_P/g_A = 13.8 \pm 2.0$.) Although not everyone agrees with the procedure, the experimental error can be reduced by averaging over world data for OMC after correcting for the estimated ortho-para conversion using the Saclay value for λ_{op}. This yields $g_P/g_A = 6.9 \pm 1.5$, in rather good agreement with the theoretical expectation based on PCAC – c.f. Eq. (3.44) – namely $g_P/g_A \approx 6.7$.

The major difficulty with OMC as a source of information on g_P/g_A is the lack of experimental sensitivity caused by the relatively low momentum transfer. Indeed, a measurement of Λ_S to an accuracy of 4 %, which presents a difficult experimental challenge, leads to an error of roughly 40 % in g_P/g_A. Clearly it will be extremely difficult to improve the determination of g_P/g_A in OMC and this has led to attempts to measure radiative muon capture, where the error in g_P/g_A should be approximately the same as the error in the measured rate.

3.4.1 Radiative Muon Capture

While radiative muon capture (RMC) is much more sensitive to the induced pseudoscalar form factor it is also enormously more difficult to measure. The rate for RMC is suppressed by a factor of 10^{-5} below OMC, which is in turn only one part in a thousand of the muon decay in orbit. Furthermore, the only detectable particles in the final state:

$$\mu^- + p \rightarrow n + \nu_\mu + \gamma, \tag{3.46}$$

are the neutron and the photon and below 53 MeV the photon spectrum is dominated by background related to the electrons from muon decay. The Time Projection Chamber group at TRIUMF has succeeded in measuring the spectrum [Jon 96] which is reproduced in Fig. 3.7.

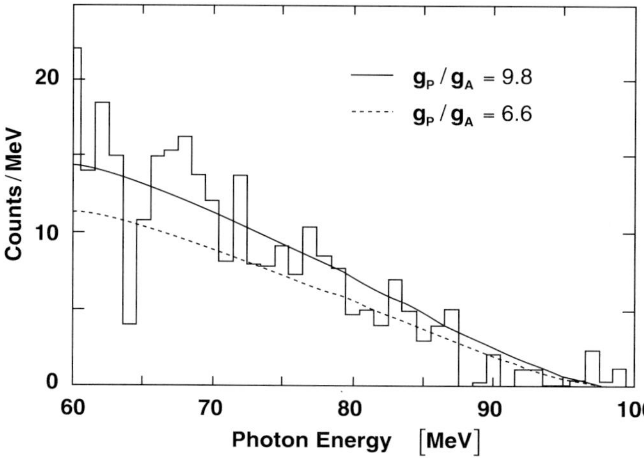

Figure 3.7: The high energy tail of the photon spectrum for radiative muon capture in hydrogen compared with a Monte Carlo estimate – from Ref.[Jon 96].

The theoretical analysis is also rather more involved. Figure 3.8 illustrates the various graphs which must be included in the calculation. For our purposes the major interest is in Figs. 3.8 (h) and (i), where the pion pole in the induced pseudoscalar form factor is shown explicitly. Unlike OMC, where the momentum transfer is fixed at the spacelike value $Q^2 = 0.88 m_\mu^2$, here the momentum transfer can be time-like with an enhancement of a factor of 4 $((m_\pi^2 + 0.88 m_\mu^2)/(m_\pi^2 - m_\mu^2))$ at the high energy end of the photon spectrum. This enhancement of the pion pole is the essential reason for studying RMC. One other technical advantage is that RMC is much less sensitive to the ortho-para conversion rate which causes so much trouble in OMC.

A careful analysis of the spectrum shown in Fig. 3.7 leads to a value of $g_P/g_A = 9.5 \pm 0.5$, which is considerably larger than the expectation based on PCAC and the value determined in OMC. This has stimulated interest in the role of Δ-excitation. However, initial estimates suggest that the Δ could give at most an 8 % correction to the PCAC prediction [BF 89, BHM 01].

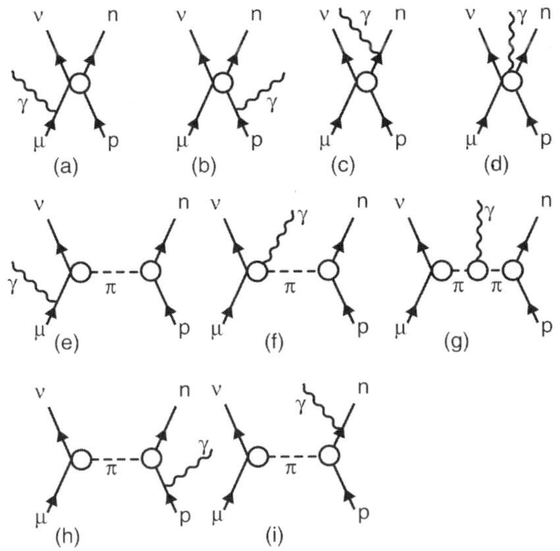

Figure 3.8: The Feynman diagrams illustrating the various contributions to radiative muon capture in hydrogen.

3.5 Near Threshold Pion Electroproduction

The general analysis of pion electroproduction in the resonance region has already been described in Section 2.3.2. We do not wish to repeat that here. Rather our interest is in the specifics of this reaction near threshold, where low energy theorems are applicable [VZ 72], to permit the *only direct determination of* G_P which has been made.

Let us first understand the importance of the kinematic flexibility afforded by the three-body final state. By adjusting both the incident energy and the scattering angle of the electron it is possible to *fix* the square of the four-momentum transfer to the nucleon, $t(\equiv -Q^2)$, the pion momentum in the πN center of mass, $|\vec{q}\,|$, *and* the invariant mass of the virtual photon, k^2, while varying the longitudinal polarization of the virtual photon, ϵ (c.f. Appendix B). Since, near threshold, there are only two multipoles involved, E_{0+} and L_{0+}, knowing the cross section for just two values of ϵ (typically $\epsilon \approx 0.2$ and $\epsilon \approx 0.7$), with k^2, Q^2 and $|\vec{q}\,|$ fixed, allows one to determine the two unknowns in the axial current matrix element, $G_A(Q^2)$ and $G_P(Q^2)$. (As usual, one assumes that the vector form factors are known from electron scattering.) The actual experiment was performed by Choi *et al.* [Cho 93] at the Saclay Linear Accelerator (ALS). The form factors $G_{A,P}(Q^2)$ were extracted at three values of Q^2, 0.0732, 0.139 and 0.179 GeV^2. Together with the values of $G_A(0)$ and $G_P(0.88m_\mu^2)$ from β-decay and OMC, respectively, this gives four points in the momentum variation of these important form factors. For G_A the data was fit in terms of the usual dipole form with the best fit being $M_A = (1.15 \pm 0.27)$ GeV – consistent with the results presented in Table 3.1. However, the really new feature of the data is shown in Fig. 3.9. This is the *only* direct determination of the induced pseudoscalar form factor and as we see it is in excellent agreement with the prediction of PCAC (or pion pole dominance) – i.e., the form given in Eq. (3.8). This is a

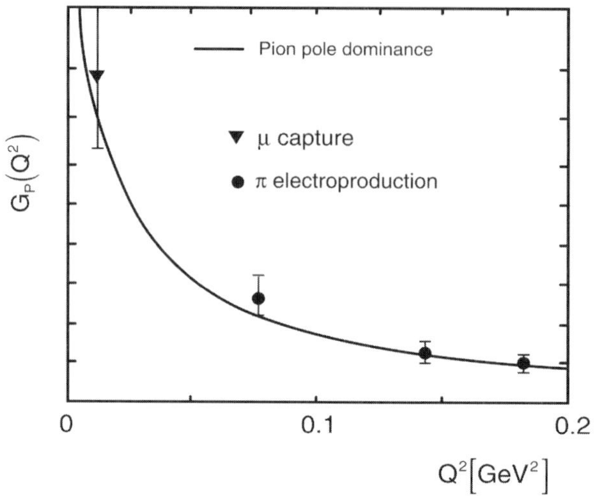

Figure 3.9: The Q^2 dependence of the induced pseudoscalar form factor, $G_P(Q^2)$, measured in ordinary muon capture ($Q^2 = 0.0112$ GeV^2) and pion electroproduction [Cho 93] – compared with the expectation based on PCAC.

very satisfying situation, although clearly one would hope for improved precision and a wider range of momentum transfers.

3.6 Neutral Current Interactions

Traditionally neutral current interactions have been investigated with beams of neutrinos. Indeed, the very existence of the neutral current was first demonstrated through neutrino scattering experiments [Has 73]. As a probe of the structure of the nucleon the weak neutral current offers two advantages. First, the weighting of strange quarks relative to non-strange quarks is different in the vector part of the current. Thus, in principle, one is able to study the distribution of s and $\bar{s}$ quarks in the nucleon – a subject of considerable interest. Second, the Z^0 also has an axial coupling to the quarks and one can therefore measure new axial form factors. These too have generated great theoretical interest in view of the suggestion of a significant, polarized, strange sea in the nucleon in the analysis of the EMC spin effect – c.f. Chapter 4.

Recently the development of a number of new, high intensity electron accelerators has added another technique for investigating these new form factors. The interference between γ and Z^0 exchange in electron scattering can give rise to parity violation at the level of 10^{-6}. Modern experiments offer the opportunity to measure this tiny signal to an accuracy of a few percent. This is a very exciting development which should yield many new results in the next decade.

By way of orientation it may be helpful to summarize the neutral current interaction before examining neutrino and electron scattering in detail. This is simply a matter of selecting the relevant terms from the summary of the Standard Model in Chapter 1. Considering first the case of electron scattering we note that the matrix element for the neutral current interaction will be proportional to

$$\frac{g^2 J^{\mu}_{e,\mathrm{NC}} J^{\mathrm{had,NC}}_{\mu}}{4\,M_Z^2\cos^2\theta_W} = G_F\sqrt{2}J^{\mu}_{e,\mathrm{NC}}(V^{\mathrm{had,NC}}_{\mu} - A^{\mathrm{had,NC}}_{\mu}), \tag{3.47}$$

where (for $Q^2 \ll M_Z^2$) we used Eqs. (1.3) and (1.4) to replace $g^2/(4\cos^2\theta_W M_Z^2)$ by $G_F\sqrt{2}$, and

$$J^{\mu}_{e,\mathrm{NC}} = \frac{1}{2}\,\bar{\psi}_e[\gamma^{\mu}(4\,\sin^2\theta_W - 1) + \gamma^{\mu}\gamma_5]\psi_e. \tag{3.48}$$

In the case of neutrino elastic scattering one finds:

$$\frac{g^2 J^{\mu}_{\nu,\mathrm{NC}} J^{\mathrm{had,NC}}_{\mu}}{4\,M_Z^2\cos^2\theta_W} = G_F\sqrt{2}J^{\mu}_{\nu,\mathrm{NC}}(V^{\mathrm{had,NC}}_{\mu} - A^{\mathrm{had,NC}}_{\mu}), \tag{3.49}$$

where (again for $Q^2 \ll M_Z^2$)

$$J^{\mu}_{\nu,\mathrm{NC}} = \frac{1}{2}\bar{\psi}_{\nu}\gamma^{\mu}(1-\gamma_5)\psi_{\nu}. \tag{3.50}$$

The appropriate, hadronic vector and axial vector currents are

$$\begin{aligned} V^{\mu}_{\mathrm{had,NC}} &= \frac{1}{2}\left(1 - \frac{8}{3}\sin^2\theta_W\right)\bar{\psi}_u\gamma^{\mu}\psi_u \\ &\quad - \frac{1}{2}\left(1 - \frac{4}{3}\,\sin^2\theta_W\right)(\bar{\psi}_d\gamma^{\mu}\psi_d + \bar{\psi}_s\gamma^{\mu}\psi_s), \end{aligned} \tag{3.51}$$

and

$$A^{\mu}_{\mathrm{had,NC}} = \frac{1}{2}(\bar{\psi}_u\gamma^{\mu}\gamma_5\psi_u - \bar{\psi}_d\gamma^{\mu}\gamma_5\psi_d - \bar{\psi}_s\gamma^{\mu}\gamma_5\psi_s + \bar{\psi}_c\gamma^{\mu}\gamma_5\psi_c). \tag{3.52}$$

The charm contribution in Eq. (3.52) will be dropped in the rest of this section as in most discussions in the literature. We include it here purely as a reminder that the Z^0 axial coupling is a weak isovector and therefore has no $U(1)$ axial anomaly.

Of course, in elastic scattering from the nucleon the information that we can obtain can be summarized as a number of form factors. These are related to the proton matrix elements of $V^{\mu}_{\mathrm{had,NC}}$ and $A^{\mu}_{\mathrm{had,NC}}$ as:

$$\langle p(P')|V^{\mu}_{\mathrm{had,NC}}(0)|p(P)\rangle = \bar{u}_p(P')\left[F_1^{\mathrm{NC}}(Q^2)\gamma^{\mu} + iF_2^{\mathrm{NC}}(Q^2)\frac{\sigma^{\mu\nu}}{2M}q_{\nu}\right]u_p(P), \tag{3.53}$$

and

$$\langle p(P')|A^{\mu}_{\mathrm{had,NC}}(0)|p(P)\rangle = \bar{u}_p(P')\left[G_A^{\mathrm{NC}}(Q^2)\gamma^{\mu}\gamma_5 + \frac{G_P^{\mathrm{NC}}(Q^2)}{2M}q^{\mu}\gamma_5\right]u_p(P), \tag{3.54}$$

and as always $q = P' - P$. Each of the vector form factors can be expressed as a linear combination of flavor triplet, octet and singlet combinations

$$F_i^{\rm NC}(Q^2) = (1 - 2\sin^2\theta_W)\left[F_i^3(Q^2) + \frac{1}{6}F_i^8(Q^2)\right] - \frac{1}{6}F_i^0(Q^2), \tag{3.55}$$

corresponding to the proton matrix elements of the appropriate linear combinations of quark vector currents ($[\bar{u}\gamma^\mu u - \bar{d}\gamma^\mu d]/2$, $[\bar{u}\gamma^\mu u + \bar{d}\gamma^\mu d - 2\bar{s}\gamma^\mu s]$ and $[\bar{u}\gamma^\mu u + \bar{d}\gamma^\mu d + \bar{s}\gamma^\mu s]$, respectively). Assuming charge independence the neutron form factor is also given by Eq. (3.55) but with the sign of F_i^3 changed. We choose to write the axial form factor as an isovector form factor minus a strangeness form factor, e.g.,

$$G_A^{\rm NC}(Q^2) = G_A^3(Q^2) - \frac{1}{2}G_A^s(Q^2), \tag{3.56}$$

and similarly for $G_P^{\rm NC}$. Here $G_A^3(Q^2)$ corresponds to the proton matrix element of $[\bar{u}\gamma^\mu\gamma_5 u - \bar{d}\gamma^\mu\gamma_5 d]/2$ and $G_A^s(Q^2)$ to the matrix element of $\bar{s}\gamma^\mu\gamma_5 s$.

3.6.1 Elastic Neutrino Scattering

In terms of the neutral current interaction given in Eq. (3.49), the differential cross section for elastic neutrino proton scattering can be calculated in a straightforward way. The result looks very much like Eq. (3.16):

$$\frac{d\sigma_{\nu(\bar{\nu})}}{dQ^2} = \frac{G_F^2 M^2}{8\pi E^2}\left[\mathcal{A} \pm \mathcal{B}\frac{s-u}{M^2} + \mathcal{C}\frac{(s-u)^2}{M^4}\right], \tag{3.57}$$

where $s - u$ is given by Eq. (3.14) with $m = 0$, $\tau = \frac{Q^2}{4M^2}$ and

$$\begin{aligned}\mathcal{A} &= 4\,\tau\left[(1+\tau)(G_A^{\rm NC})^2 - (1-\tau)(F_1^{\rm NC})^2 + \tau\,(1-\tau)(F_2^{\rm NC})^2\right] \\ &\quad + 16\,\tau^2 F_1^{\rm NC}F_2^{\rm NC}, \end{aligned}\tag{3.58}$$

$$\mathcal{B} = 4\tau G_A^{\rm NC}(F_1^{\rm NC} + F_2^{\rm NC}), \tag{3.59}$$

$$\mathcal{C} = \frac{1}{4}\left[(G_A^{\rm NC})^2 + (F_1^{\rm NC})^2 + \tau(F_2^{\rm NC})^2\right]. \tag{3.60}$$

In writing Eqs. (3.57) to (3.60) we have dropped contributions involving the induced pseudoscalar term. (By comparison with Eq. (3.17) it is clear that such terms involving G_P are suppressed by powers of the neutrino mass.)

The simplest procedure for analysing the elastic neutrino-proton data is to drop all terms involving strangeness, so that

$$F_1^{\rm NC}(Q^2) + F_2^{\rm NC}(Q^2) = \frac{\frac{1}{2}(1-2\sin^2\theta_W)(1+\kappa_p-\kappa_n) - \sin^2\theta_W(1+\kappa_p+\kappa_n)}{\left(1+\frac{Q^2}{M_V^2}\right)^2}, \tag{3.61}$$

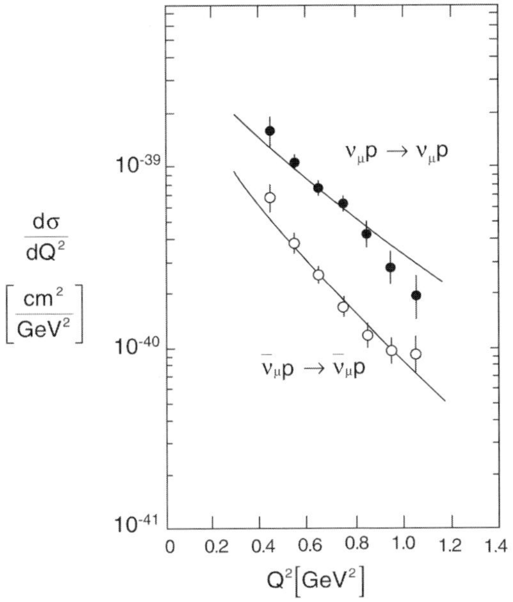

Figure 3.10: The differential cross sections for νp and $\bar{\nu}$p elastic scattering obtained at BNL [Ahr 88], in comparison with the best fits based on an axial form factor with dipole mass $M_A = 1.06$ GeV.

and

$$F_2^{\mathrm{NC}}(Q^2) = \frac{\frac{1}{2}(1 - 2\sin^2\theta_W)(\kappa_p - \kappa_n) - \sin^2\theta_W(\kappa_p + \kappa_n)}{(1+\tau)\left(1 + \frac{Q^2}{M_V^2}\right)^2}, \tag{3.62}$$

with $M_V = 0.84$ GeV – from elastic electron scattering. One then typically drops the strange quark contribution and uses a dipole parametrization for the axial form factor:

$$G_A^{\mathrm{NC}}(Q^2) = \frac{g_A}{2\left(1 + \frac{Q^2}{M_A^2}\right)^2}. \tag{3.63}$$

The best fit to ν_μp and $\bar{\nu}_\mu$p data obtained by Ahrens *et al.* [Ahr 88], which is shown in Fig. 3.10, yielded a best fit value $M_A = 1.06 \pm 0.05$ GeV (and $\sin^2\theta_W$ in agreement with other data). This agrees very well with the values quoted in Table 3.1, coming from charged current neutrino scattering.

In order to test for some strangeness (or heavier) quark contribution the data was re-analyzed by fixing both $\sin^2\theta_W$ and M_A at their respective world averages and searching on the parameter η, where g_A in Eq. (3.63) was replaced by $(1+\eta)g_A$. The result, $\eta = 0.12 \pm 0.07$ (or $0 < \eta < 0.25$ at 90 % confidence level) gives just a hint of the new physics one might hope to explore with neutral current interactions.

3.6.2 Parity Violation in Elastic Electron Scattering

As explained in the introduction, improvements in both accelerator and detector technology mean that we now have an alternative method of measuring the matrix elements of the neutral current. The parity violating asymmetry, defined as the difference over the sum of the differential cross sections for the scattering of positive and negative helicity electrons:

$$A = \frac{d\sigma_+ - d\sigma_-}{d\sigma_+ + d\sigma_-}, \tag{3.64}$$

is determined by the interference of γ and Z^0 exchange. The difference of cross sections is thus of order αG_F and while this is a small number it is much bigger than G_F^2. In addition, electron accelerators have much higher intensity than beams of neutrinos.

The asymmetry arises from terms involving one vector and one axial vector coupling in the Z^0 exchange amplitude [McK 89, Mus 94]. Thus the structure of the asymmetry is

$$A = -\frac{G_F Q^2}{2\sqrt{2}\pi\alpha}\left[\mathcal{A}_E + \mathcal{A}_M + \mathcal{A}_A\right]. \tag{3.65}$$

Both $\mathcal{A}_E$ and $\mathcal{A}_M$ arise from axial coupling to the electron and vector coupling to the proton (electric and magnetic respectively). Clearly these are most naturally expressed in terms of the Sachs form factors which for the photon were defined in Chapter 2 and for the Z^0 are defined below:

$$\mathcal{A}_E = \frac{\varepsilon\, G_E\, G_E^Z}{\varepsilon\, G_E^2 + \tau\, G_M^2}, \tag{3.66}$$

$$\mathcal{A}_M = \frac{\tau\, G_M\, G_M^Z}{\varepsilon\, G_E^2 + \tau\, G_M^2}. \tag{3.67}$$

The general structure of Eqs. (3.65 - 3.67) is clear. The factor $G_F/\sqrt{2}$ arises because of the Z^0 exchange in the interference term. We divide by α/Q^2 because the differential cross section in the denominator is purely electromagnetic and the combination $\varepsilon G_E^2 + \tau G_M^2$ is the usual combination appearing in elastic electron-proton scattering, with the virtual photon polarization parameter $\varepsilon = [1 + 2\,(1+\tau)\tan^2\theta/2]^{-1}$ (with $\tau = Q^2/4M^2$). The third term in Eq. (3.65), $\mathcal{A}_A$, arises from the vector coupling of the Z^0 to the electron (which is proportional to $(1-4\sin^2\theta_W)$) together with the neutral current axial coupling to the nucleon (c.f. Eq. (3.63)):

$$\mathcal{A}_A = -\frac{\frac{1}{2}(1-4\sin^2\theta_W)\sqrt{1-\varepsilon^2}\,\sqrt{\tau(1+\tau)}\,G_M\,G_A^{\mathrm{NC}}}{\varepsilon\, G_E^2 + \tau G_M^2}. \tag{3.68}$$

Finally, in terms of the vector form factors defined in Eqs. (3.51) and (3.53), the Sachs form factors associated with Z^0 exchange, $G_{E,M}^Z$, are:

$$\begin{aligned} G_E^Z &= F_1^{\mathrm{NC}} - \tau F_2^{\mathrm{NC}}, \\ G_M^Z &= F_1^{\mathrm{NC}} + F_2^{\mathrm{NC}}. \end{aligned} \tag{3.69}$$

Because $\sin^2\theta_W$ is so close to $\frac{1}{4}$, the factor $(1 - 4\sin^2\theta_W)$ in $\mathcal{A}_A$ is very small. It will therefore be extremely difficult to extract $G_A^{\rm NC}$, and hence G_A^s, from parity violating electron scattering. Estimates of the parity-violating asymmetry at 850 MeV (appropriate to the MAMI accelerator at Mainz) suggest that the electric term, $\mathcal{A}_E$, is even smaller – except for very small scattering angles (of order 10°). There is also some sensitivity to G_E^s under the conditions of the HAPPEX experiment [An 99] at Jefferson Lab (at a small scattering angle, 12.5°). At $Q^2 \simeq 0.5$ GeV^2 HAPPEX is sensitive to a linear combination of the strangeness electric and magnetic form factors, $G_E^s + 0.4G_M^s$. The result of their first measurement was $G_E^s + 0.4G_M^s = 0.023 \pm 0.034$ (with only statistical errors given here).

Away from the forward direction $\mathcal{A}_M$ is expected to give the dominant contribution. From a knowledge of G_M^p, G_M^n and G_M^Z it is possible to isolate the strange quark contribution to the magnetic current in the nucleon and it is instructive to briefly explore this idea. At the quark level the proton matrix elements of the vector current are the same whether one is evaluating $G_M^{p,n}$ or G_M^Z. Denoting the u, d and s magnetic matrix elements in the proton as $G_M^{u,d,s}$ and assuming charge independence we find

$$\begin{aligned}
G_M^p &= \frac{2}{3}G_M^u - \frac{1}{3}G_M^d - \frac{1}{3}G_M^s, \\
G_M^n &= \frac{2}{3}G_M^d - \frac{1}{3}G_M^u - \frac{1}{3}G_M^s, \\
G_M^Z &= \frac{1}{2}(1 - \frac{8}{3}\sin^2\theta_W)G_M^u - \frac{1}{2}(1 - \frac{4}{3}\sin^2\theta_W)G_M^d \\
&\quad - \frac{1}{2}(1 - \frac{4}{3}\sin^2\theta_W)G_M^s.
\end{aligned} \tag{3.70}$$

These simultaneous equations can be solved for G_M^s, a quantity in which there is a great deal of theoretical interest. As just one indication of the size of the strangeness contribution that one might expect, we note that initial estimates of $G_M^s(0)$ based on lattice QCD suggested a value of order -0.3 in units of nuclear magnetons, μ_N [Lei 95]. More recently it has been realized that this result is quite sensitive to chiral corrections in the extrapolation of the lattice data to the physical quark mass. An improved chiral extrapolation leads to the expectation $G_M^s(0) = -0.16 \pm 0.18$ [LT 00].

The first measurement of G_M^s alone, following this approach, was made by the SAMPLE collaboration at MIT-Bates [Spa 00]. Using a 200 MeV beam of polarized electrons and working at backward angles they were sensitive to G_M^s at Q^2 of order 0.1 GeV^2. If G_M^s is zero, their expected signal is a parity violating asymmetry of about -7 ppm. The measured asymmetry of $-4.92 \pm 0.61 \pm 0.73$ ppm led to the reported value, $G_M^s(0.1) = +0.61 \pm 0.71 \pm 0.21$. This is very far from most theoretical expectations. The situation is complicated by the fact that the isovector radiative corrections [MH 90] do not seem to be totally under control. An interesting suggestion, which may help resolve the situation experimentally, is that these corrections can be essentially removed by combining data for parity violating scattering on the deuteron with that on the proton. With more work underway at Mainz, Jefferson Lab and MIT-Bates, we can look forward to rapid progress in this field in the next few years.

4 Deep-Inelastic Lepton Scattering

The simplest experiment one can perform with a beam of leptons is to allow it to scatter from a hadronic target and to detect only the energy and scattering angle of the outgoing lepton. Consider, for example, the interaction of an electron beam with a hadronic target through single photon exchange as illustrated in Fig. 4.1. As usual the momentum transfer to

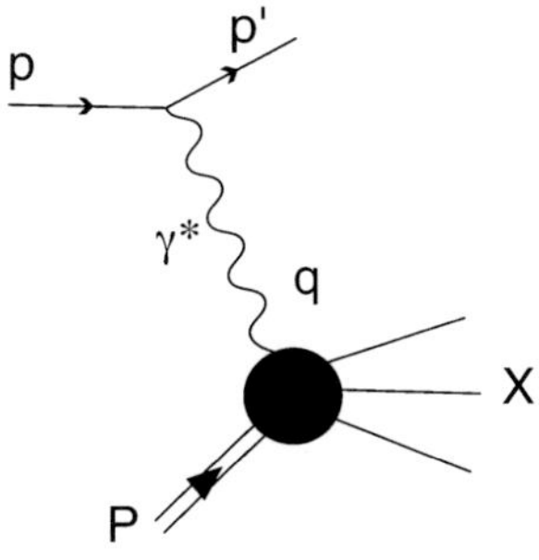

Figure 4.1: Illustration of a typical deep-inelastic scattering event in which an incident lepton transfers momentum q to the target and the hadronic final states, X, are not distinguished.

the hadronic target will be denoted q, with $Q^2 \equiv -q^2 \geq 0$, and in the laboratory frame of reference it is usual to write q^μ as $(\nu, \vec{q}\,)$. The target four-momentum is P (with $P^\mu = (M, \vec{0}\,)$ in the laboratory frame) and so the invariant mass squared of the final, hadronic state is

$$W^2 = (P+q)^2 = M^2 + 2P \cdot q - Q^2. \tag{4.1}$$

If, as in Section 2.3, we were to fix W^2 at some resonance, say M_R^2, the relation between Q^2 and ν would become:

$$Q^2 = (M^2 - M_R^2) + 2M\nu. \tag{4.2}$$

(Note that $M\nu$ is a Lorentz invariant, which is more obvious if it is written as $P \cdot q$.)

As the invariant mass of the resonance produced (M_R) increases, with Q^2 and ν both large, the number of resonances per unit energy becomes very large and their widths increase so that it is difficult to separate them individually. In addition, the fact that Q^2 is large means that form-factors suppress the contribution from any single resonance. It is this region, $Q^2 > 1-2$ GeV^2 and $\nu > 1$ GeV, that is referred to as the deep-inelastic regime. We shall see that this regime is of vital importance to our understanding of nucleon structure. In particular, the very large momentum transfer means that we can resolve objects with a size ($\lambda \sim 1/\sqrt{Q^2}$) of order $\frac{1}{10}$ fm at $Q^2 = 4$ GeV^2 and a few times $\frac{1}{100}$ fm at $Q^2 = 100$ GeV^2. Thus, by going into this kinematic region, we directly probe the energy and momentum distribution of the quarks, which are the fundamental constituents of hadronic matter.

As shown in detail in Appendix B, the inclusive cross section for unpolarized electron scattering is proportional to the product of a lepton tensor, $L_{\mu\nu}$, and a tensor $W^{\mu\nu}$ describing the response of the nucleon:

$$\begin{aligned} L_{\mu\nu} &= \frac{1}{2}\sum_{s,s'} \bar{u}_s(p)\gamma_\mu u_{s'}(p')\,\bar{u}_{s'}(p')\gamma_\nu u_s(p) \\ &= 2[p_\mu p'_\nu + p'_\mu p_\nu - g_{\mu\nu} p\cdot p'], \end{aligned} \tag{4.3}$$

and

$$\begin{aligned} W^{\mu\nu}(P,q) &= \frac{1}{4M}\sum_{s,X} < N_s(P)|J^\mu(0)|X(P')> \\ &\quad < X(P')|J^\nu(0)|N_s(P)> (2\,\pi)^3\delta^4(q+P-P'), \end{aligned} \tag{4.4}$$

$$\begin{aligned} &= W_1(\nu,Q^2)\left(\frac{q^\mu q^\nu}{q^2} - g^{\mu\nu}\right) \\ &\quad + \frac{W_2(\nu,Q^2)}{M^2}\left(P^\mu - \frac{P\cdot q}{q^2}q^\mu\right)\left(P^\nu - \frac{P\cdot q}{q^2}q^\nu\right). \end{aligned} \tag{4.5}$$

Here J^μ is the electromagnetic current given in Eq. (1.7). The resulting cross section is then

$$\frac{d^2\sigma}{d\Omega\, d\epsilon'} = \frac{\alpha^2}{4\epsilon^2 \sin^4\frac{\theta}{2}}\left(W_2(\nu,Q^2)\cos^2\frac{\theta}{2} + 2W_1(\nu,Q^2)\sin^2\frac{\theta}{2}\right), \tag{4.6}$$

where ϵ and ϵ' are the energies of the incoming and outgoing electrons ($p = (\epsilon, \vec{p})$ and $p' = (\epsilon', \vec{p}\,')$), θ is the laboratory scattering angle and $d\Omega$ the solid angle into which the electron is scattered.

Scaling

Although a quick exploration of the deep-inelastic region was one of the simplest initial experiments that one could perform with the new linear accelerator at Stanford, it turned out to be one of its most important – analogous to the famous back-angle α-particle scattering experiments which led Rutherford to the discovery of the atomic nucleus. The phenomenon discovered in these initial experiments is known as scaling and can be stated very simply. Suppose one defines the dimensionless structure functions $F_{1,2}$ from $W_{1,2}$ as:

$$\begin{aligned} F_1(x,Q^2) &= MW_1(\nu,Q^2), \\ F_2(x,Q^2) &= \nu W_2(\nu,Q^2) \equiv \frac{P\cdot q}{M}W_2(\nu,Q^2), \end{aligned} \tag{4.7}$$

where the Bjorken variable x is

$$x = \frac{Q^2}{2P\cdot q}. \tag{4.8}$$

Then it turns out that to a very good approximation these new structure functions are independent of Q^2 and depend *only* on x. For orientation, in Fig. 4.2 we display some of the original SLAC data which showed scaling [Ken 91].

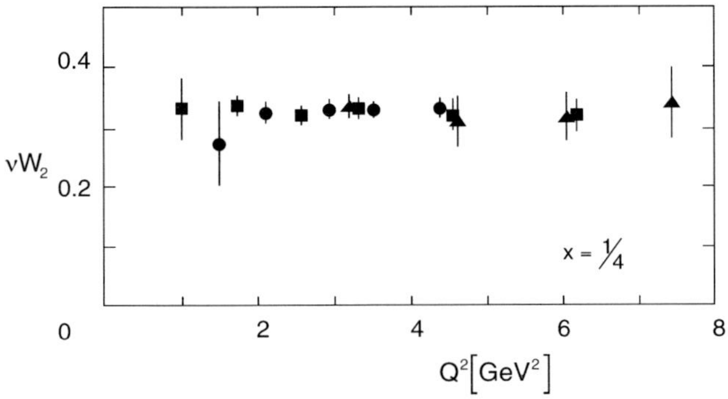

Figure 4.2: A sample of the original SLAC data which first indicated that deep-inelastic scattering yielded structure functions that depended only on x not Q^2 – i.e., that exhibited scaling.

4.1 Parton Model

While the phenomenon of scaling had been anticipated by Bjorken [Bjo 67], the parton picture of Feynman [Fey 69] gives the clearest physical idea of what scaling really means. Any object with a finite size must have a form factor and hence introduce some Q^2 dependence. Thus the first essential point in the parton model is that the proton must contain point-like objects, the "partons", which absorb the virtual photon. With years of experimental work since the original discovery [Pan 68] (the fruits of which we shall summarize below) it has become clear that these partons are quarks, but at the time there was considerable doubt whether quarks had such a solid existence.

Because the structure functions are Lorentz invariant one can formulate the parton model in any frame and it turns out to be simplest in the frame where the nucleon is moving with momentum approaching ∞ along the z-direction – a particular example of the many possible choices of an "infinite momentum frame". That is, to order M^2/P^2, we take the nucleon momentum to be $\tilde{P}^\mu = (P_z + M^2/2P_z; \vec{0}_T, P_z)$, while the photon momentum, $\tilde{q}$, is fixed by requiring that $\tilde{q}^2 = -Q^2$ and $\tilde{P} \cdot \tilde{q} = P \cdot q$ (i.e., $\tilde{q}^\mu = (-xP_z + M\nu/2P_z; \vec{0}_T, -xP_z - M\nu/2P_z)$). Time dilatation implies that in this frame there is no time for interaction between the nucleon constituents during the absorption of the virtual photon and we may therefore treat the process in impulse approximation. The partons each have a momentum which is some fraction, say $y \in [0, 1]$, of the momentum of the nucleon. After absorbing the virtual photon the parton should be on its mass shell – remember that in this frame the partons are non-interacting. As a consequence we find that

$$(y\tilde{P} + \tilde{q})^2 = y^2M^2 - Q^2 + 2y\tilde{P} \cdot \tilde{q} \approx 0, \tag{4.9}$$

and neglecting the small quantity y^2M^2 (compared with Q^2 and ν which are both very large) we find:

$$y = \frac{Q^2}{2\tilde{P}\cdot\tilde{q}} = \frac{Q^2}{2P\cdot q} = x. \tag{4.10}$$

Thus we see that in this frame the Bjorken variable x is just the fraction of the momentum of the nucleon that a parton must have in order to absorb the virtual photon.

To summarize, we have seen that in the parton model the nucleon may be viewed, in an infinite momentum frame, as a collection of non-interacting, point-like constituents, one of which must have fraction x of the momentum of the nucleon in order to absorb the virtual photon. Since they are point-like and non-interacting, the nucleon cross section is clearly just the sum of the cross sections for scattering from individual partons, i, weighted by the number density of partons of type i with momentum fraction x, which in the proton we denote $f_i(x)$. Thus it is intuitively clear that the structure functions, F_1 and F_2, must be proportional to $\sum_i e_i^2 f_i(x)$ (where e_i is the charge of the parton in units of the proton charge). A detailed calculation (which may be found in Section 5.5) yields the result:

$$F_1(x) = \frac{1}{2}\sum_i e_i^2 f_i(x). \tag{4.11}$$

As shown in Appendix B, the structure function F_1 is proportional to σ_T, the cross section for the proton to absorb a transversely polarized photon. F_2 is proportional to $\sigma_T+\sigma_L$. However, helicity conservation in the electromagnetic coupling of a high-energy, spin-$\frac{1}{2}$ fermion implies (see below for more detail) that σ_L must vanish and therefore F_2 is related to F_1 by

$$F_2(x) = 2xF_1(x) = x\sum_i e_i^2 f_i(x). \tag{4.12}$$

This relation, known as the Callan-Gross relation, is quite famous because its experimental verification showed that the partons have spin-$\frac{1}{2}$ – as expected for quarks.

In fact, it is common to use a more descriptive notation in which $f_i(x)$ is replaced by the flavor labels $u(x), d(x), s(x), \bar{u}(x)\ldots$, where, for example, $u(x)dx$ (or $\bar{u}(x)dx$) is the number of up-quarks (or anti-up-quarks) in the proton with momentum between $x\tilde{P}$ and $(x+dx)\tilde{P}$ in the infinite momentum frame. In terms of these number densities, or parton distributions, the proton structure function is

$$F_{2p}(x) = x\left[\frac{4}{9}\big(u(x)+\bar{u}(x)\big) + \frac{1}{9}\big(d(x)+\bar{d}(x)\big) + \frac{1}{9}\big(s(x)+\bar{s}(x)\big)\right], \tag{4.13}$$

while assuming charge symmetry [Lon 98, Bor 99], so that:

$$d^n = u^p \equiv u; \qquad u^n = d^p \equiv d \qquad \text{etc.,} \tag{4.14}$$

the structure function of the neutron is

$$F_{2n}(x) = x\left[\frac{4}{9}\big(d(x)+\bar{d}(x)\big) + \frac{1}{9}\big(u(x)+\bar{u}(x)\big) + \frac{1}{9}\big(s(x)+\bar{s}(x)\big)\right]. \tag{4.15}$$

The differences of the quark and anti-quark distributions for the up and down quark distributions, u_v and d_v respectively, are known as valence distributions

$$u_v = u - \bar{u}; \qquad d_v = d - \bar{d}. \tag{4.16}$$

While it is clear from Eqs. (4.13) and (4.15) that one cannot separate quark and anti-quark distributions using purely electromagnetic interactions, neutrino scattering does allow such a separation – as we shall explain later.

Parton Model with Spin

Spin dependent measurements almost always expose new features of our theoretical understanding. As we shall see this is certainly the case for deep-inelastic scattering, where many exciting developments involve spin physics. In order to appreciate these developments we need to extend the definitions of our parton densities to include polarization.

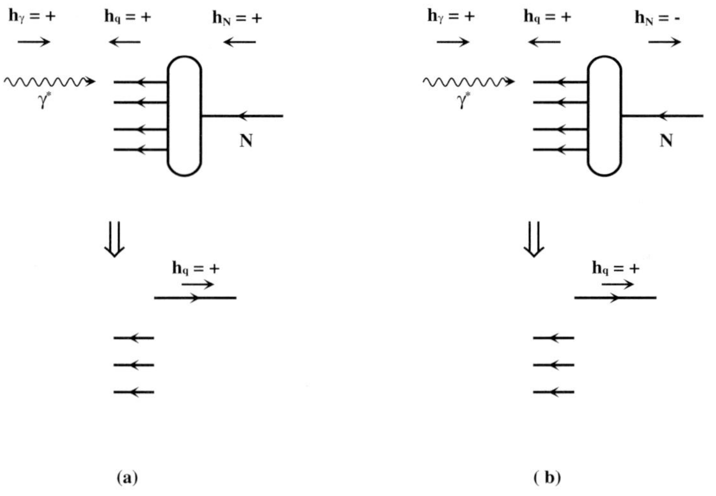

Figure 4.3: The spin dependent amplitudes for a positive helicity, virtual photon ($h_\gamma = +$) to be absorbed by a polarized quark (in a polarized proton) in the photon-parton Breit frame: (a) $\sigma_{1/2}$ and (b) $\sigma_{3/2}$.

Consider a proton with positive helicity in an infinite momentum frame. We define $f_i^\uparrow$ and $f_i^\downarrow$ to be the number densities of quarks of type i with helicity parallel and anti-parallel to that of the target proton, respectively. In the universally used, descriptive notation these become $u^\uparrow, u^\downarrow, d^\uparrow, d^\downarrow$ etc. Clearly our unpolarized distributions are related by $u = u^\uparrow + u^\downarrow$, and so on.

To appreciate the physics involved in measuring a spin structure function it is convenient to examine the absorption of the virtual photon by a parton in a particular infinite momentum frame, namely the Breit frame for the photon and the parton which absorbs it – as illustrated in Fig. 4.3. (We recall that the Breit frame is defined by the parton having equal energy before and after absorbing the photon.) The cross section corresponding to the process represented in Fig. 4.3(a) is denoted $\sigma_{1/2}$, with the subscript indicating the z-component of the total spin of the photon-proton system – i.e., the photon and target proton helicities are the same. In order to absorb the photon the parton must have positive helicity (the same as the proton) *if* it is a quark with spin $\frac{1}{2}$. (If it had negative helicity, angular momentum conservation would require the final quark to have spin $\frac{3}{2}$.) Thus $\sigma_{1/2}$ involves only $f_i^{\uparrow}$. Similarly, Fig. 4.3(b) illustrates the physics underlying $\sigma_{3/2}$, in which case only a quark with helicity opposite to that of the proton can absorb the transverse photon. That is, $\sigma_{3/2}$ involves only $f_i^{\downarrow}$.

The sum of the cross sections for the two processes shown in Fig. 4.3 just yields the unpolarized structure function given above ($F_1 \propto \sigma_{1/2} + \sigma_{3/2}$). On the other hand, the difference yields a new spin-dependent structure function, g_1:

$$\sigma_{1/2} - \sigma_{3/2} \propto g_1 = \frac{1}{2}\sum_i e_i^2 (f_i^{\uparrow} - f_i^{\downarrow}). \tag{4.17}$$

Note that there is an additional spin structure function, g_2, which has been measured and will be discussed below. However, its interpretation within the parton model is not so simple and we therefore do not discuss it further at this stage.

Incidentally, we see now the origin of the Callan-Gross relation (c.f. Eq. (4.12) above). As shown in Appendix B, $W_1 \propto \sigma_T$ and $W_2 \propto \sigma_T + \sigma_L$, where σ_T is the cross section to absorb a transversely polarized photon and σ_L is the cross section to absorb a longitudinally polarized photon. It is easy to show that the electromagnetic current for a high-energy, elementary fermion ($\sim \bar{\psi}\gamma^\mu\psi$) conserves helicity. However, it is clear from Fig. 4.3 that a longitudinal photon (with helicity zero) can only be absorbed by flipping the helicity of the absorbing quark – as it cannot change the spin and in the Breit frame the quark momentum is reversed. If σ_L is suppressed we clearly have $W_1 \propto W_2$.

Before we can actually show some data for $F_{1,2}$ and g_1 it is necessary to expand on our earlier statement that the SLAC data showed scaling "to a very good approximation". This requires an introduction to scaling violations, which have turned out to provide one of the best tests of perturbative QCD. For now our discussion will be more intuitive, with the more rigorous discussion, based on QCD and the operator product expansion, deferred to Chapter 5.

4.2 Scaling Violations

Within the Standard Model the strong interaction sector is a local gauge theory based on color $SU(3)$ and known as quantum chromodynamics (QCD). For the present all we need to know is that the quarks interact by the exchange of spin-1 gluons – at short distances very like photon exchange in QED. Thus, as in QED, when we calculate a cross section for charged particle scattering there are radiative corrections to be made. These correspond to the processes shown in Fig. 4.4. In particular, one cannot separate electron scattering from a quark (Fig. 4.4(a)) from the scattering process in which a soft gluon is also radiated (Fig. 4.4(b)). Just as in QED this soft radiation gives rise to an infinite cross section which can only be treated consistently when all other graphs at the same order – i.e., vertex and wave function renormalization – are included.

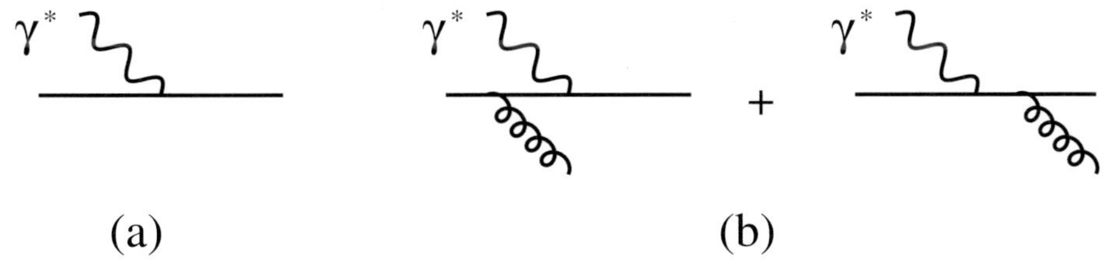

Figure 4.4: The leading order radiative corrections which give rise to a $\ln Q^2$ variation of the parton cross sections.

As a consequence of these radiative corrections the cross section for scattering on a parton develops a logarithmic dependence on Q^2. This relatively weak, logarithmic variation is precisely what we mean by saying that the data scale "to a good approximation". By incorporating the Q^2-dependence into the definition of the parton distributions one can generalize Eqs. (4.13) and (4.15) so that, e.g.,

$$\begin{aligned} F_{2p}(x,Q^2) &= x\left[\frac{4}{9}(u(x,Q^2)+\bar{u}(x,Q^2))+\frac{1}{9}(d(x,Q^2)+\bar{d}(x,Q^2))\right.\\ &\left.+\frac{1}{9}(s(x,Q^2)+\bar{s}(x,Q^2))\right]. \end{aligned} \tag{4.18}$$

Now $u(x,Q^2)dx$ should be interpreted as the number of up quarks in the proton with momentum between $x\tilde{P}$ and $(x+dx)\tilde{P}$ *when viewed with a resolution determined by* Q^2.

The physical picture associated with this definition is very simple. If one looks at the proton with low Q^2 and hence with a large wavelength ($\lambda \sim 1/\sqrt{Q^2}$) and poor spatial resolution, one cannot resolve much of its internal structure. Thus one expects the proton structure function at low Q^2 to be dominated by its valence quarks (which are required to give the correct quantum numbers). However, as Q^2 increases so does the resolving power and one expects to see more and more of the $q\bar{q}$ pairs and gluons which make up the proton in QCD.

Non-singlet Evolution Equations

The variation of the structure functions with Q^2, commonly known as QCD evolution, can be calculated, in leading order, from perturbative QCD using the Feynman rules to evaluate the diagrams shown in Fig. 4.4. By far the clearest physical interpretation is made possible by the formulation of the problem by Gribov and Lipatov [GL 72], Dokshitzer [Dok 77] and Altarelli and Parisi [AP 77], known as the Altarelli-Parisi or the DGLAP equations. These are very simple for the valence quarks, which are the prime example of a "non-singlet" quark distribution. In general terms, a flavor non-singlet distribution is one which cannot mix with gluons – e.g., for the valence distribution, $q_v \equiv q - \bar{q}$, there are always three valence quarks. In this case the DGLAP equations take the form

$$\frac{\partial q_v(x,Q^2)}{\partial lnQ^2} = \frac{\alpha_s(Q^2)}{2\pi}\int_x^1 \frac{dy}{y} P_{qq}(\frac{x}{y}) q_v(y,Q^2). \tag{4.19}$$

Here q_v is either u_v or d_v and $P_{qq}(x/y)$ is the $q-q$ splitting function which may be interpreted as the probability for a quark with momentum fraction y (in the infinite momentum frame) to radiate a gluon, leaving the quark with momentum fraction x. Because the momentum fraction carried by the radiated gluon is positive, only quarks with $y > x$ can contribute. In Eq. (4.19) the QCD coupling constant (which is itself scale-dependent) is $\alpha_s(Q^2)$, which in leading order is (c.f. Eq. (5.94)):

$$\alpha_s(Q^2) = \frac{4\pi}{\beta_0 \ln \frac{Q^2}{\Lambda^2}}; \qquad \beta_0 = 11 - \frac{2}{3}N_f, \tag{4.20}$$

with Λ a constant to be determined from experiment (c.f. Section 5.4.1, $\Lambda \sim 200 - 300$ MeV) and N_f (3 for u, d and s) the number of quark flavors which are energetically allowed at this scale.

The splitting function takes the form (with $C_2(F) = \frac{4}{3}$, corresponding to $N_f = 3$ – see Eq. (5.73))

$$P_{qq}(z) = C_2(F)\left[\frac{1+z^2}{(1-z)_+} + \frac{3}{2}\delta(1-z)\right], \tag{4.21}$$

where the first term, which increases the quark number density at lower values of x, as we just explained, involves the regularization:

$$\int_0^1 dz\, \frac{F(z)}{(1-z)_+} \equiv \int_0^1 dz\, \frac{F(z) - F(1)}{1-z}. \tag{4.22}$$

We note that because

$$\int_0^1 dz\, P_{qq}(z) = 0, \tag{4.23}$$

the total number of *valence* quarks is independent of the resolution as one might expect

$$\frac{\partial}{\partial \ln Q^2} \int_0^1 dx\, q_v(x, Q^2) = \frac{\alpha_s(Q^2)}{2\pi} \int_0^1 dy\, q_v(y, Q^2) \int_0^1 dz P_{qq}(z) = 0, \tag{4.24}$$

where we changed variables to $z = x/y$. Thus we always have two valence up quarks and one valence down quark in the proton.

More generally we may consider the n'th Mellin moment of the valence quark distribution, $M_v^n(Q^2)$, defined by

$$M_v^n(Q^2) = \int_0^1 dx\, x^{n-1} q_v(x, Q^2). \tag{4.25}$$

Taking the n'th moment of Eq. (4.19) we find

$$\frac{\partial M_v^n(Q^2)}{\partial \ln Q^2} = \frac{\alpha_s(Q^2)}{2\pi}\left(-\frac{\gamma_{\mathrm{NS}}^{(0),n}}{4}\right) M_v^n(Q^2), \tag{4.26}$$

where the new constants $\gamma_{\rm NS}^{(0),n}$ are defined through

$$\frac{\gamma_{\rm NS}^{(0),n}}{4} = -\int_0^1 dz\, z^{n-1} P_{qq}(z). \tag{4.27}$$

Using Eq. (4.20) for $\alpha_s(Q^2)$ we may rearrange Eq. (4.26) in the form

$$\partial[\ln M_v^n(Q^2)] = -d_{\rm NS}^{(0),n}\partial\left[\ln\,\ln\left(\frac{Q^2}{\Lambda^2}\right)\right], \tag{4.28}$$

with the so-called (leading order) anomalous dimensions defined as

$$d_{\rm NS}^{(0),n} = \frac{\gamma_{\rm NS}^{(0),n}}{2\beta_0}. \tag{4.29}$$

Finally we may solve Eq. (4.28) to obtain the n'th moment at arbitrary Q^2 in terms of its value at some initial point, Q_0^2:

$$M_v^n(Q^2) = M_v^n(Q_0^2)\left(\frac{\alpha_s(Q^2)}{\alpha_s(Q_0^2)}\right)^{d_{\rm NS}^{(0),n}}, \tag{4.30}$$

and as $d_{\rm NS}^{(0),n} \geq 0$ and $\alpha_s(Q^2)$ decreases as Q^2 increases (the property of QCD known as asymptotic freedom) we see that the moments all vanish logarithmically *unless* $d_{\rm NS}^{(0),n}$ actually is zero – as we just saw for the special case $n = 1$, corresponding to the conserved quantity, baryon number.

Equation (4.19) is characteristic for what are generally known as non-singlet distributions – hence the subscript "NS" on the quantities $\gamma_{\rm NS}^{(0),n}$ and $d_{\rm NS}^{(0),n}$. Perturbative QCD involves the emission or absorption of gluons by quarks and gluons. However, a gluon going to a quark-anti-quark pair ($g \to q\bar{q}$) never changes the number of valence quarks. On the other hand, the electromagnetic structure functions, $F_{1,2}$, involve $q + \bar{q}$ and the total number of $q\bar{q}$ pairs will increase as Q^2 increases. A distribution like F_1 or F_2 which mixes with gluons is known as a "singlet" distribution and in order to describe its evolution we need the singlet Altarelli-Parisi (or DGLAP) equations.

Singlet Altarelli-Parisi Equations

The singlet Altarelli-Parisi (or DGLAP) equations are simplest for a single type of quark with distribution $f_i(x, Q^2)$. Before writing the equations we need to generalize the parton number densities to include gluons. Thus $g(x, Q^2)dx$ is the number of gluons in the momentum interval $[xP, (x + dx)P]$, in a proton (or neutron) with momentum P ($P \to \infty$), when viewed with a resolution determined by Q^2. The relevant splitting functions needed in the singlet case are illustrated in Fig. 4.5:

$$\begin{aligned}
\frac{\partial}{\partial \ln Q^2} f_i(x, Q^2) &= \frac{\alpha_s(Q^2)}{2\pi}\int_x^1 \frac{dy}{y}\left(P_{qq}\left(\frac{x}{y}\right) f_i(y, Q^2) + P_{qg}\left(\frac{x}{y}\right) g(y, Q^2)\right), \\
\frac{\partial}{\partial \ln Q^2} g(x, Q^2) &= \frac{\alpha_s(Q^2)}{2\pi}\int_x^1 \frac{dy}{y}\left(P_{gq}\left(\frac{x}{y}\right) f_i(y, Q^2) + P_{gg}\left(\frac{x}{y}\right) g(y, Q^2)\right).
\end{aligned} \tag{4.31}$$

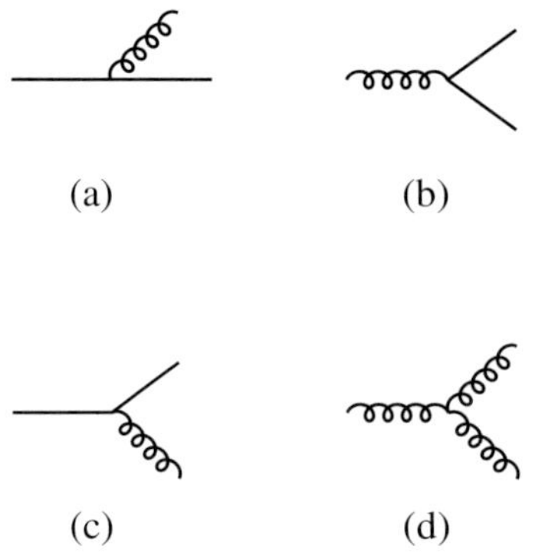

Figure 4.5: Illustration of the four processes which give rise to the splitting functions for the evolution of singlet parton distribution functions: (a) $P_{qq}(z)$, (b) $P_{gq}(z)$, (c) $P_{qg}(z)$, (d) $P_{gg}(z)$.

Note that the splitting functions do not depend on i (for light quarks) and satisfy the symmetry relations:

$$P_{qq}(z) = P_{gq}(1-z); \qquad P_{qg}(z) = P_{qg}(1-z); \qquad P_{gg}(z) = P_{gg}(1-z), \tag{4.32}$$

which are intuitively obvious from Figs. 4.5(a) and (b), Fig. 4.5(c) and Fig. 4.5(d), respectively. The explicit forms of these functions are

$$\begin{aligned} P_{qg}(z) &= \frac{1}{2}\left(z^2 + (1-z)^2\right), \\ P_{gg}(z) &= 2\,C_2(A)\left(\frac{z}{(1-z)_+} + \frac{1-z}{z} + z\,(1-z)\right) + \frac{1}{2}\,\beta_0\delta(1-z), \end{aligned} \tag{4.33}$$

where $C_2(A) = 3$ (c.f. Eq. (5.75)). P_{qq} is the same as in the non-singlet case, Eq. (4.21), and P_{gq} is given by the symmetry relation (4.32).

Because the DGLAP equations are first-order integro-differential equations, once we know the parton distributions at some scale, Q_0^2, they can be calculated at any other scale where leading order, perturbative QCD applies. Figure 4.6 shows the experimental variation of the proton electromagnetic structure function, as a function of Q^2, for a range of values of x, in comparison with the predictions of these equations. Note particularly the enormous range of x and Q^2 covered by the data. The variation with increasing Q^2 is especially rapid at small x, where the increase in resolution allows us to see more $q\bar{q}$ pairs and hence F_2 increases. There is also a significant decrease at larger values of x where momentum is lost as the resolution increases. One must also observe that for x between 0.07 and 0.3 the structure function is remarkably constant over an enormous range of Q^2 and this is what originally led to the idea of scaling. In conclusion, we observe that the beautiful description of the Q^2 variation of the data, evident in Fig. 4.6, is a tremendous success for the approach and most importantly for perturbative QCD.

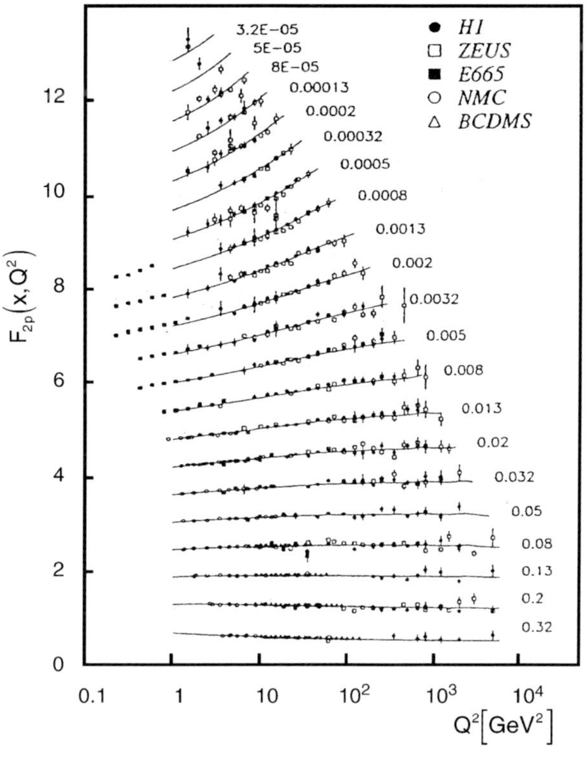

Figure 4.6: The scale dependence of the proton electromagnetic structure function, $F_{2p}(x, Q^2)$, at a range of values of Bjorken x – from Ref.[PDG 98/00]. (The vertical scale for F_{2p} has been adjusted in order to separate the plots for different x.)

4.3 Neutrino Deep-Inelastic Scattering

The differential cross section for unpolarized electron (or muon) scattering from a nucleon has already been given in Eq. (4.6). Such a measurement allows us to determine the two structure functions F_1 and F_2. However, it does not permit the separation of quark and anti-quark distributions. Since, in order to appreciate what the data means, this separation is crucial, we shall first extend the discussion of deep-inelastic scattering to include neutrino structure functions and to show how these do allow one to separate $q(x, Q^2)$ and $\bar{q}(x, Q^2)$.

To measure any cross section involving a neutrino beam presents an enormous experimental challenge. With $10^7 \nu_\mu$ incident per second on a 1,000 tonne detector one can expect typically one deep-inelastic event per second! Nevertheless, we now have data on hydrogen, deuterium and heavier nuclei, which have given important insight into the structure of the nucleon. The essential, new feature of the weak interaction is that parity is violated. For simplicity we shall consider the charged current case (i.e., $N(\nu_\mu, \mu^-)X$ and $N(\bar{\nu}_\mu, \mu^+)X$) in detail and simply summarize the corresponding neutral current results (for which there is much less data).

Suppose we begin with the reaction

$$\nu_\mu + p \to \mu^- + X. \tag{4.34}$$

The incident neutrino necessarily carries negative helicity. Therefore, in constructing the corresponding lepton tensor, $L_{\mu\nu}^{\nu,\bar{\nu}}$, we must deal with an initially polarized state. As shown in Appendix B, this adds an anti-symmetric piece, so that

$$\begin{aligned} L_{\mu\nu}^{\nu,\bar{\nu}} &= 2\left(p_\mu p'_\nu + p'_\mu p_\nu - g_{\mu\nu} p\cdot p' \pm i\epsilon_{\mu\nu\alpha\beta}\, p^\alpha p'^\beta\right), \\ &= L_{\mu\nu}^{S} + L_{\mu\nu}^{A(\nu,\bar{\nu})}. \end{aligned} \tag{4.35}$$

(Note that, for $\epsilon_{0123} = +1$, the upper and lower signs refer to ν_μ and $\bar{\nu}_\mu$ respectively, while the superscripts S and A refer to the symmetry under interchange of the Lorentz indices μ and ν.)

The hadronic tensor is much more complicated in the case of the weak interaction because we no longer have gauge invariance or parity conservation as constraints. In this case, the most general tensor representing the structure of the nucleon is

$$\begin{aligned} W^{\mu\nu} = \frac{1}{4M} \sum_{s,X} &< N_s(P)|J^\mu_{\rm CC}(0)|X(P')> \\ &< X(P')|J^\nu_{\rm CC}(0)|N_s(P)> (2\pi)^3 \delta^4(q+P-P') \end{aligned} \tag{4.36}$$

$$\begin{aligned} = -W_1 g^{\mu\nu} + W_2 \frac{P^\mu P^\nu}{M^2} + W_3 i\epsilon^{\mu\nu\alpha\beta} \frac{P_\alpha q_\beta}{2M^2} + W_4 \frac{q^\mu q^\nu}{M^2} \\ + W_5 \frac{P^\mu q^\nu + q^\mu P^\nu}{M^2} + W_6 \frac{P^\mu q^\nu - q^\mu P^\nu}{M^2}, \end{aligned} \tag{4.37}$$

with $J^\mu_{\rm CC}$ the charged weak current of Eq. (1.20). Note that we do not attach extra labels (ν or $\bar{\nu}$) to the Lorentz invariant functions, W_i, to distinguish them from the electromagnetic case, unless it is unclear from the context which is intended. In comparison with the electromagnetic case, W_3 and W_6 are allowed here because parity is violated and W_4 and W_5 are not related to W_1 and W_2 by gauge invariance. Clearly the only non-zero contributions to the cross section involve the symmetric parts of $W^{\mu\nu}$ ($W_{1,2,4,5}$) with $L^S_{\mu\nu}$ and the anti-symmetric terms, $W_{3,6}$ with $L^A_{\mu\nu}$. After working through the algebra one finds that the coefficients of the terms involving $W_{4,5,6}$ are proportional to the muon mass and are therefore negligible for the kinematics of the deep-inelastic region. Thus the laboratory differential cross section has the form (see Appendix B for details)

$$\begin{aligned} \frac{d^2\sigma^{(\nu,\bar{\nu})}}{d\Omega d\epsilon'} = \frac{G_F^2}{2\pi^2}\epsilon'^2 \left(\frac{M_W^2}{M_W^2 + Q^2}\right)^2 \\ \left[2W_1 \sin^2\frac{\theta}{2} + W_2 \cos^2\frac{\theta}{2} \pm W_3 \frac{\epsilon+\epsilon'}{M}\sin^2\frac{\theta}{2}\right]. \end{aligned} \tag{4.38}$$

In analogy with the electromagnetic case it is conventional to define the structure functions F_i ($i = 1, 2, 3$), which scale, as

$$F_1 = MW_1; \qquad F_2 = \nu W_2; \qquad F_3 = \nu W_3. \tag{4.39}$$

Invariant Cross Section

It is quite common in the experimental literature to present the data in terms of invariant cross sections, using the Lorentz invariant variables x and y. Here y is the fraction of the incident neutrino energy transferred to the hadronic target:

$$y = \frac{\nu}{\epsilon} = \frac{p \cdot q}{p \cdot k}. \tag{4.40}$$

Using

$$Q^2 = 4\epsilon\epsilon' \sin^2\frac{\theta}{2} = 4\epsilon^2(1-y)\sin^2\frac{\theta}{2} = 2M\nu x, \tag{4.41}$$

(where the lepton mass squared has been neglected) and integrating over the azimuthal angle we find

$$dQ^2 = 2\,M\epsilon y\,dx = \epsilon^2(1-y)\,\frac{d\Omega}{\pi}, \tag{4.42}$$

so

$$d\Omega\,d\epsilon' = dQ^2\,d\nu\,\frac{\pi}{\epsilon^2(1-y)}, \tag{4.43}$$

and

$$dQ^2\,d\nu = 2M\epsilon^2 y\,dx\,dy. \tag{4.44}$$

Thus

$$\frac{d^2\sigma^{(\nu,\bar{\nu})}}{dx\,dy} = \frac{G_F^2(s-M^2)}{2\,\pi}\left(\frac{M_W^2}{M_W^2+Q^2}\right)^2 \left[x\,y^2F_1 + (1-y)F_2 \pm y\left(1-\frac{y}{2}\right)xF_3\right], \tag{4.45}$$

where we note that the factor $(1-\frac{y}{2})$ arises because

$$\epsilon + \epsilon' = 2\epsilon\left(1-\frac{y}{2}\right). \tag{4.46}$$

In Eq. (4.45), s is the invariant mass squared of the neutrino target collision:

$$s = M^2 + 2M\epsilon, \tag{4.47}$$

so that the neutrino cross section is linear in the incident neutrino energy.

Parton Model

The parity conserving pieces of the hadronic tensor $W^{\mu\nu}$ are essentially identical to those in the electromagnetic case. (This is trivial for the vector-vector piece of the product of currents and not much less obvious for the axial-axial piece, as $\gamma_5^2 = 1$.) Only the weak charges are different, with

$$(Q_i^{\rm CC})^2 = 1: \quad d, s, \bar{u}, \bar{c} \\ \quad\quad = 0: \quad u, c, \bar{d}, \bar{s}, \tag{4.48}$$

in the case of $\nu_\mu \to \mu^-$, and vice-versa for $\bar{\nu}_\mu \to \mu^+$. (A more precise analysis must include the CKM matrix elements - c.f., Eq. (1.22) - which we set to unity for simplicity.) Thus, within the parton model

$$F_2(x) = 2xF_1(x) = 2x\sum_i (Q_i^{\rm CC})^2 f_i(x). \tag{4.49}$$

On the other hand, the parity violating structure function arises from the vector-axial-vector product and has the opposite sign for q (left-handed) and $\bar{q}$ (right-handed). Therefore, within the parton model we find

$$xF_3 = 2x\sum_i (Q_i^{\rm CC})^2 (-)^{\frac{1}{2}-\frac{3}{2}b_i} f_i(x), \tag{4.50}$$

where $b_i = +\frac{1}{3}(-\frac{1}{3})$ is the baryon number of the struck $q(\bar{q})$. For example,

$$xF_3^{\nu p} = 2x[d(x) + s(x) - \bar{u}(x)], \tag{4.51}$$

for three flavors. Putting all this together we find, again within the parton model, and for $Q^2 \ll M_W^2$ (and $s \gg M^2$):

$$\frac{d^2\sigma^{\nu p}}{dx\,dy} = \frac{G_F^2 s}{\pi}\left[x(d(x) + s(x)) + x\,(1-y)^2 \bar{u}(x)\right], \tag{4.52}$$

$$\frac{d^2\sigma^{\bar{\nu} p}}{dx\,dy} = \frac{G_F^2 s}{\pi}\left[x(\bar{d}(x) + \bar{s}(x)) + x\,(1-y)^2 u(x)\right]. \tag{4.53}$$

The physical origin of the coefficient 1 for νq and $\bar{\nu}\bar{q}$, and $(1-y)^2$ for $\nu\bar{q}$ and $\bar{\nu}q$, is easy to understand. In the neutrino-quark c.m. system the total spin is zero in the former case, so the scattering is isotropic. However, in the latter case the total spin projection is ± 1 (respectively, if the quark travels in the +z-direction), so the cross section is proportional to $|d_{11}^1(\cos\theta^*)|^2$, with d_{11}^1 the Wigner rotation matrix and θ^* the c.m. scattering angle. Using the fact that $1 - y = (1 + \cos\theta^*)/2 = d_{11}^1(\cos\theta^*)$ completes the explanation. Clearly the different dependence of the quark and anti-quark contributions on y allows one to separate these distributions. The results from various experiments will be summarized in the next section. It is evident from our discussion that neutrino experiments offer the chance to separate information on quarks and anti-quarks and hence to more or less completely determine the parton distributions. While there are many interesting combinations of cross sections that we could explore we finish with just two examples. Consider first the neutrino cross section on an isoscalar target (such as the deuteron) – where we neglect any nuclear corrections for the present:

$$\frac{d^2\sigma^{\nu D}}{dx\,dy} = \frac{G_F^2 s}{\pi}\Big[x\big(u(x) + d(x) + 2s(x)\big) + x\,(1-y)^2\big(\bar{u}(x) + \bar{d}(x)\big)\Big]. \tag{4.54}$$

Provided that $s(x) = \bar{s}(x)$, as is commonly assumed, we see that in the region $y \sim 0$ such a measurement provides a direct determination of the combination $(u + d + s + \bar{u} + \bar{d} + \bar{s})$, which is required for the momentum sum rule, Eq. (4.73). This sum also appears directly in the isoscalar combination of structure functions:

$$F_2^{\nu p} + F_2^{\nu n} = 2x\left[u + \bar{u} + d + \bar{d} + 2s\right]. \tag{4.55}$$

Secondly, it is easily seen from Eqs. (4.50) and (4.51) that the combination $F_3^{\nu p} + F_3^{\bar{\nu} p}$ measures the valence quark distribution directly (again assuming $s(x) = \bar{s}(x)$)

$$\frac{F_3^{\nu p}(x) + F_3^{\bar{\nu} p}(x)}{2} = \big(u(x) - \bar{u}(x)\big) + \big(d(x) - \bar{d}(x)\big). \tag{4.56}$$

Neutral Current Interactions

At sufficiently high momentum transfer, for example at HERA, the exchange of a Z^0 boson can begin to make a measurable contribution to the deep inelastic cross section. Indeed, although the photon exchange still dominates in the region of momentum transfers accessible at HERA, the $\gamma - Z^0$ interference is clearly visible in the data. It is convenient to summarize the cross section for the scattering of a polarized electron from a proton target. The subscript L or R refers to a left- or right-handed particle [PDG 98/00, IR 88, Rob 90]:

$$\frac{d^2\sigma}{dx\,dy} = \frac{\pi\alpha^2}{sx^2y^2}\left[x\sum_q \left(q(x,Q^2) + \bar{q}(x,Q^2)\right)\right] \times \left[\mathcal{A}_q + (1-y)^2\mathcal{B}_q\right]. \tag{4.57}$$

The coefficients $\mathcal{A}_q$ and $\mathcal{B}_q$ involve interference between photon and Z^0 exchange:

$$\begin{aligned}\mathcal{A}_q &= \left[e_q - g_{Lq}g_{Le}\frac{Q^2}{Q^2+M_Z^2}\right]^2 + \left[e_q - g_{Rq}g_{Re}\frac{Q^2}{Q^2+M_Z^2}\right]^2,\\ \mathcal{B}_q &= \left[e_q - g_{Rq}g_{Le}\frac{Q^2}{Q^2+M_Z^2}\right]^2 + \left[e_q - g_{Lq}g_{Re}\frac{Q^2}{Q^2+M_Z^2}\right]^2.\end{aligned} \tag{4.58}$$

As usual, e_q is the quark charge and for a left-handed electron:

$$g_{Re} = 0; \qquad g_{Le} = \frac{-\frac{1}{2} + \sin^2\theta_\mathrm{W}}{\sin\theta_\mathrm{W}\cos\theta_\mathrm{W}}, \tag{4.59}$$

while for a right handed electron:

$$g_{Le} = 0; \qquad g_{Re} = \frac{\sin^2\theta_\mathrm{W}}{\sin\theta_\mathrm{W}\cos\theta_\mathrm{W}}, \tag{4.60}$$

where $\sin\theta_\mathrm{W}$ is the sine of the Weinberg angle, given in Eq. (1.2). Finally, for the quarks the weak charges are:

$$\begin{aligned}g_{Lq} &= \frac{t_3 - e_q\sin^2\theta_\mathrm{W}}{\sin\theta_\mathrm{W}\cos\theta_\mathrm{W}},\\ g_{Rq} &= \frac{-e_q\sin^2\theta_\mathrm{W}}{\sin\theta_\mathrm{W}\cos\theta_\mathrm{W}}.\end{aligned} \tag{4.61}$$

The corresponding cross sections for left-handed (right-handed) positrons are obtained by replacing g_{Le} by g_{Re} (g_{Re} by g_{Le}).

Although it is a much more difficult experiment, Eq. (4.57) also describes neutral current scattering of neutrinos (anti-neutrinos), provided we replace g_{Le} by $g_{L\nu}$ (g_{Re} by $g_{R\bar{\nu}}$). Of course, we set $g_{L\bar{\nu}} = 0$ and $g_{R\nu} = 0$ and finally:

$$\begin{aligned} g_{L\nu} &= +\frac{1}{\sin 2\theta_{\mathrm{W}}}, \\ g_{R\bar{\nu}} &= -\frac{1}{\sin 2\theta_{\mathrm{W}}}. \end{aligned} \tag{4.62}$$

4.4 Sum Rules

As the parton distributions are number densities defined on the momentum fraction $x \in [0, 1]$, the integrals over this range are just the total numbers of each type of quark

$$n_i = \int_0^1 dx\, f_i(x). \tag{4.63}$$

Simply from the quantum numbers of the proton (strangeness zero, baryon number one, charge one) we therefore have the obvious sum rules

$$\int_0^1 dx\, u_v(x, Q^2) \equiv \int_0^1 dx\, [u(x, Q^2) - \bar{u}(x, Q^2)] = 2, \tag{4.64}$$

$$\int_0^1 dx\, d_v(x, Q^2) \equiv \int_0^1 dx\, [d(x, Q^2) - \bar{d}(x, Q^2)] = 1, \tag{4.65}$$

$$\int_0^1 dx\, [s(x, Q^2) - \bar{s}(x, Q^2)] = 0. \tag{4.66}$$

Of course, what one determines experimentally are various structure functions, not the parton distributions. Thus the sum rules must be expressed in terms of the measured structure functions. In this form they are also subject to QCD corrections involving powers of $\alpha_s(Q^2)$, as well as "higher-twist" corrections (rigorously defined in Chapter 5) which involve powers of $1/Q^2$. In practice one can expect to remove the higher-twist corrections by careful measurements over a suitable range of momentum transfer, whereas for the corrections of $O(\alpha_s)$ one usually needs to rely on perturbative QCD. Although in Section 4.2 we discussed only the simplest, or "leading twist", variation of the parton distributions we shall quote the QCD corrections to higher order if they are known.

4.4.1 Gross-Llewellyn-Smith Sum Rule

From Eqs. (4.64) and (4.65) one sees that the total number of valence quarks must be three:

$$\int_0^1 dx\, \left[u(x, Q^2) - \bar{u}(x, Q^2) + d(x, Q^2) - \bar{d}(x, Q^2)\right] = 3. \tag{4.67}$$

In terms of measured distributions, neutrino deep-inelastic scattering has a third, parity-violating structure function, F_3 (c.f. Eqs. (4.50) and (4.56)), which involves q and $\bar{q}$ with opposite sign. Therefore the sum rule associated with Eq. (4.67), known as the Gross-Llewellyn-Smith sum rule [GLS 69], is:

$$\begin{aligned} S_{\text{GLS}}(Q^2) &= \frac{1}{2}\int_0^1 dx \left(F_3^{\nu p}(x,Q^2) + F_3^{\bar{\nu} p}(x,Q^2)\right) \\ &= 3\left[1 - \frac{\alpha_s(Q^2)}{\pi} - 3.58\left(\frac{\alpha_s(Q^2)}{\pi}\right)^2 - 20.22\left(\frac{\alpha_s(Q^2)}{\pi}\right)^3\right], \end{aligned} \tag{4.68}$$

where the QCD correction corresponds to $N_f = 3$ [Lar 94]. The CCFR collaboration at Fermilab has reported a value of

$$S_{\text{GLS}}(Q^2 = 3\,\text{GeV}^2) = 2.50 \pm 0.018\,(\text{stat}) \;\pm 0.078\,(\text{syst}), \tag{4.69}$$

which is in excellent agreement with the theoretical value of 2.66 ± 0.04, expected at this scale [CTEQ 95]. In Fig. 4.7 we show a summary of the Q^2-dependence of this sum rule

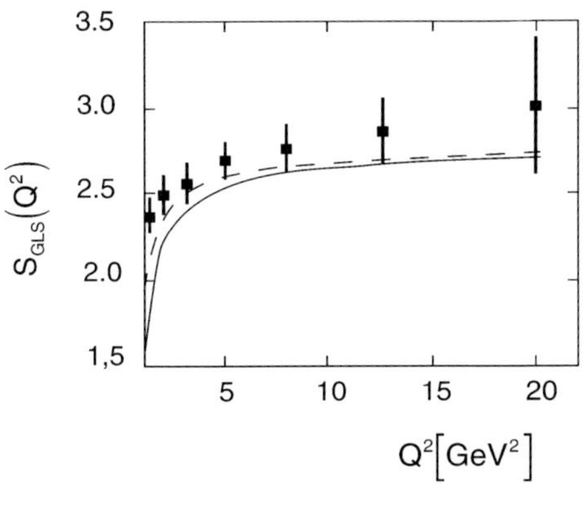

Figure 4.7: Experimental determination of the Gross-Llewellyn-Smith sum rule as a function of Q^2, from the work of the CCFR collaboration [CCFR 93]. The curves allow for QCD radiative corrections which vanish logarithmically as $Q^2 \to \infty$.

together with curves illustrating the effect of QCD radiative corrections. Within the errors it certainly seems reasonable to conclude that the number of valence quarks in the nucleon is indeed three.

4.4.2 Adler Sum Rule

The Adler sum rule [Ad 66] involves the difference of Eqs. (4.64) and (4.65):

$$\int_0^1 dx \left[\left(u(x,Q^2) - \bar{u}(x,Q^2)\right) - \left(d(x,Q^2) - \bar{d}(x,Q^2)\right)\right] = 1, \tag{4.70}$$

which can be obtained by taking the difference of the ν and $\bar{\nu}$ structure functions on the proton (c.f. Eq. (4.49)):

$$S_A = \int_0^1 dx \left(F_1^{\bar{\nu}p}(x, Q^2) - F_1^{\nu p}(x, Q^2) \right) = 1. \tag{4.71}$$

In this case there are no perturbative QCD corrections because Eq. (4.71) can be derived from the commutator of conserved currents. From the point of view of QCD evolution, the change in the integral of $u - \bar{u}$ is exactly cancelled by the change in the integral of $d - \bar{d}$.

Unfortunately this sum rule is very hard to test experimentally because of the difficulty of controlling systematic errors, especially at small-x. The value obtained by the WA25 group working at the (former) big European bubble chamber (BEBC) is [All 85]:

$$S_A = 1.08 \pm 0.08 \text{ (stat) } \pm 0.18 \text{ (syst)}, \tag{4.72}$$

which agrees with the theoretical expectation. However, one would clearly like to see a much more precise test of such a fundamental sum rule.

4.4.3 Momentum Sum Rule

By weighting the parton number densities by the momentum fraction, x, one can determine the proportion of the momentum of the nucleon carried by quarks, S_{quark}. In a naive quark model one might expect this to be one. Omitting any explicit mention of heavier quarks for the present, we define

$$S_{\text{quark}}(Q^2) = \int_0^1 dx x \Big[u(x, Q^2) + \bar{u}(x, Q^2) + d(x, Q^2) \\ + \bar{d}(x, Q^2) + s(x, Q^2) + \bar{s}(x, Q^2) \Big]. \tag{4.73}$$

As we have seen in Section 4.3, especially Eq. (4.55), this quantity can be measured directly in neutrino deep-inelastic scattering. At $Q^2 \sim 2\ \text{GeV}^2$ the data yields a value for S_{quark} of about 61 % [MRS 98], which is clearly less than our naive expectation. The rest of the momentum of the target must be carried by strongly interacting constituents that do not interact weakly or electromagnetically. Within the Standard Model these constituents must be the gluons, and using the definition of the gluon distribution in the infinite momentum frame, $g(x, Q^2)$, the fraction of the momentum of the proton carried by gluons is therefore

$$S_g = \int_0^1 dx x g(x, Q^2). \tag{4.74}$$

Thus, the momentum sum rule is

$$S_{\text{mom}} = S_{\text{quark}}(Q^2) + S_g(Q^2) = 1. \tag{4.75}$$

In practice it is not possible to test the momentum sum rule accurately because $g(x, Q^2)$ can only be determined indirectly. For example, one can use the singlet DGLAP equations, Eqs. (4.31), to self-consistently extract the gluon distribution from the Q^2-dependence of $F_2(x, Q^2)$. More generally, Eq. (4.75) is applied as a constraint on $g(x, Q^2)$, once the quark distributions have been obtained.

4.4.4 Gottfried Sum Rule

Suppose we take the difference of the proton and neutron electromagnetic structure functions:

$$\begin{aligned} F_{2p} - F_{2n} &= \frac{x}{3}\left[\left(u(x,Q^2) + \bar{u}(x,Q^2)\right) - \left(d(x,Q^2) + \bar{d}(x,Q^2)\right)\right], \\ &= \frac{x}{3}\left[\left(u(x,Q^2) - \bar{u}(x,Q^2)\right) - \left(d(x,Q^2) - \bar{d}(x,Q^2)\right)\right] \\ &\quad - \frac{2x}{3}\left(\bar{d}(x,Q^2) - \bar{u}(x,Q^2)\right). \end{aligned} \tag{4.76}$$

In the first two terms we recognize the same combination as in the Adler sum rule. To the extent that the sea of $q\bar{q}$ pairs in the nucleon is generated purely perturbatively (as described in Section 4.2) we expect $\bar{d} = \bar{u}$. That is, regardless of which valence quark (u or d) radiates a gluon, the gluon has an equal probability to make a $u\bar{u}$ or a $d\bar{d}$ pair ($s\bar{s}$ may be suppressed a little because of its greater mass). (The asymmetry arising from anti-symmetrization with the two valence u and one valence d quarks in the proton has been shown to give a very small correction to this perturbative sea [RS 79].) Under the assumption that $\bar{d} = \bar{u}$ we find the Gottfried sum rule

$$S_G = \int_0^1 \frac{dx}{x}\left(F_{2p}(x,Q^2) - F_{2n}(x,Q^2)\right) = \frac{1}{3}. \tag{4.77}$$

In a very real sense this sum rule is the first to offer genuine insight into the structure of the nucleon. This is because there is no symmetry constraint forcing $\bar{d} = \bar{u}$, it is a dynamical assumption. Any deviation of the measured value of S_G from $1/3$ would imply that the sea of the nucleon must involve physics beyond perturbative QCD.

The New Muon Collaboration (NMC) at CERN provided the first clear evidence of a violation of the Gottfried sum rule [NMC 90]. Fig. 4.8 shows the integral in Eq. (4.77) as a function of the lower cut-off in x. Because the data is taken on the deuteron there are various small corrections, illustrated by the different theoretical curves [MT 93]. Nevertheless, it is clear that even taking into account such minor uncertainties, the integral is well below $\frac{1}{3}$. The final analysis of the NMC data gave a value for the Gottfried integral over the range (0.004,0.8) equal to 0.2281 ± 0.0065 (stat) at $Q^2 = 4$ GeV^2 [NMC 97], which suggests a value for the integral of $(\bar{d} - \bar{u})$ of the order of 0.15.

Since the original NMC measurement, the deviation of $\bar{d}(x) - \bar{u}(x)$ from zero has been confirmed by two Drell-Yan experiments, NA51 at CERN [NA51 94] and E866 at Fermilab [NuSea 98]. The latter has in fact mapped out the x-dependence of $\bar{d} - \bar{u}$ in a spectacular fashion over the region $x \in (0, 0.3)$, as shown in Fig. 4.9. In the Drell-Yan reaction a quark and an anti-quark annihilate to a virtual photon, which decays into a $\mu^+\mu^-$ pair with invariant mass above the J/Ψ resonance (and not near the Υ). Energy momentum conservation means that one can reconstruct the momentum fraction of the beam (x_1) and target (x_2) carried by the partons which annihilate. By choosing the kinematics one can ensure that $x_1 > x_2$ and hence the beam parton is predominantly a valence quark. Because the process is electromagnetic the cross section involves the square of the quark charge, so the dominant contribution is a valence u in the beam proton annihilating with a $\bar{u}$ in the target. Hence the ratio of pp to pD

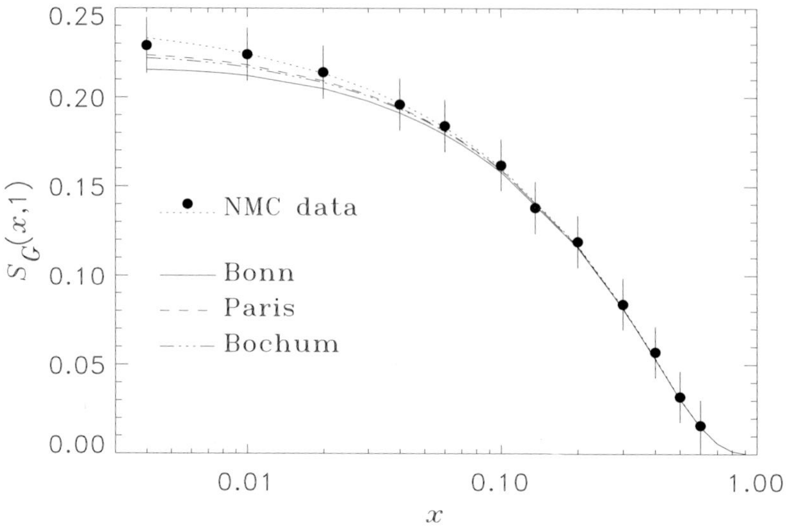

Figure 4.8: Convergence of the experimental determination of the Gottfried sum rule [NMC 90], as a function of the lower limit of the integral over x – from Ref.[MT 93].

Drell-Yan cross sections is approximately:

$$\begin{aligned}\frac{\sigma_{pD}}{\sigma_{pp}} &\approx \frac{u_v^p(x_1)\,\bar{u}^p(x_2) + u_v^p(x_1)\,\bar{u}^n(x_2)}{u_v^p(x_1)\,\bar{u}^p(x_2)} \\ &= 1 + \frac{\bar{d}(x_2)}{\bar{u}(x_2)},\end{aligned} \tag{4.78}$$

where the last step follows from charge symmetry (c.f. Eq. (4.14)). We stress that Eq. (4.78) is approximate and the complete expressions were used in extracting the $\bar{d}/\bar{u}$ ratio shown in Fig. 4.9.

While the theoretical explanation of the observed violation of the Gottfried sum rule is still a matter for research, we note that the role of the pion cloud of the nucleon [Tho 83, ST 97] is vital. The physical idea is very simple; the dominant long-range structure of the proton is given by the virtual process $p \to n\pi^+$ and, in a simple valence picture of the nucleon and the pion, this Fock component of the wavefunction involves only $\bar{d}$ quarks, not $\bar{u}$. From a more sophisticated point of view, as we shall explain in Chapter 7, Goldstone boson loops give rise to a characteristic, non-analytic behavior as a function of the quark mass ($\bar{m}$). The realization that the moments of $\bar{d} - \bar{u}$ are non-analytic in $\bar{m}$ [TFM 00] and, in particular:

$$\int_0^1 \left(\bar{d}(x) - \bar{u}(x)\right)_{\text{LNA}} dx = \frac{2g_A^2}{16\pi^2 f_\pi^2}\, m_\pi^2 \ln m_\pi^2, \tag{4.79}$$

has established the unambiguous role of dynamical symmetry breaking in this phenomenon. (From Eq. (6.54), the right hand side of Eq. (4.79) is proportional to $\bar{m}\ln\bar{m}$, which has a branch point starting at $\bar{m} = 0$. This is the "leading non-analytic contribution" and hence the subscript "LNA".)

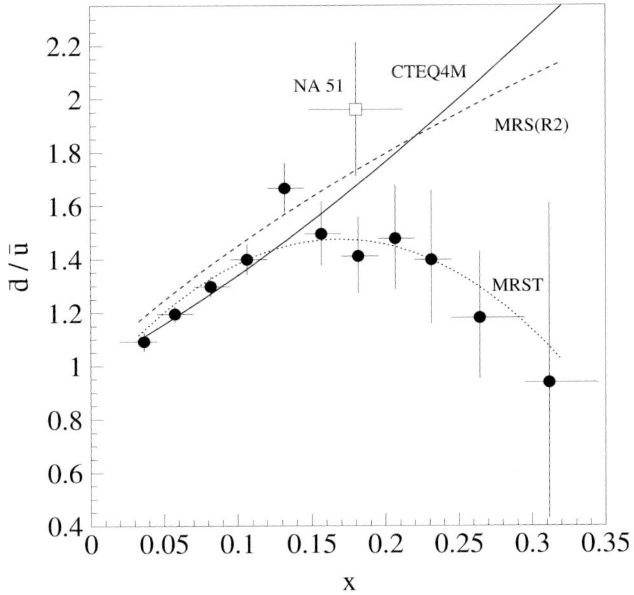

Figure 4.9: x-dependence of the difference $\bar{d} - \bar{u}$ extracted from the E866 Drell-Yan data [NuSea 98]. The curves labeled CTEQ4M, MRS(R2) and MRST are phenomenological fits to the data before and after the E866 measurements.

4.5 Experimental Results for Unpolarized Targets

Since the structure function of the nucleon depends on Q^2 as well as x it is common to display the data at some reference scale as a function of x. This is typically taken to be $5, 10$ or 15 GeV^2. Of course, in any given experiment the data are collected over a range of values of x and Q^2 which are usually highly correlated. In fact, as the spectrometer for the final lepton accepts typically quite a narrow range of energy transfers, ν, and as x is $Q^2/2M\nu$, the smaller the value of x in a given experiment the smaller the corresponding momentum transfer is likely to be. The procedure for producing $F_{2p}(x, Q^2)$ at fixed Q^2 therefore involves quite a complicated fitting process involving the DGLAP equations and overlapping data from a number of different experiments – each with its own systematic errors. Indeed, it is never possible to extract $F_{2p}(x, Q^2)$ for *all* x at a given value of Q^2. With all these difficulties in the analysis there is a remarkable consistency between the various data sets, as we have already seen in Fig. 4.6.

Figure 4.10 summarizes the key features of an enormous amount of data from both electromagnetic (BCDMS, BFP and EMC) and neutrino (CCFRR, CDHSW and CHARM) experiments[1]. Since the data are taken on nuclei, typically Ca or Fe, they are presented as averages of neutron and proton structure functions. Thus for the neutrino data F_2 is actually

[1]For a compilation of structure function data we refer to Ref.[GRM 99].

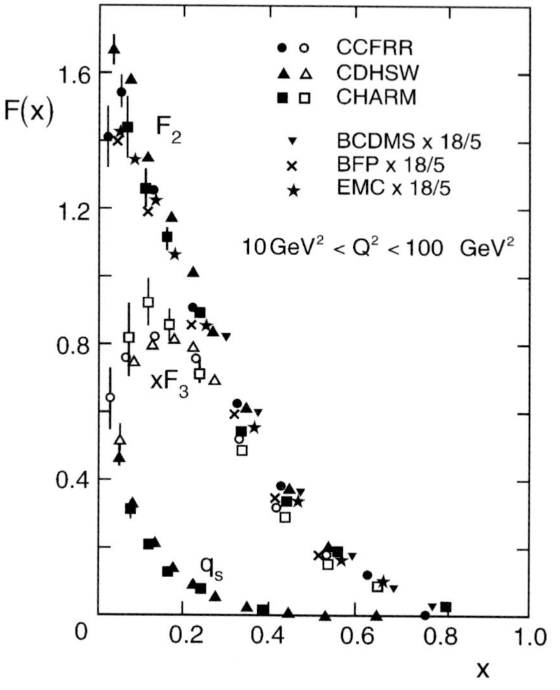

Figure 4.10: Summary of deep-inelastic scattering data from both weak and electromagnetic interactions on medium mass nuclei.

$(F_2^{\nu p} + F_2^{\nu n})/2$ – c.f. Eq. (4.55). The corresponding electromagnetic structure function is

$$F_2^{eD} \equiv \frac{F_2^{ep} + F_2^{en}}{2} = \frac{5x}{18}\left[u + \bar{u} + d + \bar{d} + \frac{2}{5}(s + \bar{s})\right]. \tag{4.80}$$

Provided that the strange quark distribution is suppressed, this leads one to expect $F_2 \approx \frac{18}{5} F_2^{eD}$. This is clearly consistent with the measurements and constitutes unambiguous evidence that deep-inelastic scattering is probing quarks whose electro-weak charges are precisely as expected in the Standard Model.

By comparing the valence quark distribution $F_3(x)$ – c.f. Eq. (4.56) – with F_2, we see that the valence quarks dominate for x beyond 0.3. On the other hand, as $x \rightarrow 0$ the sea of additional $q - \bar{q}$ pairs is dominant. This tendency grows with increasing Q^2, as we can see from Fig. 4.6.

Since the parton number densities diverge as $x \rightarrow 0$, it is convenient to use momentum fraction, rather than absolute number, to characterize the various flavors. The valence quarks carry about 35 % of the momentum of the nucleon at $Q^2 = 20$ GeV^2, while the sea has about 21 % [MRS 98]. As discussed around Eq. (4.75), the missing momentum is attributed to gluons. Within the sea $SU_F(3)$ (flavor $SU(3)$ symmetry) is badly broken. Indeed, by looking explicitly for the decay of the charm quarks produced in a deep-inelastic event involving an s-quark ($\nu_\mu + s \rightarrow \mu^- + c; c \rightarrow \mu^+ + \nu_\mu + s$), so-called opposite-sign di-muon events, the

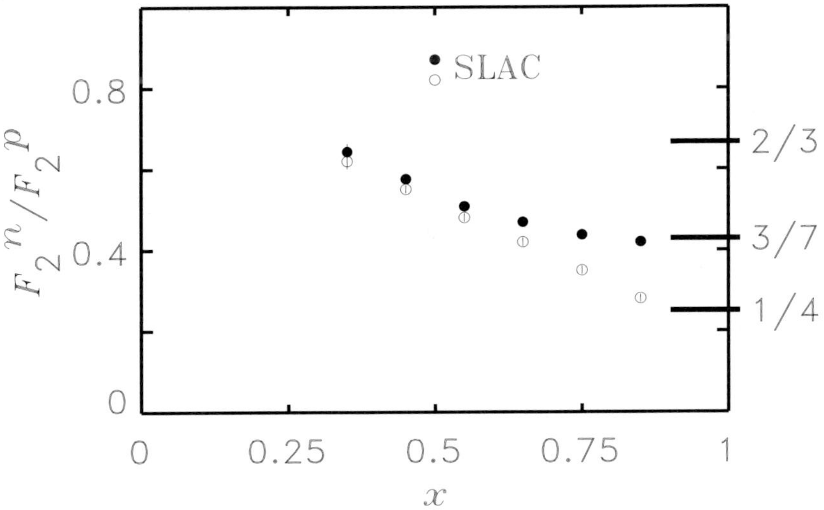

Figure 4.11: The ratio of n to p electromagnetic structure functions, extracted from SLAC data for p and D targets, using two theoretical treatments of the Fermi motion and binding corrections. The more recent analysis, the solid circles, is consistent with the expectation of perturbative QCD that $\frac{d}{u} \to \frac{1}{5}$ as $x \to 1$ – from Ref.[MT 96].

CCFR collaboration finds [CCFR 99]:

$$\frac{2 < x\bar{s} >}{< x(\bar{u} + \bar{d}) >} = 0.42 \pm 0.07 \pm 0.06. \tag{4.81}$$

With regard to $SU_F(2)$, we have already noted the clear evidence for a violation of the Gottfried sum rule. In the analysis of Martin, Roberts and Stirling [MRS 98] this leads to

$$\frac{< x\bar{d} - x\bar{u} >}{< x\bar{d} + x\bar{u} >} \sim 10\ \%, \tag{4.82}$$

at $Q^2 = 20\ \mathrm{GeV}^2$.

Returning to the valence distributions, it is clear that the minority, valence quark distribution, $d_v(x)$, is suppressed relative to $u_v(x)$. Traditionally it has been assumed that:

$$\frac{d_v(x)}{u_v(x)} \sim 0.57(1 - x). \tag{4.83}$$

If this ratio vanishes as $x \to 1$ we have the famous result that $F_2^{en}/F_2^{ep} \to \frac{1}{4}$ as $x \to 1$. Of course, in order to extract F_2^{en} one needs a deuterium target and, as $x = 1$ is a kinematic boundary for a free nucleon (corresponding to elastic scattering), the extraction is very sensitive to Fermi motion and binding corrections in the deuteron. Fig. 4.11 shows the ratio F_2^{en}/F_2^{ep} extracted for two choices of how to deal with Fermi motion and binding [MT 96].

The more recent analysis (the solid dots), clearly suggests that the ratio tends to $\frac{3}{7}$ rather than $\frac{1}{4}$. The former actually corresponds to $d_v(1)/u_v(1) = \frac{1}{5}$, as predicted by perturbative QCD [FJ 75]. It is therefore vital that this ambiguity be resolved.

While, as we have explained earlier, the gluon distribution can only be obtained somewhat indirectly, the range of x and Q^2 over which the structure functions have been determined is such that $g(x, Q^2)$ is now rather well known. Figure 4.12, which summarizes the situation at $Q^2 = 5$ and 20 GeV^2, was obtained by a global fit to the combined data from the H1, NMC and BCDMS experiments on p and D targets.

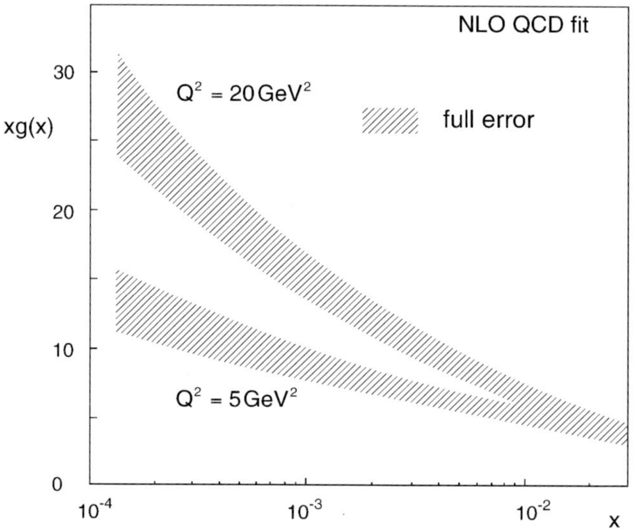

Figure 4.12: The gluon distribution in the nucleon at $Q^2 = 5$ and 20 GeV^2. (Adapted from Ref.[Lev 97].)

4.6 Spin Dependent Structure Functions

Suppose that the incident lepton beam has definite helicity ($h \equiv \vec{\sigma} \cdot \vec{p}/|\vec{p}| = \pm 1$). Then, as shown in Appendix B (and as already discussed for a neutrino beam which is always polarized), the lepton tensor has a piece that is anti-symmetric in μ and ν ($L^{A\pm}_{\mu\nu}$):

$$L^{\pm}_{\mu\nu} = L^{S}_{\mu\nu} + L^{A\pm}_{\mu\nu}. \tag{4.84}$$

Here $L^{S}_{\mu\nu}$ is the tensor already presented in Eq. (4.3) – without the superscript S – and the anti-symmetric term is:

$$L^{A\pm}_{\mu\nu} = \mp 2i\epsilon_{\mu\nu\alpha\beta} p^{\alpha} p'^{\beta} \tag{4.85}$$

(with $\epsilon_{0123} = +1$). Of course, if we have an unpolarized target and parity is conserved, the hadronic tensor $W^{\mu\nu}$ is symmetric in μ and ν and so the cross section is the same as if the beam were unpolarized. Thus we need both the beam and target polarized in order to measure spin dependent quantities.

Let us consider the case where the target is polarized with spin vector S^μ ($S^2 = -1, P \cdot S = 0$). In this case, $W^{\mu\nu}$ has an anti-symmetric piece which must involve $\epsilon_{\mu\nu\alpha\beta}$ in order to preserve parity:

$$W^{\mu\nu} = W_S^{\mu\nu} + W_A^{\mu\nu}, \tag{4.86}$$

with

$$W_A^{\mu\nu} = i\frac{\epsilon^{\mu\nu\alpha\beta}q_\alpha}{P \cdot q}\left[g_1(x,Q^2)S_\beta + g_2(x,Q^2)\left(S_\beta - \frac{S \cdot q}{P \cdot q}P_\beta\right)\right], \tag{4.87}$$

and we have chosen to express $W_A^{\mu\nu}$ directly in terms of the structure functions $g_{1,2}$ which are expected to exhibit scaling.

The differential cross section for the scattering of a polarized e or μ from a target with polarization S^μ is therefore

$$d\sigma \propto L^S_{\mu\nu}W_S^{\mu\nu} + L^A_{\mu\nu}W_A^{\mu\nu}. \tag{4.88}$$

We shall consider just two orientations of the spin of the target with respect to the beam. Following our discussion in Section 4.1 and especially Eq. (4.17), we consider first the difference of the cross sections in which a lepton beam with positive helicity, denoted "→", scatters from a nucleon target with spin parallel or anti-parallel to the direction of the incident lepton, denoted "⇒" and "⇐" respectively. The symmetric piece of Eq. (4.86) cancels in the difference of cross sections and one finds:

$$\frac{d^2\sigma^{\overset{\rightarrow}{\Leftarrow}}}{d\Omega d\epsilon'} - \frac{d^2\sigma^{\overset{\rightarrow}{\Rightarrow}}}{d\Omega d\epsilon'} = \frac{4\alpha^2\epsilon'}{M\nu Q^2\epsilon}\left[(\epsilon + \epsilon'\cos\theta)g_1(x,Q^2) - 2Mxg_2(x,Q^2)\right]. \tag{4.89}$$

It is immediately evident from Eq. (4.89) that (because $\epsilon \gg M$) one is not sensitive to the structure function g_2 in such an experiment. In order to measure that it is simplest to take the polarization of the target at right angles to the incident beam – in the scattering plane. We use "⇑" to denote the target spin pointing to the opposite side of the beam from the outgoing lepton and "⇓" to denote the spin pointing to the same side. Then we find that

$$\frac{d^2\sigma^{\rightarrow\Uparrow}}{d\Omega d\epsilon'} - \frac{d^2\sigma^{\rightarrow\Downarrow}}{d\Omega d\epsilon'} = \frac{4\alpha^2\epsilon'^2}{M\nu Q^2\epsilon}\left[g_1(x,Q^2) + \frac{2\epsilon}{\nu}g_2(x,Q^2)\right]\sin\theta. \tag{4.90}$$

Clearly such a measurement should allow one to determine $g_2(x,Q^2)$ if $g_1(x,Q^2)$ is already known.

While the two cases we have just described are sufficient to determine $g_{1,2}$, there are an infinite variety of target spin configurations. In addition, one could determine the spin of the outgoing lepton – although that offers no new theoretical information. For a complete summary of these possibilities we refer to the review of Anselmino *et al.* [AEL 95].

4.6.1 Spin Structure Function g_1

Using Eq. (4.17) for the proton and generalizing it to include scaling violations, we find

$$g_{1p}(x,Q^2) = \frac{4}{18}\Delta u(x,Q^2) + \frac{1}{18}\Delta d(x,Q^2) + \frac{1}{18}\Delta s(x,Q^2), \tag{4.91}$$

where

$$\begin{aligned}\Delta u(x,Q^2) &= u^{\uparrow}(x,Q^2) - u^{\downarrow}(x,Q^2) + \bar{u}^{\uparrow}(x,Q^2) - \bar{u}^{\downarrow}(x,Q^2),\\ \Delta d(x,Q^2) &= d^{\uparrow}(x,Q^2) - d^{\downarrow}(x,Q^2) + \bar{d}^{\uparrow}(x,Q^2) - \bar{d}^{\downarrow}(x,Q^2),\\ \Delta s(x,Q^2) &= s^{\uparrow}(x,Q^2) - s^{\downarrow}(x,Q^2) + \bar{s}^{\uparrow}(x,Q^2) - \bar{s}^{\downarrow}(x,Q^2).\end{aligned} \tag{4.92}$$

Assuming charge symmetry we can also write the spin structure function of the neutron as:

$$g_{1n}(x,Q^2) = \frac{4}{18}\Delta d(x,Q^2) + \frac{1}{18}\Delta u(x,Q^2) + \frac{1}{18}\Delta s(x,Q^2). \tag{4.93}$$

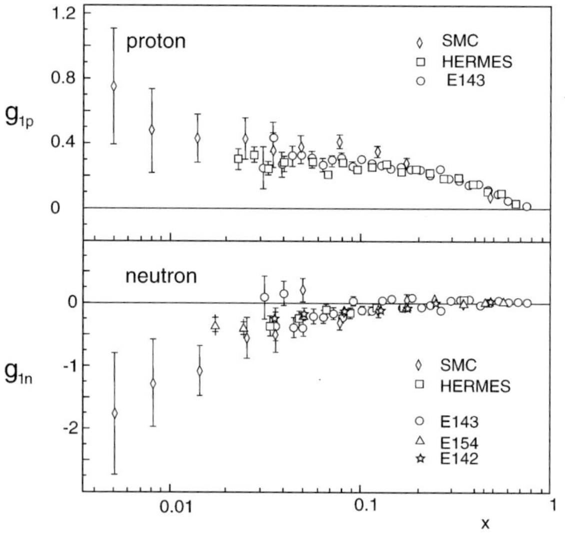

Figure 4.13: The proton and neutron spin structure functions, g_{1p} and g_{1n} – from Ref.[PW 00]; we thank U. Stoesslein for the compilation of data.

Proton

As a result of very demanding experimental work at a number of laboratories, notably SLAC and CERN, we have quite an impressive body of data for g_{1p}, which is summarized in Fig. 4.13. For completeness we note that what is actually measured is the asymmetry,

$A_{//}(x, Q^2)$, defined by

$$\begin{aligned} A_{//}(x, Q^2) &= \frac{d^2\sigma^{\rightleftarrows} - d^2\sigma^{\rightrightarrows}}{d^2\sigma^{\rightleftarrows} + d^2\sigma^{\rightrightarrows}}, \\ &\propto \frac{g_1(x, Q^2)}{F_1(x, Q^2)} + \mathcal{O}\left(\frac{1}{\epsilon}\right). \end{aligned} \tag{4.94}$$

The spin structure function, g_1, can then be extracted by using a parametrization of world data for $F_1(x, Q^2)$.

Neutron

As always, the neutron presents even more experimental difficulty. One can only determine g_{1n} from deep inelastic scattering on polarized deuteron or ^{3}He targets after correcting for various nuclear effects. For example, although the deuteron with spin projection $+1$ is primarily a neutron or a proton in an s-state each with spin $\frac{1}{2}$, there is a small probability ($P_D \sim 5\ \%$) to find them in a relative d-state in which case the spins can be up or down. In addition, one must correct for Fermi motion, binding and the fact that the bound nucleons are slightly off mass-shell. These small corrections are exceptionally important because g_{1n} is considerably smaller than g_{1p} and one must subtract g_{1p} from g_{1d} to get it. Nevertheless, the best estimates [MPT 95, CSP 92] are that for x below 0.8 g_{1d} may be approximated by

$$g_{1d} = \frac{g_{1p} + g_{1n}}{2}\left(1 - \frac{3}{2}P_D\right), \tag{4.95}$$

and so one can extract g_{1n}.

In the case of polarized ^{3}He the nuclear physics is simpler, as the largest component of the ^{3}He wave function consists of two protons with spins opposite, so that the nuclear spin is carried entirely by the neutron. On the other hand, the Fermi motion and binding corrections will be bigger than they are for the deuteron and one must again correct for the D-state component of the ^{3}He wave function.

We summarize the data for g_{1n} in Fig. 4.13. Note that the relative errors are much greater than for g_{1p} for the reasons just explained. Thus, for example, it is not yet clear whether g_{1n} becomes positive in the valence region ($x \geq 0.4$).

4.6.2 Bjorken Sum Rule

The Bjorken sum rule was derived well before either scaling was discovered or the parton model invented [Bjo 66]. Apart from well understood QCD corrections, which will be discussed in Chapter 5, this sum rule can be derived from current algebra – c.f. Section 6.1. It therefore constitutes a very fundamental test of the structure of QCD. For $N_f = 3$ the Bjorken sum rule takes the form

$$\begin{aligned} S_{Bj} &= \int_0^1 dx[g_{1p}(x, Q^2) - g_{1n}(x, Q^2)] \\ &= \left[1 - \frac{\alpha_s(Q^2)}{\pi} - 3.583\left(\frac{\alpha_s(Q^2)}{\pi}\right)^2 - 20.215\left(\frac{\alpha_s(Q^2)}{\pi}\right)^3\right]\frac{g_A}{6}, \end{aligned} \tag{4.96}$$

where g_A is the isovector, axial charge of the nucleon, measured in neutron β-decay – c.f. Section 3.3.

The appearance of $g_A/6$ can be immediately understood from the parton model expressions for g_{1p} and g_{1n} (Eqs. (4.91) and (4.93) respectively):

$$\int_0^1 dx\,[g_{1p}(x,Q^2) - g_{1n}(x,Q^2)] = \frac{1}{6}\int_0^1 dx\,[\Delta u(x,Q^2) - \Delta d(x,Q^2)]. \tag{4.97}$$

The connection of the right hand side of Eq. (4.97) with g_A follows from Eqs. (3.4), (3.11) and (3.29), which imply that

$$g_A = \frac{\langle p\uparrow|\bar{\psi}(0)\gamma^z\gamma_5\tau_3\psi(0)|p\uparrow\rangle}{\langle p\uparrow|p\uparrow\rangle}, \tag{4.98}$$

for a proton at rest.

By combining the data for g_{1p} and g_{1n} that we have just presented, and assuming that the small-x behavior of $g_{1p} - g_{1n}$ is given by isovector Regge exchange, it is possible to test the Bjorken sum rule at an impressive level of precision. Fig. 4.14 illustrates the situation beautifully. After allowing for perturbative QCD corrections, that seem to be well under control, the Bjorken sum rule has certainly been confirmed at the level of 10 %.

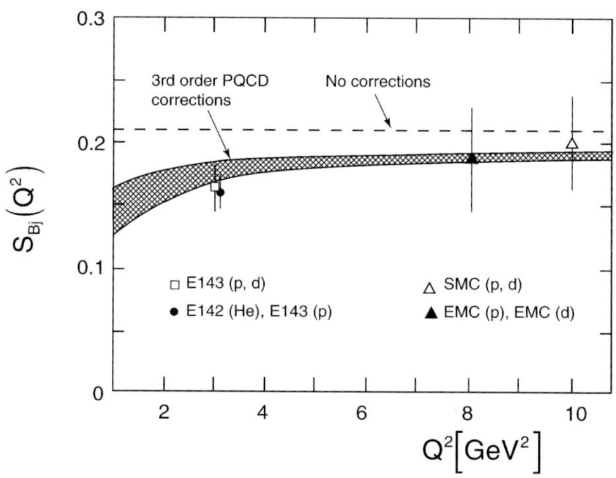

Figure 4.14: Comparison between the measured value of the Bjorken sum rule at several values of Q^2 and the theoretical expectations including third-order QCD corrections and no allowance for higher-twist corrections. (Adapted from Ref.[Rit 97].)

4.6.3 Ellis-Jaffe Sum Rule

The theoretical simplicity of the Bjorken sum rule arises because flavor singlet and octet contributions to the spin structure functions cancel in the difference between g_{1p} and g_{1n}. This leaves only the isovector axial current which is partially conserved and for which we can exploit the success of chiral $SU(2) \times SU(2)$. To derive a rigorous sum rule for the proton or neutron alone presents far more difficulty.

Working with three flavors, we find from Eq. (4.91) that the integral of g_{1p} has the form

$$\begin{aligned} S_p &= \int_0^1 dx\, g_{1p}(x, Q^2) \\ &= \int_0^1 dx \left[\frac{4}{18}\Delta u(x, Q^2) + \frac{1}{18}\Delta d(x, Q^2) + \frac{1}{18}\Delta s(x, Q^2)\right], \\ &\equiv \frac{4}{18}\Delta u + \frac{1}{18}\Delta d + \frac{1}{18}\Delta s. \end{aligned} \tag{4.99}$$

In Eq. (4.99) we have adopted the conventional notation for the first moments of the polarized u, d and s distributions, namely $\Delta u, \Delta d$ and Δs. By rewriting Eq. (4.99) in terms of the isovector, octet and singlet combinations:

$$\begin{aligned} \Delta q_3 &= \Delta u - \Delta d = g_A, \\ \Delta q_8 &= \Delta u + \Delta d - 2\Delta s, \\ \Delta q_0 &= \Delta u + \Delta d + \Delta s, \end{aligned} \tag{4.100}$$

we find

$$S_p = \frac{\Delta q_3}{12} + \frac{\Delta q_8}{36} + \frac{\Delta q_0}{9}. \tag{4.101}$$

In the case of the neutron only the isovector term changes sign

$$S_n = -\frac{\Delta q_3}{12} + \frac{\Delta q_8}{36} + \frac{\Delta q_0}{9}. \tag{4.102}$$

We have already explained that Δq_3 is obtained from neutron β-decay. Provided one accepts $SU(3)$ symmetry, Δq_8 is similarly obtained from hyperon β-decay. As there are well documented cases of $SU(3)$ symmetry breaking, there is some controversy over the precise value and error associated with it, but $\Delta q_8 = 0.57 \pm 0.06$ is a conservative choice. The real difficulty in establishing a rigorous sum rule is associated with Δq_0, which has not been determined from any other experiment. One therefore needs some physics input. Assuming that the polarized strange quark distribution in the proton is very small, $\Delta s \sim 0$, one has $\Delta q_0 = \Delta q_8$ and obtains the Ellis-Jaffe sum rule [EJ 74]:

$$S_{p,n} = \pm\frac{\Delta q_3}{12} + \frac{5}{36}\Delta q_8. \tag{4.103}$$

The QCD radiative corrections, which involve extremely complicated calculations involving a great deal of computer algebra, were derived [Lar 97] after the original work of Ellis and

Jaffe. They modify these sum rules so that[2]:

$$S_{p,n}(Q^2) = \left[1 - \frac{\alpha_s(Q^2)}{\pi} - 3.583\left(\frac{\alpha_s(Q^2)}{\pi}\right)^2 \right.$$
$$\left. -20.215\left(\frac{\alpha_s(Q^2)}{\pi}\right)^3\right]\left(\pm\frac{\Delta q_3}{12} + \frac{\Delta q_8}{36}\right) \tag{4.104}$$
$$+\left[1 - \frac{\alpha_s(Q^2)}{3\,\pi} - 0.55\left(\frac{\alpha_s(Q^2)}{\pi}\right)^2\right]\frac{\Delta q_0}{9}.$$

If we work to just first order in $\alpha_s(Q^2)$ and adopt the Ellis-Jaffe hypothesis, we obtain

$$S^{\rm EJ}_{p,n}(Q^2) = \pm\frac{\Delta q_3}{12}\left[1 - \frac{\alpha_s(Q^2)}{\pi}\right] + \frac{5}{36}\Delta q_8\left[1 - \frac{7}{15}\frac{\alpha_s(Q^2)}{\pi}\right]. \tag{4.105}$$

Using the standard value for g_A and the value quoted earlier for Δq_8, this leads to the expectation

$$S^{\rm EJ}_p = 0.167 \pm 0.008, \tag{4.106}$$

at 10 GeV^2. Much excitement was generated by the first measurement of this quantity by the European Muon Collaboration (EMC), which reported a value of only $0.114 \pm 0.012 \pm 0.026$ in 1988 [EMC 88]. This is clearly incompatible with the Ellis-Jaffe hypothesis. Returning to the original expression, Eq. (4.104), one easily sees that this implies a value for Δq_0 very near zero. On the other hand, within the parton model, Eq. (4.100) suggests that Δq_0 is the fraction of the spin of the proton carried by its quark constituents. Given the success of the quark model in describing magnetic moments, it is difficult to understand how almost none of the spin could reside on its quarks. This issue became known as the "proton spin crisis".

Without considerable discussion of key aspects of non-perturbative QCD, we cannot give a complete discussion of the resolution of this apparent crisis. In particular, the $U(1)$ axial anomaly – c.f. Chapter 6 – plays a vital role in the flavor singlet sector. As the minimal consequence of this anomaly, if the proton contains gluons with non-zero polarization, Δg, then the naive parton model value of Δq_0 should be replaced by [AR 88, BQ 90]

$$\Delta q_0 = \Delta q_0|_{\rm CI} - 3\frac{\alpha_s(Q^2)}{2\,\pi}\Delta g(Q^2), \tag{4.107}$$

where the term $\Delta q_0|_{\rm CI}$ (CI for chiral invariant) corresponds to a renormalization scheme where the axial current satisfies the usual commutation relations of spin [Kod 80, Cre 75]. The QCD evolution of Δg is such that $\alpha_s(Q^2)\Delta g(Q^2)$ is independent of Q^2 (at high Q^2), so that the correction term in Eq. (4.107) survives even in the Bjorken limit. We shall return to the fact that Δg would need to be uncomfortably large in order to explain the observed value of the sum rule when we come to discuss nucleon models.

[2] The flavor singlet quantity, Δq_0, appearing in Eq. (4.104) is strictly renormalization group invariant. For an introduction to the renormalization group we refer to Section 5.4 and for a complete discussion in relation to this sum rule to Ref.[Lar 97]. For the present purposes it is enough to note that the invariant form of Δq_0 is equivalent to $\Delta q_0(Q^2)$, appearing in Eq. (4.107), in the limit $Q^2 \to \infty$.

Table 4.1: The integrals over $g_1(x, Q^2)$ for the Ellis-Jaffe sum rule (for the proton and neutron) and the Bjorken sum rule (for the difference) measured at SLAC (E155 [E155 99] and E143 [E143 98]), CERN (SMC [SMC 97]) and HERA (Hermes [Hermes 97]).

Experiment	Q^2 (GeV^2)	$\int_0^1 dx g_1(x, Q^2)$
p (E155)	5	0.118 ± 0.008
(E143)	5	0.129 ± 0.010
(SMC)	5	0.140 ± 0.011
n (E155)	5	-0.058 ± 0.010
(E143)	5	-0.034 ± 0.018
(Hermes)	2.5	-0.037 ± 0.015
$p - n$ (E155)	5	0.176 ± 0.008
(E143)	5	0.164 ± 0.022

The experimental situation has improved dramatically since the original EMC measurement, with more accurate data over a wider range of x and Q^2 for the proton and even accurate data for the neutron. Table 4.1 summarizes the results of measurements for both nucleons. The corresponding values of Δq_0 are all around 0.3, which is unambiguously smaller than the Ellis-Jaffe expectation. This is a vital piece of information in our quest to understand the structure of the nucleon.

4.6.4 Transverse Spin Structure Function g_2

When it comes to the transverse spin structure function, $g_2(x, Q^2)$, the parton model faces a major challenge. Until now we have dealt with deep inelastic cross sections which can be factorized as the hard photon-quark cross section times the number density of free quarks moving collinearly with the target proton in an infinite momentum frame. However, for an on-shell quark the asymmetry corresponding to g_2 is a difference of cross sections which vanishes identically. As a consequence, a non-zero value for g_2 for a composite system like the nucleon is a direct measure of the degree to which its constituents are off-shell – a concept foreign to the parton model.

To be more specific we consider the spin dependent piece of the tensor $W^{\mu\nu}$ for a quark of momentum p and spin s, which is taken to be off-mass-shell before absorbing the virtual photon of momentum q (i.e., $p^2 = m^2 \neq m_q^2$). One finds easily that [LA 88]:

$$W_A^{\mu\nu} \sim \epsilon^{\mu\nu\alpha\beta} s_\alpha \left[\left(1 - \frac{m}{m_q}\right) p_\beta - \frac{m}{m_q} q_\beta \right]. \tag{4.108}$$

Clearly this is not even gauge invariant unless the struck quark is on-shell. (In that case the coefficient of p^β vanishes, as does $q^\mu \epsilon_{\mu\nu\alpha\beta} q^\beta$.) If the struck quark is on-shell $W_A^{\mu\nu}$ becomes $-\epsilon^{\mu\nu\alpha\beta} s_\alpha q_\beta$ and we see by comparison with Eq. (4.87) that g_2 is zero (as stated earlier).

A natural extension of the naive parton model is to allow the quarks inside the nucleon to have some intrinsic transverse momentum, $\vec{p}_\perp$. In this case one finds

$$g_2(x) = \frac{1}{2} \sum_q e_q^2 \left(\frac{m_q}{xM} - 1 \right) \Delta q(x), \tag{4.109}$$

with $\Delta q(x)$ the usual, spin dependent distributions. Whereas in the naive parton model, with $\vec{p}_\perp = 0$, consistency requires that we choose $m_q = xM$ and hence $g_2 = 0$, the extra degree of freedom provided by the intrinsic transverse momentum allows a non-trivial result. On the other hand, it is disturbing that the result (4.109) should depend totally on the quark mass.

For a definitive treatment of g_2 one must have recourse to the operator product expansion

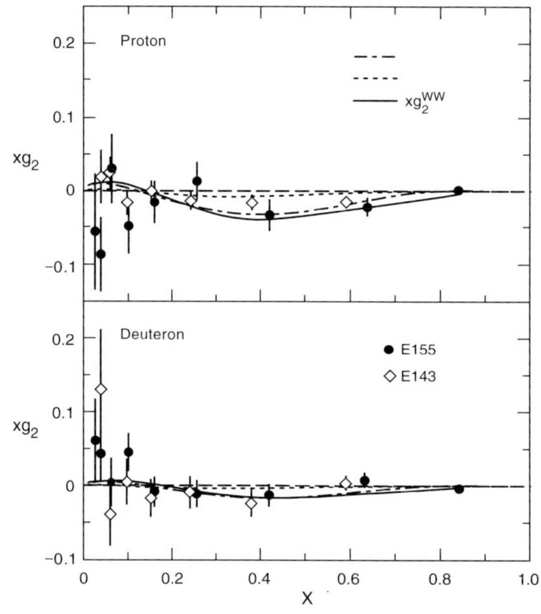

Figure 4.15: The structure function $xg_2(x, Q^2)$ for the proton and the deuteron measured at SLAC [E155 99]. The solid line shows the prediction of the Wandzura-Wilczek term, Eq. (4.111), alone. The dashed and dash-dotted lines refer to other theoretical models.

which will be introduced in the next chapter. In particular, one can show that $g_2(x, Q^2)$ can be written as the sum of two distinct contributions [AEL 95]:

$$g_2(x, Q^2) = g_2^{WW}(x, Q^2) + g_2^{(3)}(x, Q^2). \tag{4.110}$$

The first term, labelled WW for Wandzura and Wilczek [WW 77], is entirely determined by $g_1(x, Q^2)$:

$$g_2^{WW}(x, Q^2) = \int_x^1 \frac{dy}{y} g_1(y, Q^2) - g_1(x, Q^2). \tag{4.111}$$

This is the leading twist contribution. The real excitement over $g_2(x, Q^2)$ is the second term in Eq. (4.110), which is formally of twist-3. It involves matrix elements of quark *and* gluon operators such as $\langle p \uparrow | \psi^\dagger \vec{B} \psi | p \uparrow \rangle$ and $\langle p \uparrow | \psi^\dagger \vec{\alpha} \times \vec{B} \psi | p \uparrow \rangle$, with $\vec{B}$ the color magnetic field inside the proton [JJ 91]. The opportunity to make a direct study of the correlations between quarks and gluons promises new insight into the structure of the nucleon. However, as we see from Fig. 4.15, it will require an order of magnitude improvement over the already remarkable efforts at SLAC to have sufficient sensitivity to extract $g_2^{(3)}(x, Q^2)$ with confidence. For the present the data is consistent with the Wandzura-Wilczek result.

4.7 Qualitative Understanding of Parton Distributions

Thus far in our investigation of deep-inelastic scattering as a probe of nucleon structure we have summarized the experimental facts that seem most relevant and presented the ideas of the parton model. However, we have not indicated the shapes that one might expect for the various parton distributions. To the extent that a detailed comparison with any given model is concerned it would not be appropriate here. Nevertheless, there are a number of results, which are either model independent or likely to hold rather generally, which provide valuable insight into the observed distributions. It is these which we briefly outline here.

4.7.1 Counting Rules

As $x \to 1$ is the elastic limit, it is not surprising that one can relate the asymptotic (high-Q^2) behavior of the elastic form factors of a composite system to the behavior of its structure function as x approaches unity. It is clear that the minimum Q^2-dependence of the elastic form factor of a composite system with N strongly interacting constituents is a single gluon exchange between each of them. As a result we expect

$$F_{\text{el}} \sim \left(\frac{1}{Q^2} \right)^{N_s - 1}, \tag{4.112}$$

where $N_s = (N - 1)$ is the number of spectators to the constituent which interacts directly with the probe. On the basis of reasonable behavior of the light-cone wave function, Drell, Yan and West [DYW 70] were able to show that the corresponding parton distribution should satisfy

$$q(x) \sim (1 - x)^{2N_s - 1}, \qquad \text{as } x \to 1. \tag{4.113}$$

Suppose we consider the minimum number of constituents which can have nucleon quantum numbers. For a valence quark the minimum number is three (qqq), while for a gluon it is four ($qqqg$) and for a sea-quark five ($qqq\bar{q}q$). Thus one expects

$$q_v(x) \sim (1 - x)^3, \qquad \text{as } x \to 1, \tag{4.114}$$

($N_s = 2$), while

$$g(x) \sim (1 - x)^5, \qquad \text{as } x \to 1, \tag{4.115}$$

($N_s = 3$) and finally

$$\bar{q}(x) \sim (1-x)^7 , \qquad \text{as } x \to 1. \tag{4.116}$$

It is remarkable how close these predictions are to the values extracted from experiment. For example, at $Q^2 \sim 1$ GeV^2, Martin, Roberts and Stirling (MRS) find the leading power of $(1-x)$ to be respectively 3.4, 6.1 and 7.5 for the valence (up quark), gluon and sea distributions [MRS 98].

4.7.2 Small-x Behaviour

As $x \to 0$, with Q^2 large, the invariant mass squared of the virtual-photon-nucleon system ($s_{\gamma p} \equiv W^2$ in the notation of Eq. (4.1)) tends to infinity. Thus this region corresponds to high-energy electro-production. Such reactions have been studied extensively and are well understood in terms of Regge-pole exchange [Hei 73, Squ 64].

On the basis of Regge-pole phenomenology one knows that the structure functions W_1 and W_2 should behave as $s_{\gamma p}^{\alpha} \sim \nu^{\alpha}$ and $s_{\gamma p}^{\alpha-2} \sim \nu^{\alpha-2}$ respectively, times appropriate form factors. (Here α is the leading Regge intercept, for example 1 for the Pomeron and $\frac{1}{2}$ for the ρ, f, ω or A_2.) If the structure functions are to exhibit even approximate scaling behavior, the Q^2-dependence of these form-factors must be such as to convert these powers of ν to inverse powers of x. Thus we expect that as $x \to 0$

$$F_1(x) \sim x^{-\alpha}, \tag{4.117}$$

and

$$F_2(x) = \nu W_2 \sim x^{1-\alpha}, \tag{4.118}$$

as $x \to 0$. In the spin-dependent case the trajectory is less well known, but once again in the limit $x \to 0$, one usually expects

$$g_1(x) \sim x^{-\alpha}; \qquad -0.5 \leq \alpha \leq 0. \tag{4.119}$$

These expectations are once again in fair agreement with the phenomenological fits to data, which typically go like $x^{\alpha}(1-x)^{\beta}$ times a polynomial in x. The MRS analysis yields a valence quark distribution which behaves as $x^{-0.5}$ as $x \to 0$, while the gluon and sea distributions go as $x^{-1.3}$. Of course, the interpretation in terms of Regge-poles is complicated by the possible existence of Regge-cuts, our lack of understanding of the nature of the Pomeron, and so on. The enormous kinematic range made available by HERA has created a major industry for the study of the small-x region. This work is vital to our understanding of QCD and, in particular, the QCD vacuum. To see this, we note that from Eq. (4.120) (below), in the usual sense of a Fourier transform, small-x corresponds to large coordinate space distances (on the light-cone) z – in the laboratory frame. Thus at very small values of x one is sensitive to physics over distance scales way beyond the nucleon size – i.e., in the QCD vacuum – and we do not expect that such data should be particularly relevant to our theme of the structure of the nucleon. Therefore, we simply refer to some helpful reviews on the subject [MP 94, BLW 95, Lip 97].

4.7.3 Formal Parton Model

The components of the nucleon structure functions which survive at large-Q^2 (i.e., those which evolve only logarithmically with Q^2) are those described by the DGLAP equations. As we shall explain in detail in the next Chapter, these distributions are determined by the matrix elements of a certain set of operators known as "twist-2" (or leading twist) operators. Within QCD one can show that the leading twist quark distribution of a target $|N>$ is given (in the laboratory frame, for $x > 0$) by the following *connected* matrix element (hence the subscript "C"):

$$q(x) = \frac{1}{4\pi} \int dz\, e^{-iMxz} \langle N|\psi_+^\dagger(\xi_-)\psi_+(0)|N\rangle_C. \tag{4.120}$$

In Eq. (4.120) M is the target (nucleon) mass in terms of which x is defined, ψ is the quark field operator, ξ_- is the light-like separation $(z; \vec{0}_T, -z)$, the subscript $+$ indicates multiplication by $(1+\gamma^0\gamma^3)/2$ and finally we are working in the light-cone gauge, $A_+ = 0$ (axial gauge). (The latter is normal for the parton model. Indeed, in other gauges one would need to add a factor between $\psi_+^\dagger(\xi_-)$ and $\psi_+(0)$ to ensure gauge invariance.)

Inserting a formal, complete set of energy-momentum eigenstates, $|n(\vec{p})\rangle$, between the field operators in Eq. (4.120) and using the translational properties of the field operator $\psi_+(\xi_-)$ to make the z-dependence explicit, we can carry out the z-integration [Jaf 85]:

$$q(x) = \frac{1}{2}\sum_n \int \frac{d^3p}{2E_p(2\pi)^3} |\langle n(\vec{p})|\psi_+(0)|N\rangle|^2 \delta\left(M(1-x) - (\sqrt{m_n^2+\vec{p}^{\,2}} + p_z)\right). \tag{4.121}$$

Equation (4.121) clearly illustrates that the support of the parton distribution is limited to $x \leq 1$, with $q(1) = 0$. (This is because $\sqrt{m_n^2+\vec{p}^{\,2}} + p_z \equiv p_n^+$ is necessarily positive and hence the δ-function can only be satisfied for $x < 1$.) The corresponding equation for the anti-quark distribution is

$$\bar{q}(x) = \frac{1}{2}\sum_n \int \frac{d^3p}{2E_p(2\pi)^3} |\langle n(\vec{p})|\psi_+^\dagger(0)|N\rangle|^2 \delta\left(M(1-x) - (\sqrt{m_n^2+\vec{p}^{\,2}} + p_z)\right). \tag{4.122}$$

Although Eq. (4.121) is quite formal, it does permit some insight into the valence quark distribution. In particular, if the nucleon were to consist of just three constituent quarks in the s-state of a potential well, the most probable momentum of the quark (in the nucleon rest frame) would be zero. In that case $p_n^+ = m_n$ and one sees that the δ-function in Eq. (4.121) leads to a maximum in $q(x)$ at $x = 1 - m_n/M$, where m_n is the mass of the intermediate state. Clearly then the major contribution to the valence distribution at large-x will come from the lightest mass intermediate state, namely two constituent quarks. (Note that the mass m_n should be understood here in the same sense as the mass of a constituent quark.) Thus we expect $q_v(x)$ to peak at about $\frac{1}{3}$ in the constituent quark model, or $\frac{1}{4}$ in the bag model (c.f. Chapter 8), where the relativistic (confining) potential carries $\frac{1}{4}$ of the mass of the nucleon.

This approach also illustrates very clearly why a hyperfine interaction, which makes a spin-0 pair of quarks less massive than a spin-1 pair, implies that the former will dominate in the valence region [CT 88]. (The peak value, $1 - m_n/M$, will be larger for the lighter mass spectator pair.) In particular, for an $SU(6)$ proton wave function, only a u-quark comes with a pair of spectators in spin-0 and therefore $\frac{d}{u}$ should be small at large-x. Within the simple parametrizations based on counting rules this argument was used to increase the exponent in the leading power of $1 - x$ (c.f. Eq. (4.114)) by one, so that $d_v(x)/u_v(x) \sim (1 - x)$. Of course, this argument must be revisited in the extreme limit, $x \to 1$, where we see from Eq. (4.121) that the struck quark must have infinite momentum ($p_z \to -\infty$). Such momenta can only come from hard collisions and hence are a proper subject for perturbative QCD, to which we turn in the next chapter.

4.8 Off-Forward Parton Distributions

Through the optical theorem the structure functions presented in this chapter, which are derived from inclusive lepton scattering, can be related to the imaginary part of the forward Compton amplitude for a high energy, virtual photon. Although the experiments are challenging it has become clear that it is now possible to investigate a new set of generalized parton distributions by direct measurement of deeply virtual Compton scattering (DVCS) – i.e., virtual Compton scattering in the Bjorken regime. The cross section for DVCS can be expressed in terms of generalized parton distributions, known as off-forward parton distributions (OFPD's) [Rad 96, Ji 97, Ji 98, Gui 98]. (One may also find the terms "off-diagonal", "skewed" or "non-forward" used to describe these distributions.)

Unlike the usual parton distributions, the OFPD's are not number densities but probability amplitudes. However, they do also have a straightforward interpretation in an infinite momentum frame. While they reduce to the usual parton distributions in the forward limit, they can also be related to the nucleon elastic form factors. In general, they probe the structure of the nucleon in new ways and promise considerable insight. From the experimental point of view, a direct determination of the OFPD's seems most likely at very high energies – for example, the region covered by COMPASS at CERN. At lower energies one would need to separate it from the Bethe-Heitler process (c.f. the determination of the generalized polarizabilities discussed in Chapter 2).

In Fig. 4.16, we show the two diagrams expected to dominate the OFPD's in the Bjorken region. Apart from Q^2 and $x \equiv k^+/p^+$, there are two other variables necessary to specify the OFPD's. The invariant four-momentum squared transferred to the target proton is $t = \Delta^2$ and $\xi \equiv -\Delta^+/2p^+$ is essentially the light-cone momentum fraction transferred to the target. It is most convenient to define the OFPD's in the Bjorken limit in terms of the sort of light-cone

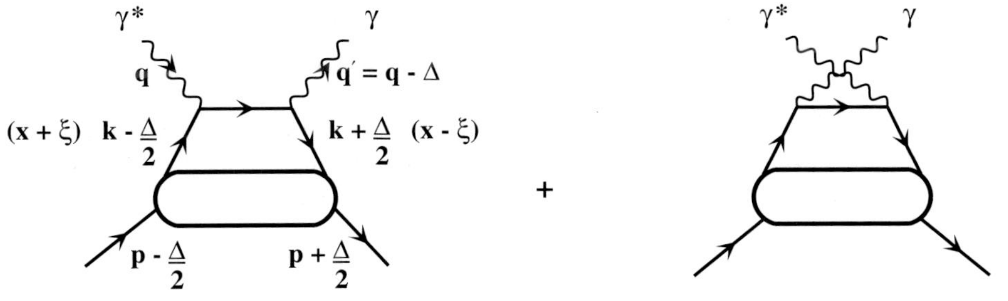

Figure 4.16: Leading order, perturbative QCD contributions to deeply virtual Compton scattering.

Fourier transform introduced for the parton distributions in the preceding section. That is:

$$\begin{aligned}
&\frac{p^+}{2\pi}\int dy^- \exp\left(-ixp^+y^-\right)\langle p'|\bar{\psi}_\alpha(y)\psi_\beta(0)|p\rangle_{y^+=y_\perp=0} \\
&= \frac{1}{4}\gamma^-_{\alpha\beta}\left[H(x,\xi,\Delta^2)\bar{u}(p')\gamma^+u(p) + E(x,\xi,\Delta^2)\bar{u}(p')\sigma^{+\mu}\frac{\Delta_\mu}{2M}u(p)\right] \\
&\quad + \frac{1}{4}\left(\gamma_5\gamma^-\right)_{\alpha\beta}\left[\tilde{H}(x,\xi,\Delta^2)\bar{u}(p')\gamma^+\gamma_5 u(p)\right. \\
&\qquad\qquad \left. +\tilde{E}(x,\xi,\Delta^2)\bar{u}(p')\gamma_5\frac{\Delta^+}{2M}u(p)\right],
\end{aligned} \tag{4.123}$$

where we have chosen to work in light-cone gauge ($A_+ = 0$).

Whereas the distributions H and $\tilde{H}$ are related to the usual parton distributions as:

$$\begin{aligned}
H(x,\xi=0,\Delta^2=0) &= q(x), \\
\tilde{H}(x,\xi=0,\Delta^2=0) &= \Delta q(x),
\end{aligned} \tag{4.124}$$

the distributions E and $\tilde{E}$ have no analogue in this limit. On the other hand, after integrating the OFPD's over x one obtains the standard vector and axial vector form factors:

$$\begin{aligned}
&\int_{-1}^{+1} H(x,\xi,\Delta^2)\,dx = F_1(\Delta^2), \\
&\int_{-1}^{+1} E(x,\xi,\Delta^2)\,dx = F_2(\Delta^2), \\
&\int_{-1}^{+1} \tilde{H}(x,\xi,\Delta^2)\,dx = G_A(\Delta^2)
\end{aligned} \tag{4.125}$$

and

$$\int_{-1}^{+1} \tilde{E}(x,\xi,\Delta^2)\,dx = G_P(\Delta^2). \tag{4.126}$$

Of course, one should interpret these equations as the contribution of the quark of flavor q to the Dirac, Pauli, axial and induced-pseudoscalar form factors, respectively. (Note that the dependence on ξ drops out after integration over x.)

In view of the interest in the failure of the Ellis-Jaffe sum rule, described earlier in this Chapter, a great deal of interest in the possibility of measuring the OFPD's has been generated by the realization that the total angular momentum (spin plus orbital angular momentum) carried by quark q is given by the sum rule [Ji 97, HJL 99]:

$$J_q = \frac{1}{2}\int_{-1}^{+1} dx x \left[H(x,\xi,\Delta^2=0) + E(x,\xi,\Delta^2=0)\right] . \tag{4.127}$$

(As in Eq. (4.126), the ξ dependence of H and E is removed by the integration over x.) The total spin of the nucleon is then given by:

$$\sum_q J_q + J_g = \frac{1}{2}, \tag{4.128}$$

where J_g is the total angular momentum carried by gluons.

The physical interpretation of the OFPD's can be derived by expanding the field operators in Eq. (4.123) in terms of light-cone quantized creation and annihilation operators (in $A_+ = 0$ gauge). In the region $x > \xi$ ($x < \xi$) the OFPD has an interpretation which is a simple generalization of the usual parton distribution. That is, it is the amplitude to take a quark (anti-quark) of momentum $k - \Delta/2$ out of the proton and reinsert a quark (anti-quark) of momentum $k + \Delta/2$ into the proton, at some later point on the light-cone, to re-form the recoiling proton. On the other hand, in the region $-\xi < x < \xi$, which is often known as the "mesonic region", one is either taking out (or inserting) a $q\bar{q}$ pair with momentum $k - \Delta/2$ and $-k - \Delta/2$ (or $k + \Delta/2$ and $-k + \Delta/2$), respectively.

While, as we have already noted, the OFPD's must usually be interpreted as probability amplitudes rather than densities, the interpretation simplifies in the special case where $\xi \to 0$. In particular, as explained by Burkardt [Bur 00], in this limit where the "off-forwardness" is only in the transverse direction, the OFPD's may be interpreted, in the infinite momentum frame, as the Fourier transforms of impact parameter dependent parton distributions, with respect to the impact parameter. The proof requires a fairly subtle construction of wave packets, but the result is analogous to the well-known identification of the charge form factor as the Fourier transform of the charge distribution in the Breit frame, discussed in Chapter 2.

5 Elements of QCD

The concept of local gauge invariance as the guiding principle in constructing renormalizable quantum field theories is one of the great discoveries of the second half of the twentieth century. In Chapter 1 we have already summarized the $SU(2) \times U(1)$ sector of the Standard Model (SM) in which electromagnetism and the weak interaction are so beautifully combined. Here we turn to the strong interaction sector which is primarily responsible for the structure of the nucleon. Within the SM, the strong interaction is described by QCD which is a local gauge theory built on the non-Abelian internal symmetry of the quarks known as color – that is, $SU(3)_c$.

The need for three colors of quarks became apparent very soon after the development of the quark model. For example, the $\Delta^{++}(J_z = +\frac{3}{2})$ has the totally symmetric spin-flavor structure $|u \uparrow u \uparrow u \uparrow\rangle$, whereas the quarks are supposed to be fermions. In order to be able to write an antisymmetric wave function for baryons one therefore needed an extra label, with at least three degrees of freedom – traditionally taken to be $i \in (R, B, G)$ (for red, blue and green). As an invariant under interchanges of the colors, the antisymmetric combination, $\epsilon_{ijk} q_i q_j q_k$, is said to be a color singlet and it is believed that all physical strongly interacting particles are color singlets.

There are many excellent texts available from which the dedicated student can learn QCD, including the original review of Abers and Lee [AL 73] and the books of Muta, Bailin and Love, Ellis, Stirling and Webber and Peskin and Schroeder [Mut 98, BL 86, ESW 96, PS 95]. While this is not a text book on QCD, it does seem worthwhile to outline here the main physical and mathematical ideas and to review the application of perturbative QCD to deep inelastic scattering – the latter being a major source of information on nucleon structure. In doing so we shall be led to a somewhat deeper understanding of the parton model, including the Q^2-dependence of the parton distributions and the QCD radiative corrections to various cross sections and sum-rules.

5.1 Basic Lagrangian

We shall write the quark fields as:

$$\psi(x) = \begin{pmatrix} \psi_R(x) \\ \psi_B(x) \\ \psi_G(x) \end{pmatrix}, \tag{5.1}$$

where each of the ψ_i is a Dirac field operator for color i and the pattern is repeated for each active flavor. (Thus, in the case of 6 flavors, $\psi(x)$ is a 72-component object.) The unitary transformations which mix quarks of different colors are generated by the elements of the Lie algebra $SU(3)$, namely $\lambda_a,\ a = 1, \ldots, 8$, satisfying

$$[\lambda_a, \lambda_b] = 2if_{abc}, \qquad \mathrm{Tr}(\lambda_a\lambda_b) = 2\delta_{ab}, \qquad \mathrm{Tr}\,\lambda_a = 0, \tag{5.2}$$

where the structure constants f_{abc} are totally anti-symmetric in the color labels a, b and c. The standard 3×3, matrix representation of the λ_a is given in Appendix A.2. In terms of these generators the most general transformation amongst the colors is given by the operator U

$$U = \exp\left(-i\,\frac{\lambda_a\theta_a}{2}\right), \tag{5.3}$$

where $(\theta_1, \theta_2, \ldots, \theta_8)$ are eight real parameters. Under this gauge transformation, ψ transforms as:

$$\psi \to \tilde{\psi} = U\psi. \tag{5.4}$$

Local gauge invariance requires that the Lagrangian density be invariant under an arbitrary transformation amongst the colors at each point in space – i.e., with $\theta_a = \theta_a(x)$. As in QED, this requires the introduction of a covariant derivative involving a new Lorentz-vector field, in this case eight of them, $A_a^\mu (\equiv A^{a\mu})$ [1]. These are the gluon fields and their transformation properties are defined so that the covariant derivative of the quark field (ψ), namely $D_\mu\psi$:

$$\begin{aligned} D_\mu\psi &= (\partial_\mu - igA_\mu)\,\psi, \\ A_\mu &\equiv \sum_{a=1}^{8} t_a A_\mu^a; \qquad t_a \equiv \frac{\lambda_a}{2}, \end{aligned} \tag{5.5}$$

transforms in exactly the same way as ψ itself. Once this is achieved, $\bar{\psi}i\gamma^\mu D_\mu\psi$ is gauge invariant.

Suppose that indeed

$$\begin{aligned} \widetilde{D_\mu\psi} &\equiv \partial_\mu\tilde{\psi} - ig\tilde{A}_\mu\tilde{\psi} \\ &= U\,(D_\mu\psi)\,, \end{aligned} \tag{5.6}$$

then clearly

$$\tilde{A}_\mu U = UA_\mu - \frac{i}{g}\partial_\mu U, \tag{5.7}$$

and hence

$$\tilde{A}_\mu = U\left[A_\mu - \frac{i}{g}U^{-1}\partial_\mu U\right]U^{-1}. \tag{5.8}$$

In terms of an infinitesimal gauge transformation, $U = 1 - it_a\theta_a (U^{-1} = 1 + it_a\theta_a)$, using Eq. (5.4) we find that the gluon fields transform as

$$A_a^\mu \to A_a^\mu - \frac{1}{g}\partial^\mu\theta_a + f_{abc}\theta_b A_c^\mu. \tag{5.9}$$

[1] Unlike the Lorentz indices, μ, there is no difference between color indices, a, that are raised or lowered and the choice is dictated by appearance.

The appearance of the term $f_{abc}\theta_b A_c^\mu$ on the right hand side of Eq. (5.9) is characteristic of a non-Abelian gauge theory. It is a direct consequence of the fact that the generators of the gauge transformations, $t_a \equiv \frac{\lambda_a}{2}$, do not commute with each other. It is generally believed that from this simple fact follow the most spectacular properties of the strong interaction, notably asymptotic freedom and confinement.

Having established the gauge invariant Lagrangian density for the quarks, $\bar{\psi}(i\gamma^\mu D_\mu - m)\psi$, we now examine the gluons. Clearly a gluon mass term, involving $A_\mu A^\mu$, is not invariant under Eq. (5.8), so the gluons must be massless. This is as in QED. However, unlike QED, the analogue of the usual kinetic energy density, $F^a_{\mu\nu}F^{\mu\nu}_a$, with:

$$F^a_{\mu\nu} = \partial_\mu A^a_\nu - \partial_\nu A^a_\mu, \tag{5.10}$$

is not gauge invariant. Instead we define

$$G^{\mu\nu} \equiv t_a G^{\mu\nu}_a, \tag{5.11}$$

with

$$G^{\mu\nu}_a \equiv \partial^\mu A^\nu_a - \partial^\nu A^\mu_a + g f_{abc} A^\mu_b A^\nu_c, \tag{5.12}$$

and it is straightforward to show that

$$G^{\mu\nu} = \frac{i}{g}\left[D^\mu, D^\nu\right]. \tag{5.13}$$

Since D^μ transforms as

$$D^\mu \rightarrow U D^\mu U^{-1} \tag{5.14}$$

(from Eqs. (5.4) and (5.6)), we see easily that

$$\mathrm{Tr}\left(G^{\mu\nu}G_{\mu\nu}\right) \equiv \frac{1}{2} G^{\mu\nu}_a G^a_{\mu\nu} \tag{5.15}$$

is invariant under gauge transformations. (Here "Tr" is a trace over the 3×3 color matrices.) We are therefore led naturally to the QCD Lagrangian density

$$\mathcal{L}_{\mathrm{QCD}} = \bar{\psi}\left(i\gamma^\mu D_\mu - m\right)\psi - \frac{1}{2}\,\mathrm{Tr}\left(G^{\mu\nu}G_{\mu\nu}\right). \tag{5.16}$$

(Note that the coupling constant, g, is dimensionless, so we could redefine $A_\mu \rightarrow gA_\mu$, in which case:

$$\mathcal{L} \rightarrow \bar{\psi}\left(i\gamma^\mu\left[\partial_\mu - iA_\mu\right] - m\right)\psi - \frac{1}{2g^2}\,\mathrm{Tr}\left(G^{\mu\nu}G_{\mu\nu}\right). \tag{5.17}$$

As this depends only on g^2, the sign of g is physically irrelevant.)

As we have seen, Eq. (5.16) is the natural consequence of local color symmetry. It looks deceptively like QED, but expanding the gluon field energy yields a completely new feature. The gluons self interact, with a tri-linear coupling proportional to g

$$\mathcal{L}^{(3)} = -\frac{g}{2} f_{abc}\left(\partial_\mu A_{a\nu} - \partial_\nu A_{a\mu}\right) A^\mu_b A^\nu_c, \tag{5.18}$$

and a four-gluon vertex proportional to

$$\mathcal{L}^{(4)} = -\frac{g^2}{4} f_{abe} f_{cde} A_{a\nu} A_{b\nu} A_c^{\mu} A_d^{\nu}. \tag{5.19}$$

In order to appreciate the consequences of these terms we need to obtain the Feynman rules appropriate to Eq. (5.16).

5.2 Feynman Rules

The derivation of the Feynman rules for QCD is a complicated task for which the appropriate formalism is the Feynman path integral [FH 65]. A thorough grounding in path integral methods is beyond the scope of this book. Nevertheless, the method is conceptually simple and we aim to set out the steps in the derivation in a manner which should be broadly accessible. For those wanting complete derivations we refer to the specialized texts already cited. We believe that the overview presented here will enable some students to understand those texts rather more quickly than would otherwise be the case. Appendix E summarizes the main formulas arising in the path integral formalism.

In essence the Feynman rules are the expressions for the irreducible, connected 2-, 3- and 4-particle Green functions. Within the path integral formalism the Green functions are the naturally occurring objects. For example, in a simple scalar field theory with action

$$S = \int d^4x \mathcal{L}(\phi, \partial_\mu \phi) \tag{5.20}$$

the n-point Green function is

$$\left\langle 0 \middle| \mathcal{T}\left[\hat{\phi}(x_1)\cdots\hat{\phi}(x_n)\right] \middle| 0 \right\rangle = \frac{\int [d\phi]\phi(x_1)\ldots\phi(x_n)\exp(iS)}{\int [d\phi]\exp(iS)}, \tag{5.21}$$

where the path integral, $\int [d\phi]$, is a formal notation for an integral over all possible values of the field at all space-time points. As a convenient formal device for generating products of fields, such as those on the right hand side of Eq. (5.21), one introduces the generating functional

$$\mathcal{Z}(J) = \int [d\phi] \exp\left(i \int d^4x \left[\mathcal{L} + \phi(x)J(x)\right]\right) \tag{5.22}$$

so that the n-point Green function is

$$\begin{aligned} \mathcal{G}^{(n)}(x_1,\ldots,x_n) &= \left\langle 0 \middle| \mathcal{T}\left[\hat{\phi}(x_1)\ldots\hat{\phi}(x_n)\right] \middle| 0 \right\rangle \\ &= \frac{(-i)^n}{\mathcal{Z}(0)} \frac{\delta \mathcal{Z}(J)}{\delta J(x_1)\cdots\delta J(x_n)}\bigg|_{J=0} \end{aligned} \tag{5.23}$$

with

$$\frac{\delta \mathcal{Z}\left[J(x)\right]}{\delta J(y)} = \lim_{\epsilon\to 0+} \frac{\mathcal{Z}\left(J(x) + \epsilon\delta(x-y)\right) - \mathcal{Z}\left(J(x)\right)}{\epsilon}. \tag{5.24}$$

In short, each functional derivative with respect to $J(x_i)$ pulls down one factor of $\phi(x_i)$ and then one removes the source, J, which is no longer needed, by setting it to zero.

5.2.1 Gauge Invariance

In evaluating these path integrals in QCD, one is integrating over infinitely many gluon field configurations which are related by gauge transformations and therefore have the same QCD action. In order to avoid this we introduce a gauge fixing condition which ensures that each set of gauge equivalent gluon fields is integrated just once.[2]

In QED the gauge fixing condition is often taken to be the Lorentz condition, $\partial_\mu A^\mu = 0$. For QCD this is usually generalized to

$$\Theta^\mu A^a_\mu(x) = B^a(x), \tag{5.25}$$

with Θ^μ some Lorentz vector operation and $B^a(x)$ some function chosen as a constraint. This condition is imposed on the generating functional of QCD by introducing unity through a product of δ-functions

$$1 = \det M \prod_a \left(\int [d\theta^a]\, \delta\big(\Theta^\mu A^a_\mu(x) - B^a(x)\big) \right). \tag{5.26}$$

Here $\det M$ is just the Jacobian required to give unity inside a path integral over the gauge fields A^a_μ:

$$M^{ab}(x,y) = \frac{\delta\big(\Theta^\mu A^a_\mu(x)\big)}{\delta\big(\theta^b(y)\big)}. \tag{5.27}$$

In the case of an Abelian gauge theory the determinant of M is simply a constant, independent of A^a_μ, and is physically irrelevant – cancelling in the ratio $\mathcal{Z}(J)/\mathcal{Z}(0)$. Even in QCD it is possible to choose conditions so that this is the case. For example, given a space-like, constant vector $n^\mu(n^2 < 0)$, the axial gauge, defined by

$$n^\mu A^a_\mu = 0, \tag{5.28}$$

satisfies this condition. Although this choice simplifies some formal demonstrations, the presence of the vector n^μ in objects like Green functions makes practical calculations very difficult indeed.

The alternative, which is much simpler in practical calculations, is to use the mathematical identity for the functional integral over Grassmann fields (see Appendix E)

$$\det M = \int [d\psi]\,[d\bar{\psi}] \exp\left(\int d^4x\, d^4y\, \bar{\psi}(x)\, M(x,y)\, \psi(y) \right). \tag{5.29}$$

At the price of introducing a set of anti-commuting boson fields one can use Eq. (5.29) to replace the Jacobian in the path integral. These fields, which are a purely mathematical artifice and quite unphysical, are known as Faddeev-Popov ghosts [FP 67]. They are there simply to ensure that the gauge fixing procedure does not alter the original path integral.

---

[2] The complications associated with so-called Gribov copies [Gri 78] are believed to play no role in perturbative QCD, which is the only use for Feynman rules.

Adopting the shorthand notation

$$\prod_a \int [d\chi^a]\,[d\chi^{a\star}] = \int [d\chi]\,[d\chi^\star], \tag{5.30}$$

we may use Eq. (5.29) to write

$$\det M = i \int [d\chi]\,[d\chi^\star] \exp\left(-i \int d^4x\, d^4y\, \chi^{a\star}(x)\, M^{ab}(x,y)\, \chi^b(y)\right). \tag{5.31}$$

For the most common choice of covariant gauge, one can integrate the exponent inEq. (5.31) to obtain

$$\det M = i \int [d\chi]\,[d\chi^\star] \exp\left(i \int d^4x\, (\partial^\mu \chi^{a\star}(x))\, D^{ab}_\mu\, \chi^b(x)\right), \tag{5.32}$$

with D^{ab}_μ the covariant derivative

$$D^{ab}_\mu = \delta^{ab}\partial_\mu - g f^{abc} A^c_\mu. \tag{5.33}$$

Returning to Eq. (5.26), it is convenient to formally eliminate the δ-functions by performing a functional integral over the functions $B^a(x)$, with the simple weight function $\exp(-\frac{i}{2\lambda}\int d^4x (B^a(x))^2)$. The δ-functions then replace B^a by $\partial^\mu A^a_\mu$ (for $\Theta^\mu = \partial^\mu$, as above). In summary then, the gauge fixing procedure given in Eq. (5.26), for a covariant gauge, involves the addition of unity in the QCD generating functional in the form[3]

$$1 = \int [d\chi]\,[d\chi^\star] \exp\left(-i \int d^4x (\mathcal{L}_{\text{GF}} + \mathcal{L}_{\text{FP}})\right), \tag{5.34}$$

where the Faddeev-Popov (FP) piece is:

$$\mathcal{L}_{\text{FP}} = (\partial^\mu \chi^{a\star}) D^{ab}_\mu \chi^b \tag{5.35}$$

and the gauge-fixing (GF) piece is:

$$\mathcal{L}_{\text{GF}} = -\frac{1}{2\lambda}(\partial^\mu A^a_\mu)^2. \tag{5.36}$$

Including Eq. (5.34) the QCD generating functional is therefore

$$\mathcal{Z}[J,\xi,\xi^\star,\eta,\bar{\eta}] = \int [dA]\,[d\chi]\,[d\chi^\star]\,[d\psi]\,[d\bar{\psi}]$$
$$\exp\left(i \int d^4x \left[\mathcal{L} + A^a_\mu J^{a\mu} + \chi^{a\star}\xi^a + \xi^{a\star}\chi^a + \bar{\psi}\eta + \bar{\eta}\psi\right]\right), \tag{5.37}$$

where the sources for the gluon, quark and ghost fields are $J^{a\mu}$, $\eta(\bar{\eta})$ and $\xi^a\,(\xi^{a\star})$, respectively. Combining Eqs. (5.16) and (5.34) we have

$$\mathcal{L}_{\text{QCD}} = \bar{\psi}(i\gamma^\mu D_\mu - m)\psi - \frac{1}{4}G^{a\mu\nu}G^a_{\mu\nu} + \mathcal{L}_{\text{GF}} + \mathcal{L}_{\text{FP}}. \tag{5.38}$$

[3]The factor of i in Eqs. (5.31) and (5.32) is irrelevant because we only ever need the ratio $\mathcal{Z}(J)/\mathcal{Z}(0)$ – c.f. Eq. (5.23).

5.2.2 Free Propagators and Interactions

The usual technique for calculating Feynman rules is to divide $\mathcal{L}_{\text{QCD}}$, in Eq. (5.38), into free and interacting pieces, $\mathcal{L}_0$ and $\mathcal{L}_I$, respectively. The structure of the free quark and gluon pieces is the same as in QED, but one has in addition a ghost term:

$$\mathcal{L}_0 = \mathcal{L}_0^{\text{quark}} + \mathcal{L}_0^{\text{glue}} + \mathcal{L}_0^{\text{ghost}}, \tag{5.39}$$

with

$$\begin{aligned}
\mathcal{L}_0^{\text{quark}} &= \bar{\psi}(i\,\not{\partial} - m)\psi, & (5.40)\\
\mathcal{L}_0^{\text{glue}} &= -\frac{1}{4}(\partial_\mu A_\nu^a - \partial_\nu A_\mu^a)(\partial^\mu A^{a\nu} - \partial^\nu A^{a\mu}) - \frac{1}{2\lambda}(\partial_\mu A_a^\mu)^2, & (5.41)\\
\mathcal{L}_0^{\text{ghost}} &= (\partial^\mu \chi^{a\star})\partial_\mu \chi^a. & (5.42)
\end{aligned}$$

By definition, the remaining terms constitute $\mathcal{L}_I$: a three-gluon coupling, $\mathcal{L}^{(3)}$ in Eq. (5.18); a four-gluon coupling, $\mathcal{L}^{(4)}$ in Eq. (5.19); the usual vector coupling of the quarks to the gauge boson and finally, through Eqs. (5.32) and (5.33), a gluon ghost interaction:

$$\begin{aligned}
\mathcal{L}_I &= \mathcal{L}_I(A_\mu^a, \chi^a, \chi^{a\star}, \psi, \bar{\psi})\\
&= -\frac{g}{2} f_{abc}(\partial_\mu A_\nu^a - \partial_\nu A_\mu^a)A^{b\mu}A^{c\nu}\\
&\quad - \frac{g^2}{4} f_{abe} f_{cde} A_\mu^a A_\nu^b A^{c\mu} A^{d\nu} + g\bar{\psi}\gamma^\mu t^a \psi A_\mu^a - g f_{abc}(\partial_\mu \chi^{a\star})\chi^b A^{c\mu}.
\end{aligned} \tag{5.43}$$

In complete analogy with QED, the generating functional corresponding to $\mathcal{L}_0$ may be written as:

$$\mathcal{Z}_0(J, \xi, \xi^\star, \eta, \bar{\eta}) = \mathcal{Z}_0^{\text{quark}}(\eta, \bar{\eta})\, \mathcal{Z}_0^{\text{glue}}(J)\, \mathcal{Z}_0^{\text{ghost}}(\xi, \xi^\star). \tag{5.44}$$

Here the quark piece is simply expressed in terms of the free fermion propagator, $S(z)$:

$$\begin{aligned}
S(z) &\equiv \langle 0|T\left(\psi(z)\bar{\psi}(0)\right)|0\rangle,\\
&= i\int \frac{d^4k}{(2\pi)^4}\frac{\exp(-ik\cdot z)}{\not{k} - m + i\epsilon},
\end{aligned} \tag{5.45}$$

with

$$\mathcal{Z}_0^{\text{quark}}(\eta, \bar{\eta}) = \exp\left(-\int d^4x\, d^4y\, \bar{\eta}(x)\, S(x-y)\, \eta(y)\right). \tag{5.46}$$

(Note that our $S(z)$ is often written as $iS_F(z)$. It has the advantage that the momentum space Feynman propagator, Eq. (5.58), is simply the Fourier transform of $S(z)$.) Similarly, the gluon piece can be written as:

$$\mathcal{Z}_0^{\text{glue}}(J) = \exp\left(-\frac{1}{2}\int d^4x\, d^4y\, J^{a\mu}(x)\, D_{\mu\nu}^{ab}(x-y)\, J^{b\nu}(y)\right), \tag{5.47}$$

in terms of the free gluon propagator

$$\begin{aligned} D^{ab}_{\mu\nu}(z) &\equiv \langle 0|\mathcal{T}\left(A^a_\mu(z)A^b_\nu(0)\right)|0\rangle, \\ &= i\delta_{ab}\int \frac{d^4k}{(2\pi)^4}\left(-g_{\mu\nu}+(1-\lambda)\frac{k_\mu k_\nu}{k^2+i\epsilon}\right)\frac{\exp(-ik\cdot z)}{k^2+i\epsilon}, \end{aligned} \tag{5.48}$$

and the ghost term is

$$\mathcal{Z}_0^{\text{ghost}}(\xi,\xi^\star) = \exp\left(-\int d^4x\,d^4y\,\xi^{a\star}(x)\,D_{ab}(x-y)\,\xi^b(y)\right), \tag{5.49}$$

with

$$D_{ab}(z) = i\delta_{ab}\int \frac{d^4k}{(2\pi)^4}\frac{\exp(-ik\cdot z)}{k^2+i\epsilon}. \tag{5.50}$$

In order to calculate the connected Green functions for the fully interacting theory it is convenient to begin with the formal result that

$$\begin{aligned} \mathcal{Z}(J,\xi,\xi^\star,\eta,\bar\eta) =& \exp\left(i\int d^4x\,\mathcal{L}_I\left(-i\frac{\delta}{\delta J},-i\frac{\delta}{\delta\xi^{a\star}},+i\frac{\delta}{\delta\xi^a},-i\frac{\delta}{\delta\bar\eta},+i\frac{\delta}{\delta\eta}\right)\right) \\ &\mathcal{Z}_0(J,\xi,\xi^\star,\eta,\bar\eta). \end{aligned} \tag{5.51}$$

This is simple to understand because $\mathcal{L}=\mathcal{L}_0+\mathcal{L}_I$ and the functional derivatives with respect to $J^{a\mu}(x),\xi^{a\star}(x),\bar\eta(x)$ (etc.) just pull a factor $A^a_\mu(x),\chi^a(x)$ or $\psi(x)$ out of $\mathcal{Z}_0$. (Note the difference in sign between $+i\frac{\delta}{\delta\eta}$ and $-i\frac{\delta}{\delta\bar\eta}$ arises because these are anti-commuting, Grassmann variables and the first acts on $+i\bar\psi\eta$ whereas the second acts on $+i\bar\eta\psi$.)

One can now obtain the Feynman rules used in perturbative QCD calculations by expanding the exponential in Eq. (5.51)

$$\mathcal{Z}(J,\xi,\xi^\star,\eta,\bar\eta) = \left[1+i\int d^4x\mathcal{L}_I+\cdots\right]\mathcal{Z}_0(J,\xi,\xi^\star,\eta,\bar\eta), \tag{5.52}$$

and using the fundamental relation, Eq. (5.23). Applying this to the various pieces of $\mathcal{Z}$ generated by the term linear in $\mathcal{L}_I$ in Eq. (5.52) we can identify the appropriate rules for constructing Feynman diagrams. For example, the three-gluon vertex is obtained by calculating the three-point Green function

$$\begin{aligned} \mathcal{G}^{(3)\mu_1\mu_2\mu_3}_{a_1a_2a_3} &= (-i)^3\frac{\delta^3}{\delta J^{a_1}_{\mu_1}(x_1)\delta J^{a_2}_{\mu_2}(x_2)\delta J^{a_3}_{\mu_3}(x_3)}\int d^4x\mathcal{L}^{(3)}\left(-i\frac{\delta}{\delta J}\right)\mathcal{Z}_0^{\text{gluon}}(J)\bigg|_{J=0} \\ &= -igf_{abc}\int d^4x\left[\partial^\mu D^{\nu\mu_1}_{aa_1}(x-x_1)-\partial^\nu D^{\mu\mu_1}_{aa_1}(x-x_1)\right] \\ &\qquad D^{ba_2\mu_2}_\mu(x-x_2)D^{ca_3\mu_3}_\nu(x-x_3) \quad + \quad \text{even permutations.} \end{aligned} \tag{5.53}$$

Using Eq. (5.48) for the free gluon propagators and realizing that the derivative becomes the gluon momentum after a Fourier transform, we obtain the expression to be used for a three-gluon vertex in a Feynman graph [4]:

$$+gf_{a_1a_2a_3}\left[g^{\mu_1\mu_2}(p_1-p_2)^{\mu_3}+g^{\mu_2\mu_3}(p_2-p_3)^{\mu_1}+g^{\mu_3\mu_1}(p_3-p_1)^{\mu_2}\right]. \tag{5.54}$$

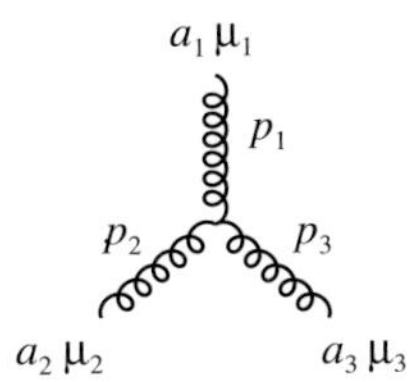

Figure 5.1: The three-gluon vertex of Eq. (5.54).

The quark-gluon, four-gluon and ghost-gluon vertices can be obtained in the same way. The quark-gluon vertex factor is:

$$ig\gamma^\mu t^a, \tag{5.55}$$

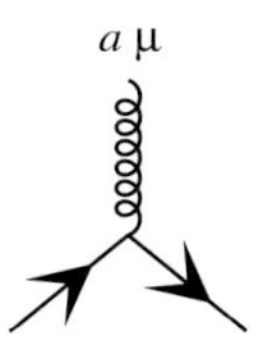

Figure 5.2: The quark-gluon vertex of Eq. (5.55).

while the four-gluon vertex is given by:

$$\begin{aligned}-ig^2\Big[&f_{a_1a_2a}f_{a_3a_4a}(g^{\mu_1\mu_3}g^{\mu_2\mu_4}-g^{\mu_1\mu_4}g^{\mu_2\mu_3})\\&+f_{a_1a_3a}f_{a_2a_4a}(g^{\mu_1\mu_2}g^{\mu_3\mu_4}-g^{\mu_1\mu_4}g^{\mu_2\mu_3})\\&+f_{a_1a_4a}f_{a_2a_3a}(g^{\mu_1\mu_2}g^{\mu_3\mu_4}-g^{\mu_1\mu_3}g^{\mu_2\mu_4})\Big],\end{aligned} \tag{5.56}$$

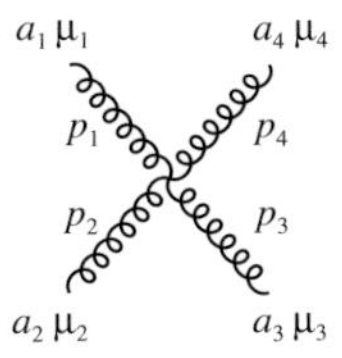

Figure 5.3: The four-gluon vertex of Eq. (5.56).

[4]This is the three point Green function truncated by removing the external gluon propagators.

and the ghost-gluon vertex introduces the factor

$$g f_{abc} p^{\mu}. \tag{5.57}$$

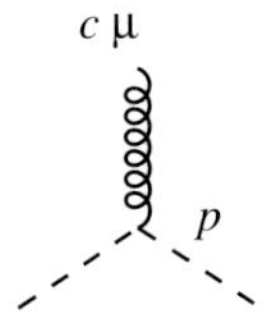

Figure 5.4: The ghost-gluon vertex of Eq. (5.57).

The momentum space rules for the free propagators in Eqs. (5.45), (5.48) and (5.50) are:

$$\delta_{ij}\delta_{ab}\frac{i}{\not{p} - m + i\epsilon} \tag{5.58}$$

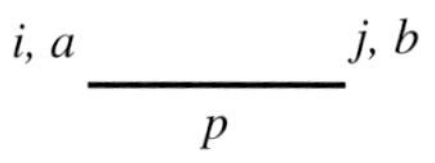

Figure 5.5: The quark propagator of Eq. (5.58).

for the quark,

$$\delta_{ab}\left[-g^{\mu\nu} + (1-\lambda)\frac{p^{\mu}p^{\nu}}{p^2 + i\epsilon}\right]\frac{i}{p^2 + i\epsilon} \tag{5.59}$$

for the gluon,

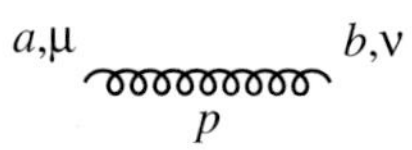

Figure 5.6: The gluon propagator of Eq. (5.59).

and

$$\delta_{ab}\frac{i}{p^2 + i\epsilon} \tag{5.60}$$

for the ghost.

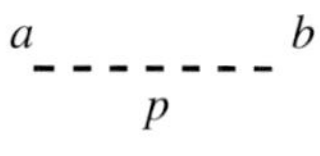

Figure 5.7: The ghost propagator of Eq. (5.60).

Finally, Fermi-Dirac statistics for quarks and ghosts lead to an extra factor of -1 for any closed loops involving these particles (as for electrons in QED). There are also some statistical factors to compensate for overcounting in gluon loops: a factor of $1/2!$ in a two gluon loop on a gluon line; $1/2!$ for a single gluon loop (tadpole) on a gluon line; and $1/3!$ for a three gluon loop on a gluon line.

5.3 Renormalization

In applying the Feynman rules developed in the last section one soon encounters divergent loop integrals. The modern technique for dealing with such divergences is to regulate them using dimensional regularization and to renormalize by adding suitable counter terms, so that the renormalized theory is finite and well defined as the regulator is removed. Dimensional regularization is thoroughly described in any recent text book on quantum field theory. In brief the idea is that an integral which is divergent in 4-dimensional space-time will be convergent in D-dimensions with $D < 4$ [HV 72a]. One evaluates the integrals for D-dimensions and then analytically continues towards $D = 4$. The divergences then show up as logarithms or inverse powers of $\epsilon = (4 - D)/2$. As a consequence, the counter terms which must be added to the Lagrangian density also involve logarithms or inverse powers of ϵ. We summarize some key results of dimensional regularization in Appendix D.

The renormalization procedure for gauge theories such as QCD has been made much easier by the proof that the renormalized Lagrangian density must be gauge invariant if the bare Lagrangian density is itself gauge invariant [HV 72b, LZ 72]. We introduce renormalized fields which are related to the bare fields by multiplicative renormalization constants:

$$\begin{aligned} A_a^\mu &= Z_A^{-\frac{1}{2}} \, A_a^\mu|_{\text{bare}} \,, \\ \psi &= Z_\psi^{-\frac{1}{2}} \, \psi|_{\text{bare}} \,, \\ \chi^a &= Z_\chi^{-\frac{1}{2}} \, \chi^a|_{\text{bare}} \,. \end{aligned} \tag{5.61}$$

Furthermore, the quark masses and the gauge parameter are also renormalized

$$\begin{aligned} m &= Z_m^{-1} Z_\psi m_{\text{bare}}, \\ \lambda &= Z_\lambda Z_A^{-1} \lambda_{\text{bare}}. \end{aligned} \tag{5.62}$$

Finally, there are four different vertices involving the gauge coupling constant, each of which must be separately renormalized, for example

$$g = Z_\chi Z_A^{\frac{1}{2}} Z_1^{-1} g_{\text{bare}} \tag{5.63}$$

for the ghost-gluon vertex and similarly for the quark-gluon vertex (with a new constant Z_2) and the 3- and 4-gluon vertices (Z_3 and Z_4). In order that gauge invariance is preserved for both the bare and renormalized Lagrangians, the constants $Z_1, \ldots, Z_4$ must be related. These relationships are the Slavnov-Taylor identities

$$Z_\chi Z_1^{-1} \;=\; Z_\psi Z_2^{-1} \;=\; Z_A Z_3^{-1} \;=\; Z_A^{\frac{1}{2}} Z_4^{-\frac{1}{2}}. \tag{5.64}$$

Using Eqs. (5.61) – (5.64), it is straightforward to write the Lagrangian density in terms of the renormalized one plus a set of counter terms

$$\mathcal{L}_{\text{bare}} = \mathcal{L} + \Delta\mathcal{L}, \tag{5.65}$$

where

$$\begin{aligned}\Delta\mathcal{L} = &-\frac{1}{4}\Delta Z_A\left(\partial_\mu A^a_\nu - \partial_\nu A^a_\mu\right)\left(\partial^\mu A^{a\nu} - \partial^\nu A^{a\mu}\right) - \frac{\Delta Z_\lambda}{2\lambda}\left(\partial_\mu A^\mu_a\right)^2 \\ &+ \Delta Z_\chi\left(\partial_\mu \chi^{a\star}\right)\left(\partial^\mu \chi^a\right) + \Delta Z_\psi \bar{\psi} i\,\partial\!\!\!/\psi - m\Delta Z_m\bar{\psi}\psi \\ &- g f_{abc}\Delta Z_1\left(\partial_\mu\chi^{a\star}\right)\chi^b A^\mu_c + g\Delta Z_2\bar{\psi}\gamma_\mu t_a A^\mu_a\psi \\ &- g\Delta Z_3 f_{abc}A^\mu_a A^\nu_b\partial_\mu A_{c\nu} - g^2\Delta Z_4 f_{abc}f_{ade}A^\mu_b A^\nu_c A_{d\mu}A_{e\nu}.\end{aligned} \tag{5.66}$$

(Here, ΔZ_α is uniformly defined as $Z_\alpha - 1$, for $\alpha = A, \ldots, m, 1, \ldots, 4$.)

5.3.1 Renormalization Scale

In order that the action be dimensionless in D-dimensional space-time, the gauge coupling, g, must acquire dimensions of mass to the power ϵ (recall $\epsilon = (4 - D)/2$). We can define a dimensionless coupling constant, $\hat{g}$, by introducing an arbitrary mass scale, μ, known as the renormalization scale, such that

$$g \equiv g(\mu) = \hat{g}\mu^\epsilon. \tag{5.67}$$

No physical observable can depend on the choice of μ.

5.3.2 Renormalization of the Gauge Coupling Constant

We saw from Eq. (5.63) and the following discussion that the renormalized coupling constant can be calculated in four equivalent ways. Although it is not the simplest, the most natural approach is to consider the renormalization of the quark-gluon vertex

$$g = Z_\psi\, Z_A^{\frac{1}{2}}\, Z_2^{-1}\, g_{\text{bare}}. \tag{5.68}$$

By way of illustration (and because of its importance) we shall evaluate the renormalization to order g^3. In that case, we may write

$$g_{\text{bare}} = \left(1 - \Delta Z_\psi - \frac{1}{2}\Delta Z_A + \Delta Z_2\right) g, \tag{5.69}$$

where each of ΔZ_ψ, ΔZ_A and ΔZ_2 is evaluated to order g^2.

Given the Feynman rules for perturbative QCD and the standard formulas for integration in $4 - 2\epsilon$ dimensions, it is a straightforward but quite technical exercise to evaluate the right hand side of Eq. (5.69). We choose to emphasize the physics by showing the Feynman graphs relevant to each piece of the calculation, together with the corresponding analytic expression.

To evaluate ΔZ_ψ we need only the self-energy graph shown in Fig. 5.8. In the minimal subtraction scheme (MS) we retain only the piece of Fig. 5.8 which behaves as $\frac{1}{\epsilon}$ as $\epsilon \to 0$:

$$\Delta Z_\psi = -\frac{2\,\hat{g}^2\,\lambda}{32\,\pi^2\,\epsilon}\,C_2(F). \tag{5.70}$$

Figure 5.8: Diagram contributing to the wave function renormalization constant for the quarks, Z_ψ.

Incidentally, the mass renormalization constant resulting from the evaluation of the same diagram is:

$$\Delta Z_m = -\frac{\hat{g}^2\,(3+\lambda)}{16\,\pi^2\,\epsilon}\,C_2(F). \tag{5.71}$$

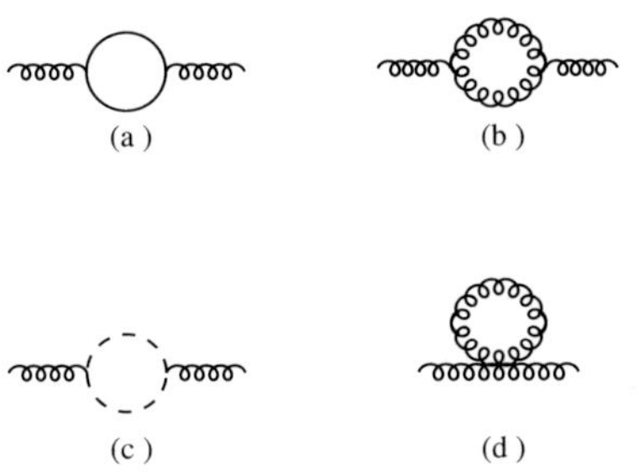

Figure 5.9: Diagrams contributing to the wave function renormalization constant for the gluon field, Z_A.

The calculation of the wave function renormalization constant for the gluon field, Z_A, is rather more complicated, involving the diagrams shown in Fig. 5.9. The last diagram there is known as a "tadpole" and vanishes in dimensional regularization. Evaluating the other three leads to

$$\Delta Z_A = -\frac{\hat{g}^2}{32\,\pi^2\,\epsilon}\left(\left[\frac{8}{3}T_2(F)N_f\right] - \left[\left(\frac{25}{6}-\lambda\right)C_2(A)\right] - \left[\frac{1}{6}C_2(A)\right]\right), \tag{5.72}$$

where the terms in square brackets correspond to Figs. 5.9 (a), (b) and (c), respectively. Note that N_f is the number of quark flavors and the group factors are defined by:

$$(t^a\,t^a)_{ij} = C_2(F)\,\delta_{ij} = \frac{N_c^2-1}{2N_c}\,\delta_{ij}, \tag{5.73}$$

$$\mathrm{Tr}(t^a\,t^b) = T_2(F)\,\delta_{ab} = \frac{\delta_{ab}}{2}, \tag{5.74}$$

$$f_{acd}\,f_{bcd} = C_2(A)\,\delta_{ab} = N_c\,\delta_{ab}, \tag{5.75}$$

where N_c is the number of colors.

Finally, in order to evaluate ΔZ_2 in Eq. (5.69) we need to calculate the diagrams shown in Fig. 5.10. The result is

$$\Delta Z_2 = -\frac{2\,\hat{g}^2}{32\,\pi^2\epsilon}\left(\left[C_2(F) - \frac{1}{2}\,C_2(A)\right]\lambda - \left[\frac{3}{4}\,C_2(A)(1-\lambda)\right]\right), \tag{5.76}$$

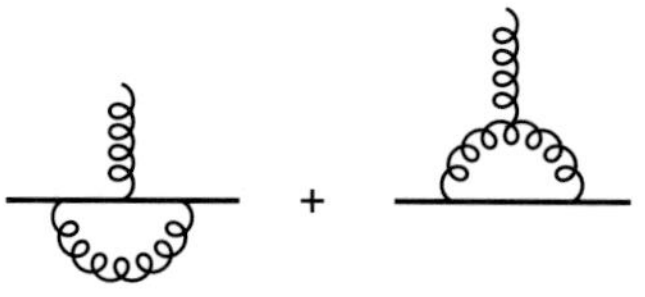

Figure 5.10: Diagrams contributing to the renormalization of the quark-gluon vertex, Z_2.

where once again the terms in square brackets correspond to Figs. 5.9 (a) and (b) in that order. Putting Eqs. (5.70), (5.72) and (5.76) into Eq. (5.69), we obtain

$$g_{\text{bare}} = g - \frac{33 - 2\,N_f}{96\,\pi^2\epsilon}\,g\,\hat{g}^2, \tag{5.77}$$

$$= g - \frac{33 - 2\,N_f}{96\,\pi^2\epsilon}\,g^3\,\mu^{-2\epsilon}, \tag{5.78}$$

where the second line follows from the definition (5.67).

From Eq. (5.78) we see that the bare coupling is divergent as $D \to 4$. Of far more interest is the variation of the renormalized coupling constant with the choice of μ, through Eq. (5.78). The introduction of μ was necessary only in order to regularize divergent integrals appearing in the renormalization procedure and hence g_{bare} is independent of μ in $D = 4$. Using this fact we can calculate the famous β-function of QCD:

$$\beta(g) = \frac{\partial g}{\partial \ln \mu}, \tag{5.79}$$

to order g^3. Acting on the right hand side of Eq. (5.78) with $\frac{\partial}{\partial \mu}$, and setting the result to zero, we can solve for $\beta(g)$ (to $\mathcal{O}(g^3)$):

$$\begin{aligned} \beta(g) &= -\frac{33 - 2\,N_f}{48\,\pi^2}\,g^3 \\ &= -\frac{\beta_0}{16\,\pi^2}\,g^3, \end{aligned} \tag{5.80}$$

with $\beta_0 = 11 - \frac{2}{3}N_f$, the constant already introduced in Chapter 4. Theories in which $\beta(g)$ is negative (for g positive – remember that the sign of g is arbitrary) are said to be **asymptotically free**. This property is crucial to our understanding of high energy interactions in QCD, including deep inelastic scattering phenomena.

5.4 Renormalization Group

As the choice of renormalization scale, μ, is completely arbitrary, no physical observable can depend on it. On the other hand, intermediate steps in a calculation, such as the renormalized Green functions, will usually depend on μ. It is important that this behavior is under mathematical control and the general framework for ensuring this is known as the renormalization group.

Our purpose in discussing the renormalization group here is three-fold. First, we shall see that, in order to ensure that physical quantities are independent of μ, it is natural to introduce the so-called **running coupling**, $\bar{g}$. This replaces the renormalized coupling constant in renormalization group improved perturbation theory. (Similarly, a running quark mass, $\bar{m}$, replaces m.) We show that $\bar{g}$ vanishes logarithmically as its argument increases, and that the famous property of **asymptotic freedom** is a direct consequence. That is, in QCD, as the external momenta increase, the effective coupling decreases and (renormalization group improved) perturbation theory converges, permitting rigorous predictions. Second, this analysis will lead to an understanding of the concept of an anomalous dimension associated with operators in QCD. Finally, through the use of the operator product expansion we will be led to a deeper understanding of the scaling violations in deep inelastic scattering, which were discussed at a phenomenological level in Chapter 4.

In order to simplify the discussion, we specialize to the Landau gauge (for which $\lambda = \lambda_{\text{bare}} = 0$ and hence – c.f. Eq. (5.70) – $\Delta Z_\psi = 0$) and to massless quarks. The latter permits us to avoid the anomalous dimension associated with the renormalization of the mass of the quark (c.f. Eqs. (5.62), (5.70) and (5.71)). We define $\Gamma(p_i, g, \mu)$ to be the one-particle irreducible, truncated Green function [5] for n_A gluons and n_ψ quarks (with momenta p_i, $i = 1, \ldots, n(\equiv n_A + n_\psi)$). In general, Γ is related to the corresponding bare Green function by:

$$\Gamma = Z\Gamma_{\text{bare}}. \tag{5.81}$$

For example, for $n_A = 1$ and $n_\psi = 2$ we would have $Z = Z_A^{\frac{1}{2}} Z_\psi Z_2^{-1}$.

We define the anomalous dimension, γ, of the renormalized Green function, through:

$$\gamma = \frac{\partial \ln Z}{\partial \ln \mu}. \tag{5.82}$$

(Note that the partial derivative indicates that g_{bare} and whatever cut-off is used in the bare theory are kept fixed while the derivative with respect to μ is calculated.) Then the invariance of all bare quantities under changes in μ:

$$\mu \frac{\partial \Gamma_{\text{bare}}}{\partial \mu} = \mu \frac{\partial}{\partial \mu}\left(Z^{-1}\Gamma\right) = 0, \tag{5.83}$$

leads to the Callan-Symanzik equation:

$$\left(\mu \frac{\partial}{\partial \mu} + \beta(g)\frac{\partial}{\partial g} - \gamma\right)\Gamma(p_i, \mu, g) = 0, \tag{5.84}$$

where $\beta(g)$ was defined in Eq. (5.79).

5.4.1 Running Coupling

It is perhaps simplest to appreciate the power of Eq. (5.84) by first setting $\gamma = 0$. Then clearly a change in μ, from μ to $\bar{\mu}$, can always be compensated by a change in g, from g to $\bar{g}$, such that

$$\Gamma(p_i, \bar{\mu}, \bar{g}) = \Gamma(p_i, \mu, g). \tag{5.85}$$

[5] This is the Green function obtained by removing the external gluon and quark propagators.

This is most simply realized by setting

$$\bar{\mu} = \exp(t)\mu; \qquad \bar{g} \equiv \bar{g}(t), \tag{5.86}$$

with $\bar{g}(t)$ defined by:

$$t = \int_{\bar{g}(0)}^{\bar{g}(t)} \frac{dg'}{\beta(g')}; \qquad \bar{g}(0) = g. \tag{5.87}$$

The new coupling, $\bar{g}(t)$, defined through Eq. (5.87), is known as the running coupling.

To leading order in QCD perturbation theory we have already established that

$$\beta(g) = -\frac{\beta_0}{16\,\pi^2}\, g^3 + \mathcal{O}(g^5), \tag{5.88}$$

with $\beta_0 = 11 - \frac{2}{3}N_f$. Then Eq. (5.87) yields

$$\begin{aligned} t &= -\frac{16\,\pi^2}{\beta_0} \int_{\bar{g}(0)}^{\bar{g}(t)} \frac{dg'}{g'^3}, \\ &= \frac{16\,\pi^2}{2\,\beta_0} \left(\frac{1}{\bar{g}^2(t)} - \frac{1}{\bar{g}^2(0)} \right). \end{aligned} \tag{5.89}$$

Although the parameter t is arbitrary, we will be interested in the limit where it becomes large. In that case, Eq. (5.89) can only be consistent if the running coupling, $\bar{g}(t)$, vanishes as $t \to \infty$ – i.e., $\bar{g}(\bar{\mu})$ should vanish logarithmically as a function of $\bar{\mu}$. (Recall that, from Eq. (5.86), t behaves like $\ln(\bar{\mu})$.)

It is easy to solve Eq. (5.89) for $\bar{g}(t)$:

$$\begin{aligned} \bar{g}^2(t) &= \frac{\bar{g}^2(0)}{1 + \frac{2\,\beta_0}{16\,\pi^2}\, \bar{g}^2(0) t}, \\ &= \frac{16\,\pi^2}{\beta_0 \ln(\frac{\bar{\mu}^2}{\Lambda^2})}, \end{aligned} \tag{5.90}$$

where

$$\Lambda = \mu \, \exp\left(-\frac{16\,\pi^2}{2\,\beta_0\, \bar{g}^2(0)} \right). \tag{5.91}$$

Of course, the same analysis can be carried out to any order in perturbation theory. For example, at next-to-leading order (NLO), which means that the perturbative renormalization includes all diagrams with one and two loops, the running coupling takes the form:

$$\bar{g}^2(t) = \frac{16\,\pi^2}{\beta_0 \ln(\frac{\bar{\mu}^2}{\Lambda^2})} \left[1 - \frac{16\,\pi^2 \beta_1}{\beta_0^2} \frac{\ln\ln(\frac{\bar{\mu}^2}{\Lambda^2})}{(\ln \frac{\bar{\mu}^2}{\Lambda^2})} \right], \tag{5.92}$$

with an appropriately complicated definition of the Λ parameter, $\Lambda_{\rm NLO}$:

$$\Lambda_{\rm NLO} = \mu \exp\left(-\frac{8\,\pi^2}{\beta_0\,g^2}\right)\left(\frac{1+\frac{\beta_1\,g^2}{(4\,\pi)^2\beta_0}}{\frac{\beta_0\,g^2}{(4\,\pi)^2}}\right)^{\frac{\beta_1}{2\,\beta_0^2}}. \tag{5.93}$$

(The parameter $\beta_1 = 102 - \frac{38N_f}{3}$ (divided by $(4\pi)^4$) is the coefficient of g^5 in the expansion of $\beta(g)$.) Λ is the only parameter in massless QCD which has to be specified from outside. It can only be determined from experiment by fitting the calculation of the cross section for some process to experiment. After that, all physical quantities should be calculable in terms of this scale parameter.

It is quite common to deal with a strong coupling constant, $\alpha_s(\mu)$, defined (by analogy with the fine structure constant in QED) by:

$$\alpha_s(\mu) = \frac{\bar{g}^2(\mu)}{4\pi}. \tag{5.94}$$

In Fig. 5.11 we show the scale dependence of $\alpha_s(\mu)$ as determined from a large number of experimental measurements. Although the technical details go beyond the scope of this book, we note that in calculations beyond NLO (i.e. at $\mathcal{O}(g^7)$ and higher) the definition of $\alpha_s(\mu)$ is renormalization scheme dependent. It is conventional to use the $\overline{MS}$ scheme, and that is what has been used in computing the theoretical curve in Fig. 5.11. One final complication that should be noted is that as μ increases the number of active flavors, N_f, increases – from 3 below charm threshold to 5 above the bottom production threshold. The central value of the QCD Λ parameter within the $\overline{MS}$ scheme, for four flavors ($\Lambda^{(4)}_{\overline{MS}}$), is of order 300 ± 50 MeV [PDG 98/00].

5.4.2 Asymptotic Freedom

The logarithmic decrease of the running coupling at large momentum scales, or short distances, has as its main consequence the asymptotic freedom of QCD. Returning to Eq. (5.85), it is now relatively easy to see the practical significance of asymptotic freedom. Suppose that $\tilde{\Gamma}$ is Γ with all necessary powers of the external momenta removed, so that it is dimensionless. Then $\tilde{\Gamma}$ can only depend on the dimensionless combination $\frac{p_i}{\mu}$ and g, which we express as

$$\tilde{\Gamma}(p_i,\,\mu,\,g) \equiv \tilde{F}\left(\frac{p_i}{\mu},g\right). \tag{5.95}$$

Using Eq. (5.85), this implies that

$$\begin{aligned}\tilde{\Gamma}(p_i,\,\mu,\,g) &= \tilde{\Gamma}(p_i,\,\bar{\mu},\,\bar{g}(t))\\ &= \tilde{\Gamma}\big(\exp(-t)p_i,\,\mu,\,\bar{g}(t)\big),\end{aligned} \tag{5.96}$$

since

$$\frac{\exp(-t)p_i}{\mu} = \frac{p_i}{\exp(t)\mu}. \tag{5.97}$$

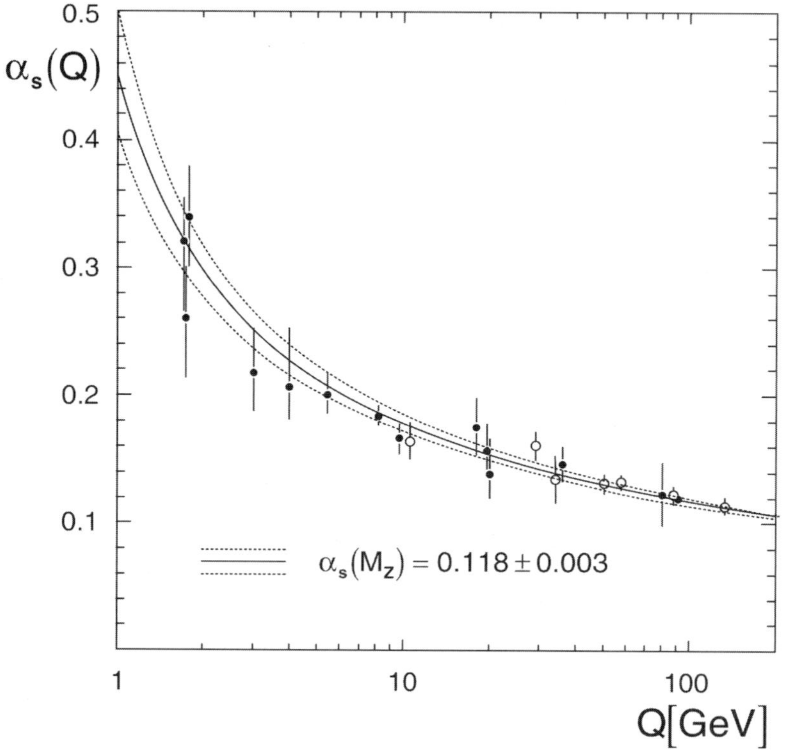

Figure 5.11: Scale dependence of the strong coupling constant $\alpha_s(Q)$ [PDG 98/00].

Replacing p_i by $\exp(t)p_i$ in Eq. (5.96) we find:

$$\tilde{\Gamma}\big(\exp(t)p_i,\, \mu,\, g\big) = \tilde{\Gamma}\big(p_i,\, \mu,\, \bar{g}(t)\big). \tag{5.98}$$

That is, the calculation of processes in QCD at high momentum can be carried out in terms of a coupling constant that vanishes logarithmically as the momentum scale increases. Thus, high energy, or "hard" processes should be calculable in perturbation theory in terms of the running coupling.

Equation (5.98) gives the essential physical importance of asymptotic freedom. Of course, in practice the situation is slightly more complicated. Suppose that the Green function has naive (or engineering) dimensions (in powers of energy) $d(\Gamma)$, then

$$\Gamma\big(\exp(t)p_i,\, \mu,\, g\big) = \exp\big(d(\Gamma)t\big)\Gamma\big(p_i,\, \mu,\, \bar{g}(t)\big). \tag{5.99}$$

Of course, in a quantum field theory the Green function will, in general, need to be renormalized and its anomalous dimension, γ – c.f. Eq. (5.82) – will not be zero. Returning to Eq. (5.84), with γ non-zero, we can replace $\mu\frac{\partial}{\partial\mu} + \beta\frac{\partial}{\partial g}$ by $\frac{d}{d\ln\mu}$ (using Eq. (5.80)), and hence

$$\Gamma(p_i,\, \bar{\mu},\, \bar{g}) = \exp\left(\int_0^t \gamma(\bar{g}(t'))\, dt'\right)\Gamma\big(p_i,\, \mu,\, g\big). \tag{5.100}$$

Using the same trick we used to obtain Eq. (5.98), we find

$$\Gamma\big(\exp(t)p_i,\, \mu,\, g\big) = \exp\left(d(\Gamma)t - \int_0^t \gamma\big(\bar{g}(t')\big)\, dt'\right)\Gamma\big(p_i,\, \mu,\, \bar{g}(t)\big). \tag{5.101}$$

Finally, it is common to use the β-function for the running coupling $\left(\beta\big(\bar{g}(t)\big) = \frac{d\bar{g}(t)}{dt}\right)$ to rewrite Eq. (5.101) as:

$$\Gamma(\exp(t)p_i,\, \mu,\, g) = \exp\left(d(\Gamma)t - \int_{\bar{g}(0)}^{\bar{g}(t)} \frac{\gamma(g')}{\beta(g')}\, dg'\right)\Gamma(p_i,\, \mu,\, \bar{g}(t)). \tag{5.102}$$

While the presence of the exponential factor involving the anomalous dimension of the Green function under consideration is vital (e.g. we shall see that it explains the scale dependence of the deep inelastic structure functions discussed in the previous chapter), it does not spoil the crucial results associated with asymptotic freedom.

With this background it is now relatively easy to see that the procedure followed for $\bar{g}$ can also be followed for the quark masses. One finds that a running quark mass, $\bar{m}(t)$, defined by:

$$\bar{m}(t) = m \exp\left(-t - \int_0^t dt' \gamma_m\left[\bar{g}(t')\right]\right) \tag{5.103}$$

appears on the right hand side of Eq. (5.99). In fact, this running mass vanishes as $t \to \infty$, so QCD at high momenta is essentially a massless theory.

The renormalization group analysis which we have just outlined is essential to the understanding of QCD phenomenology, from the study of e^+e^- annihilation into hadrons to hard scattering processes. However, our concern is nucleon structure and especially the information about that structure which can be obtained from deep inelastic scattering. We therefore turn next to the consideration of deep inelastic scattering within the framework of the renormalization group.

5.5 Deep Inelastic Scattering: Operator Product Expansion

The structure functions measured in deep inelastic scattering with an electromagnetic probe are related to the forward Compton amplitude

$$T_{\mu\nu} = i \int d^4z\, \exp(iq\cdot z)\, \langle N|\mathcal{T}(j_\mu(z) j_\nu(0))|N\rangle, \tag{5.104}$$

by

$$\begin{aligned} 2MW_{\mu\nu} &= \frac{1}{\pi}\, \mathrm{Im}\, T_{\mu\nu}, \\ &= \frac{1}{2\pi} \int d^4z\, \exp(iq\cdot z)\, \langle N|\left[j_\mu(z), j_\nu(0)\right]|N\rangle. \end{aligned} \tag{5.105}$$

(Rather than working with the full electromagnetic current of QCD, for the present introduction we consider the simpler case of a single quark flavor with unit charge – i.e., $j_\mu = \bar{\psi}\gamma_\mu\psi$.)

As we have already explained, in working with deep inelastic probes we are primarily concerned with the Bjorken limit, where $Q^2 = -q^2$ and $p \cdot q$ (with p the target four momentum) are both large – typically greater than 2 GeV^2. In this limit the Fourier transform in Eq. (5.104) is dominated by the behavior as $z^2 \to 0$. The operator product expansion is the ideal tool for dealing with such problems.

In general the product of two operators, say $A(x)B(y)$, is not well defined as $x \to y$ in a quantum field theory. The operator product expansion, or OPE [Wil 69], expresses such products as a sum over local operators, $O_i(\frac{x+y}{2})$, which can be renormalized in the usual way, and a set of (possibly singular) c-number coefficient functions, $C_i(x-y)$:

$$A(x)B(y) = \sum_i C_i(x-y)\, O_i\left(\frac{x+y}{2}\right). \tag{5.106}$$

After taking the Fourier transform required in Eq. (5.104), it is simple to see that on purely dimensional grounds the operators which are most important at high-Q^2 are those for which the coefficient functions, $C_i(z)$, are most singular as $z^2 \to 0$.

In the case of interest, the OPE for the time ordered product of $j_\mu(z)$ and $j_\nu(0)$ can be obtained from:

$$\begin{aligned}\mathcal{T}(j_\mu(z)j_\nu(0)) &= \left(g_{\mu\nu} - 2\frac{z_\mu z_\nu}{z^2}\right)\frac{z^2}{\pi^4(z^2-i\epsilon)^4}\\ &+ \sigma_{\mu\lambda\nu\rho}O_V^\rho(z,0)\frac{iz^\lambda}{2\pi^2(z^2-i\epsilon)^2}\\ &+ \epsilon_{\mu\lambda\nu\rho}O_A^\rho(z)\frac{z^\lambda}{2\pi^2(z^2-i\epsilon)^2} + O_{\mu\nu}^{\rm NS}(z,0).\end{aligned} \tag{5.107}$$

This is obtained from the Wick expansion of $\mathcal{T}(j_\mu(z)j_\nu(0))$ and we have used:

$$\sigma_{\mu\lambda\nu\rho} = g_{\mu\lambda}g_{\nu\rho} - g_{\mu\nu}g_{\lambda\rho} + g_{\mu\rho}g_{\nu\lambda} \tag{5.108}$$

$$O_V^\rho(z,0) = :\bar\psi(z)\gamma^\rho\psi(0) - \bar\psi(0)\gamma^\rho\psi(z): \tag{5.109}$$

$$O_A^\rho(z,0) = :\bar\psi(z)\gamma^\rho\gamma_5\psi(0) + \bar\psi(0)\gamma^\rho\gamma_5\psi(z): \tag{5.110}$$

The explicit form of $O_{\mu\nu}^{\rm NS}$, which involves the normal ordered product of four quark fields, will not be needed here and $\epsilon_{\mu\lambda\nu\rho}$ is the anti-symmetric tensor with $\epsilon_{0123} = +1$.

The first term in Eq. (5.107) is target independent and is known as a disconnected term. The most singular, connected contributions, which are therefore the contributions relevant to deep inelastic scattering in the Bjorken limit, come from the operators $O_V^\rho(z,0)$ and $O_A^\rho(z,0)$. These are "bi-local", in that they involve quark field operators at two different space-time points. Now, using the definition of the time ordered product and the fact that the electromagnetic current is Hermitian, it is a simple matter to check that:

$$\left[j_\mu(z), j_\nu(0)\right] = \epsilon(z_0)\left[\mathcal{T}(j_\mu(z)j_\nu(0)) - \left(\mathcal{T}(j_\mu(z)j_\nu(0))\right)^\dagger\right]. \tag{5.111}$$

It is also useful to have the identity:

$$\frac{1}{(z^2-i\epsilon)^2} = \frac{\mathcal{P}}{z^4} - i\pi\delta^{(1)}(z^2), \tag{5.112}$$

where $\delta^{(1)}(z^2)$ is the first derivative of the Dirac δ-function and $\mathcal{P}$ indicates a principal value. Then, using Eq. (5.111) for the commutator of currents in Eq. (5.105), as well as Eq. (5.112), and keeping only the terms involving $O_V^\rho(z,0)$ and $O_A^\rho(z,0)$ which dominate the connected part of the interaction with the target in the Bjorken limit, we find:

$$2MW_{\mu\nu} = \frac{1}{2\pi^2}\int d^4z z^\lambda \epsilon(z_0)\delta^{(1)}(z^2)\langle N|\sigma_{\mu\lambda\nu\rho}O_V^\rho(z,0) - i\epsilon_{\mu\lambda\nu\rho}O_A^\rho(z,0)|N\rangle. \tag{5.113}$$

It is important to stress that none of the steps leading to Eq. (5.113) have involved the choice of target. In anticipation of the need to renormalize matrix elements of the operators appearing here, it is usual to expand them in a Taylor series about $z = 0$, so that:

$$O_{V,A}^\rho(z,0) = \sum_{n=0}^{\infty}\frac{1}{n!}z^{\mu_1}\ldots z^{\mu_n}O_{V,A\mu_1\ldots\mu_n}^\rho(0), \tag{5.114}$$

where $O_{V,A\mu_1\ldots\mu_n}^\rho(0)$ are purely local operators

$$\begin{aligned} O_{V\mu_1\ldots\mu_n}^\rho(z) &= :\left[\partial_{\mu_1}\ldots\partial_{\mu_n}\bar\psi(z)\right]\gamma^\rho\psi(z) - \bar\psi(z)\gamma^\rho\left[\partial_{\mu_1}\ldots\partial_{\mu_n}\psi(z)\right]: \\ O_{A\mu_1\ldots\mu_n}^\rho(z) &= :\left[\partial_{\mu_1}\ldots\partial_{\mu_n}\bar\psi(z)\right]\gamma^\rho\gamma_5\psi(z) \\ &\quad +\bar\psi(z)\gamma^\rho\gamma_5\left[\partial_{\mu_1}\ldots\partial_{\mu_n}\psi(z)\right]: . \end{aligned} \tag{5.115}$$

Note that by a series of integrations by parts one can show that $O_{V\mu_1\ldots\mu_n}^\rho(z)$ and $O_{A\mu_1\ldots\mu_n}^\rho(z)$ are non-zero only for n odd or even, respectively.

We recall from Chapter 4 that, because of the anti-symmetric tensor accompanying $O_{A\mu_1\ldots\mu_n}^\rho(z)$, this term cannot contribute to the unpolarized deep inelastic cross section. For simplicity we carry the argument through for the unpolarized case – i.e. for $W_{\mu\nu}^{(S)}$. If we take the momentum of the target to be p, on very general grounds the matrix element of the local operator $O_{V\mu_1\ldots\mu_n}^\rho(0)$ can only depend on combinations of $p_{\mu_i}p_{\mu_j}$ and $p^2 g_{\mu_i\mu_j}$:

$$\langle N|O_{V\mu_1\ldots\mu_n}^\rho(0)|N\rangle = 2a_n p^\rho p_{\mu_1}\cdots p_{\mu_n} + \cdots . \tag{5.116}$$

The additional terms in Eq. (5.116), ..., involve at least one term in $p^2 g_{\mu_i\mu_j}$. When contracted with $z^{\mu_i}z^{\mu_j}$ and Fourier transformed, such terms behave like $\frac{p^2}{q^2}$ in comparison with $\frac{(p\cdot q)^2}{q^4}$ for the leading terms. In the Bjorken limit ($\frac{p\cdot q}{q^2} \to$ constant, with $-q^2$ and $p\cdot q \to \infty$) the latter dominate and so we can drop the additional terms in Eq. (5.116).

Suppose we define the function $\tilde f(y)$ as

$$\tilde f(y) = \sum_{n=0,2,\ldots}\frac{y^n}{n!}a_n, \tag{5.117}$$

then in the Bjorken limit we have established that $W_{\mu\nu}^{(S)}$ can be expressed as

$$2\,MW_{\mu\nu}^{(S)} = \frac{\sigma_{\mu\lambda\nu\rho}p^\rho}{\pi^2}\int d^4z\,\exp(iq\cdot z)\,\delta^{(1)}(z^2)\,\epsilon(z_0)\,z^\lambda\,\tilde f(p\cdot z). \tag{5.118}$$

One can evaluate this integral relatively easily in terms of the Fourier transform of $\tilde{f}(y)$:

$$f(\zeta) = \int_{-\infty}^{\infty} dy\, \exp(i\zeta y)\, \tilde{f}(y). \tag{5.119}$$

The result is

$$W_{\mu\nu}^{(S)} = \frac{1}{M} \int_{-\infty}^{\infty} d\zeta f(\zeta)\, \epsilon(q_0 + \zeta p_0)\, \delta\big(2\,\zeta p \cdot q + \zeta^2 p^2 - Q^2\big) \times \left(p_\mu q_\nu + p_\nu q_\mu + 2\,\zeta p_\mu p_\nu - g_{\mu\nu}(p \cdot q + \zeta p^2)\right), \tag{5.120}$$

where $Q^2 = -q^2$, as usual. In the Bjorken limit, the δ-function in Eq. (5.120) implies that $\zeta \equiv x = \frac{Q^2}{2p \cdot q}$ and hence:

$$W_{\mu\nu}^{(S)} = \frac{f(x)}{2M} \left(-g_{\mu\nu} + \frac{q_\mu q_\nu}{q^2}\right) + \frac{x f(x)}{M\, p \cdot q} \left(p_\mu - \frac{p \cdot q}{q^2} q_\mu\right) \left(p_\nu - \frac{p \cdot q}{q^2} q_\nu\right). \tag{5.121}$$

From Eq. (4.5) we recognize immediately that the usual structure functions F_1 and F_2 can be expressed as

$$F_1 = \frac{1}{2} f(x) \qquad \text{and} \qquad F_2 = x f(x). \tag{5.122}$$

These are precisely the simple parton model formulas for one flavor of quark with unit charge. Clearly the structure functions scale in the sense that they depend only on x and not at all on Q^2.

5.6 Deep Inelastic Scattering within QCD

We have accumulated almost all the techniques and concepts necessary to understand what the QCD improved parton model is and why it works. However, a few preliminaries remain. We first need to define the concept of **twist** and to replace the operators introduced in Eq. (5.115) by operators which have definite twist and are gauge invariant. It is also important to derive the relationship between the matrix elements of these gauge invariant operators and the moments of the measured structure functions. Then we will be ready to consider the renormalization of the OPE within renormalization group improved QCD and hence to rigorously derive the evolution of the structure functions.

After integration by parts the operators $O_V^{\rho\mu_1 \cdots \mu_n}$, introduced in Eq. (5.115), are essentially $\bar{\psi}\gamma^\rho \partial^{\mu_1} \ldots \partial^{\mu_n} \psi$, for n odd. While this object is appropriate for a naive parton model, QCD is based on the concept of local gauge invariance, so it is natural in QCD to consider instead the operator $\bar{\psi}\gamma^\rho i D^{\mu_1} \ldots i D^{\mu_n} \psi$, where D^μ is the covariant derivative (5.5), and the factors of i are conventionally introduced to simplify the Feynman rules for these operators. While this object has $n+1$ Lorentz indices, its Lorentz transformation properties are not simple. That is, it does not belong to an irreducible representation of the Lorentz group. It is theoretically much simpler to consider operators which do belong to an irreducible representation and the representation which dominates in the Bjorken limit is the largest – in this case the $n+1$-dimensional representation.

The simple prescription for building elements of the largest irreducible representation of the Lorentz group is to first symmetrize in all indices and then make the result traceless on all pairs of indices, by subtracting all possible terms in which any two elements, say i and j, have been contracted (multiplied by $g^{\mu_i \mu_j}$). The result, for n rather than $n+1$:

$$O_n^{\mu_1 \ldots \mu_n} = \mathcal{S} \bar{\psi} \gamma^{\mu_1} i D^{\mu_2} \ldots i D^{\mu_{n-1}} \psi - \text{traces involving } g^{\mu_i \mu_j}, \tag{5.123}$$

is said to have "spin" n. (Here $\mathcal{S}$ symmetrizes the Lorentz indices.) Since it has dimensions of energy to the power $n+2$, its "engineering" dimensions minus its spin give 2. This combination is known as **twist** and these operators are called **twist-2** or **leading twist** operators.

Although the leading twist operator of spin-n is unique, we shall see that radiative corrections in QCD lead to a violation of the Callan-Gross relation at leading order in $\bar{g}^2$. In anticipation of this, we note that the tensor structure of $W^{\mu\nu}$ involves $g^{\mu\nu}$ and $p^\mu p^\nu$ – the extra terms involving q can be written in later on the basis of current conservation. These two terms are quite naturally associated with the operators $q^2 g^{\mu\nu} O_n^{\mu_1 \ldots \mu_n}$ and $O_n^{\mu\nu\mu_3 \ldots \mu_n}$, respectively. (Recall that the matrix elements of O give products of p.)

Let us define a projection operator which picks out the piece of $W^{\mu\nu}$ corresponding to the absorption of longitudinal photons, $P_{\mu\nu}^{(L)}$, and a tensor, $P_{\mu\nu}^{(2)}$, which picks out the W_2 piece:

$$\begin{aligned} P_{\mu\nu}^{(L)} &= g_{\mu\nu} - \frac{q_\mu q_\nu}{q^2}, \\ P_{\mu\nu}^{(2)} &= -p_\mu p_\nu \frac{q^2}{(p \cdot q)^2} + \frac{p_\mu q_\nu + p_\nu q_\mu}{p \cdot q} - g_{\mu\nu}. \end{aligned} \tag{5.124}$$

Then the forward Compton amplitude can be written as

$$\begin{aligned} T_{\mu\nu}(P,q) &= i \int d^4 z \exp(iq \cdot z) \langle N(P) | \mathcal{T} \left(j_\mu(z) j_\nu(0) \right) | N(P) \rangle \\ &= \sum_{n=\text{even}} \langle N(P) | O_n^{\mu_1 \ldots \mu_n} | N(P) \rangle \frac{2^n}{(Q^2)^n} \Big(P_{\mu\nu}^{(L)} C_n^{(L)}(Q^2) q_{\mu_1} \cdots q_{\mu_n} \\ &\quad + \left[-q^2 g_{\mu\mu_1} g_{\mu_2\nu} + [g_{\mu\mu_1} q_{\mu_2} q_\nu + g_{\mu_2\nu} q_\mu q_{\mu_1}] - g_{\mu\nu} q_{\mu_1} q_{\mu_2} \right] \\ &\quad \times C_n^{(2)}(Q^2) q_{\mu_3} \cdots q_{\mu_n} \Big), \end{aligned} \tag{5.125}$$

where we recognize the tensor structure of $P_{\mu\nu}^{(2)}$ in the second term on the right hand side of the equation. In Eq. (5.125) we have defined the Fourier transform of the coordinate space coefficient functions to be $C_n^{(L,2)}(Q^2)$. Using Eq. (5.116) in Eq. (5.125) we find

$$T_{\mu\nu}(P,q) = \sum_n \left(\frac{1}{x} \right)^n \left(a_n C_n^{(L)}(Q^2) P_{\mu\nu}^{(L)} + a_n C_n^{(2)}(Q^2) P_{\mu\nu}^{(2)} \right). \tag{5.126}$$

Clearly this expression is a power series in $\frac{1}{x}$, while the physical region is $x \in [0,1]$. It is not therefore very useful as it stands. However, as we now show it can be continued to the region of interest by using the analytic properties of $T_{\mu\nu}(P,q)$.

5.6.1 Connection to Moments of the Structure Functions

For fixed Q^2 the forward Compton scattering amplitude is analytic in the complex photon energy except along the real axis when $(P+q)^2 \geq M^2$, which corresponds to $\omega \equiv \frac{1}{x} \in [1,\infty)$. In addition, crossing implies that $T_{\mu\nu}(P,q) = T_{\mu\nu}(P,-q)$ and hence there must also be a cut from $-\infty$ to 1 in ω. If we integrate over the contour $\mathcal{C}$ in the complex ω-plane, as shown in Fig. 5.12, we find:

$$\begin{aligned}\frac{1}{2\pi i}\int_{\mathcal{C}} \frac{T_{\mu\nu}(\omega)}{\omega^n} d\omega &= \int_1^\infty \frac{d\omega}{\omega^n}\frac{1}{\pi}\,\mathrm{Im}\,T_{\mu\nu}(\omega) \\ &= 2\,M\int_0^1 dx\, n^{n-2}\, W_{\mu\nu},\end{aligned} \tag{5.127}$$

where the last step follows from Eq. (5.105).

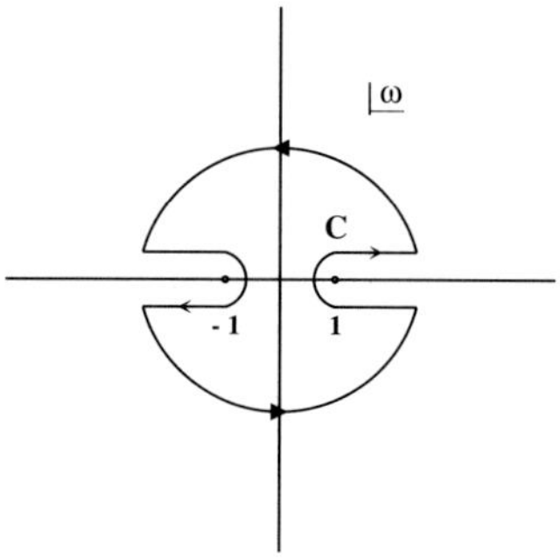

Figure 5.12: Illustration of the contour in the complex-ω ($\equiv x^{-1}$) plane, needed to relate a particular moment of the structure function to the matrix element of a leading twist operator and its coefficient function.

If we now substitute the expansion (5.126) into the left hand side of Eq. (5.127) and use Cauchy's theorem:

$$\frac{1}{2\pi i}\int_{\mathcal{C}} d\omega\, \omega^{m-n} = \delta_{n,m+1}, \tag{5.128}$$

we see that the integral just picks out the n'th term in the sum. As a result, using the definition:

$$2MW_{\mu\nu} = P^{(L)}_{\mu\nu}\frac{F_L}{x} + P^{(2)}_{\mu\nu}\frac{F_2}{x}, \tag{5.129}$$

where F_2 is the structure function, νW_2, already defined in Eq. (4.7) and

$$F_L(x,Q^2) = 2M\left(\left(1+\frac{\nu^2}{Q^2}\right)W_2 - W_1\right), \tag{5.130}$$

we find the key results:

$$a_n C^{(L)}_n(Q^2) = \int_0^1 dx\, x^{n-2} F_L(x,Q^2), \tag{5.131}$$

and

$$a_n C_n^{(2)}(Q^2) = \int_0^1 dx\, x^{n-2}\, F_2(x, Q^2). \tag{5.132}$$

That is the products of the coefficient function and the proton matrix element of the leading twist operator is directly given by the n'th moment of the structure function in the Bjorken limit. As we shall explain shortly, the Mellin transform may be used to reconstruct F_2 and F_L from their moments.

5.6.2 Flavour Structure

In our discussion in this chapter, we have so far ignored the fact that the quarks come in several flavors, with different charges. Below charm threshold the electromagnetic current operators involve a sum over up, down and strange quarks, weighted by the appropriate charge squared. By decomposing the resulting structure function into isovector, octet and singlet components, as explained in Eqs. (4.91)-(4.93), we see that rather than a single leading twist quark operator for each value of the spin, we need three operators, one containing τ_3, one λ_8 (an SU(3) flavor operator) and finally a singlet quark operator in which we sum over u, d and s with equal weight. As we explained in our introduction to the parton model in Chapter 4, the first two cases are referred to as non-singlet operators.

Formally we can summarize all this by adding an extra, flavor label, i, to the operators and summing over this label. If we define $O_{3,n} = (O_n^u - O_n^d), O_{8,n} = (O_n^u + O_n^d - 2O_n^s)$ and $O_{0,n} = (O_n^u + O_n^d + O_n^s)$, where the Lorentz indices have been suppressed for simplicity, then the forward Compton amplitude for the proton becomes:

$$T_{\mu\nu}(P, q) = \sum_n \sum_{i=3,8,0} e_i^2 \left(\frac{1}{x}\right)^n \left(a_{i,n} C_{i,n}^{(L)} P_{\mu\nu}^{(L)} + a_{i,n} C_{i,n}^{(2)} P_{\mu\nu}^{(2)}\right), \tag{5.133}$$

where $e_i^2 = \frac{1}{12}, \frac{1}{36}$ and $\frac{1}{9}$ for $i = 3, 8, 0$.

The procedure for computing the coefficient functions within QCD is now straightforward, if technically complicated. We can exploit the fact that the OPE is target independent to compute the forward Compton amplitude on a single quark in two ways: directly within QCD and also using the OPE. By equating the two we can find the coefficient functions as a power series in the renormalized QCD coupling constant g^2. These coefficient functions can then be used in the analysis of deep inelastic scattering data on any target.

From Eq. (5.126) we recognize that the forward Compton amplitude can be written in terms of its two components:

$$T_{\mu\nu}(P, q) = \sum_i e_i^2 \left(T_i^{(L)} P_{\mu\nu}^{(L)} + T_i^{(2)} P_{\mu\nu}^{(2)}\right). \tag{5.134}$$

where

$$T_i^{(L,2)} = \sum_n \left(\frac{1}{x}\right)^n T_{i,n}^{(L,2)}. \tag{5.135}$$

Evaluating the renormalized Compton scattering amplitude on a single quark to a given order in g^2, yields

$$T_{i,n}^{(L,2)} = t_{i,n}^{(L,2)}(0) + g^2 t_{i,n}^{(L,2)}(1) + \cdots \tag{5.136}$$

There is one final complication to be faced, namely that the renormalization of the singlet operators (subscript 0, above) is a little more complicated than the renormalization of the non-singlet operators (subscripts 3 and 8). The former mix under renormalization (as illustrated in Fig. 5.14) with leading twist gluon operators,

$$O_{A,n}^{\mu_1 \dots \mu_n} = \mathcal{S} G_a^{\sigma\mu_1} i D^{\mu_2} \dots i D^{\mu_{n-1}} G_{a\sigma}^{\mu_n} - \text{traces.} \tag{5.137}$$

For simplicity we consider first the non-singlet case.

5.6.3 Non-singlet Case

As we have explained the non-singlet operators involving the flavor matrices λ_3 and λ_8 cannot mix with gluon operators under renormalization. Formally we need to introduce the appropriate operator, designated $O_{i,n}$ (with other indices omitted for simplicity), with its associated source $J_{i,n}$, into the functional integral (5.51) and differentiate with respect to $\eta, \bar{\eta}$ and $J_{i,n}$ in order to obtain the matrix element $\langle N|\mathcal{T} O_{i,n}\psi(x)\bar{\psi}(y)|N\rangle$. After renormalization of this matrix element, we will have a power series in g^2 for the matrix elements of the operator on a single quark target:

$$a_{i,n} = A_{i,n}(0) + g^2 A_{i,n}(1) + \cdots \tag{5.138}$$

Finally, if we write a power series for the coefficients $C_{i,n}$:

$$C_{i,n} = c_{i,n}(0) + g^2 c_{i,n}(1) + \cdots, \tag{5.139}$$

we can determine the expansion coefficients, $c_{i,n}(j)$, order by order by equating Eq. (5.136) and the product of Eqs. (5.138) and (5.139). Thus, $t_{i,n}(0) = A_{i,n}(0)c_{i,n}(0)$ determines $c_{i,n}(0)$. Then equating the coefficients of g^2 yields $t_{i,n}(1) = A_{i,n}(0)c_{i,n}(1) + A_{i,n}(1)c_{i,n}(0)$, which we can solve for $c_{i,n}(1)$, knowing $c_{i,n}(0)$, and so on.

It may be helpful to illustrate how the calculation is performed. The overall result is gauge invariant, but for definiteness we show the intermediate results in Feynman gauge ($\lambda = 1$). Fig. 5.13 shows the processes which must be calculated. The simple insertion of the operator between external quark lines, which is proportional to $\mathcal{S}\gamma^{\mu_1} p^{\mu_2} \dots p^{\mu_n} \lambda_i$, is shown in the first graph. It also includes the counter term we are calculating, which involves the same factor times $\Delta Z_\psi + \Delta Z_O$. The second process yields a factor

$$-\frac{g^2\, C_2(F)}{32\,\pi^2 \epsilon} \frac{4}{n(n+1)} \tag{5.140}$$

times $\mathcal{S}\gamma^{\mu_1} p^{\mu_2} \dots p^{\mu_n} \lambda_i$. (Recall that $C_2(F)$ was defined in Eq. (5.73).) The third and fourth terms, which are identical, yield a factor

$$\frac{g^2\, C_2(F)}{32\,\pi^2 \epsilon} \sum_{k=2}^{n} \frac{4}{k}, \tag{5.141}$$

where the sum arises from the ordering of the covariant derivatives to which the virtual gluon couples.

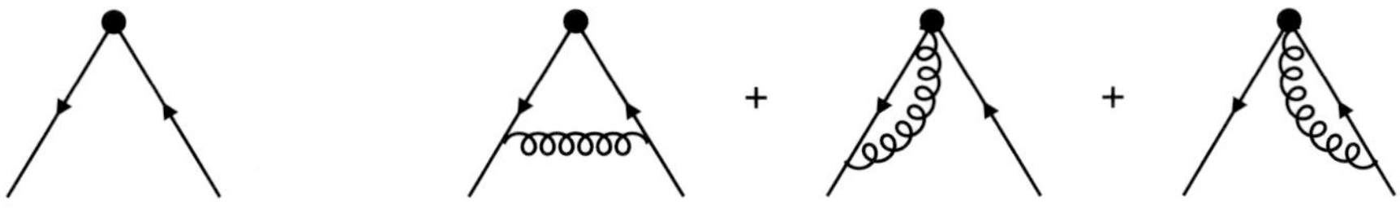

Figure 5.13: Diagrams involved in the renormalization of the flavor non-singlet operators to one loop.

Using our explicit expression for ΔZ_ψ, Eq. (5.70), we can solve for the counter term associated with the spin-n, leading twist, non-singlet operator:

$$\Delta Z_O = \frac{2\,g^2\,C_2(F)}{32\,\pi^2\epsilon}\left(1 - \frac{2}{n(n+1)} + 4\sum_{k=2}^{n}\frac{1}{k}\right). \tag{5.142}$$

Using this we can finally evaluate the anomalous dimension of this operator as:

$$\begin{aligned}\gamma_n &= \mu\,\frac{\partial}{\partial\mu}\ln(1+\Delta Z_O)\\ &= -\frac{2\,g^2\,C_2(F)}{16\,\pi^2}\left(1 - \frac{2}{n(n+1)} + 4\sum_{k=2}^{n}\frac{1}{k}\right).\end{aligned} \tag{5.143}$$

Renormalization group improvement

As usual in perturbative QCD, our expression for the forward Compton amplitude necessarily involves terms like $\ln\frac{Q^2}{\mu^2}$, which can grow rapidly. This was precisely why we introduced the renormalization group analysis earlier, leading to Eq. (5.102). Applied to our coefficients in the OPE, which are dimensionless so $d(O) = 0$ in Eq. (5.102), we find:

$$C_{i,n}(Q^2,\mu^2) = C_{i,n}\left(\mu^2,\bar{g}(t)\right)\exp\left(-\int_{\bar{g}(0)}^{\bar{g}(t))} dg'\,\frac{\gamma_n(g')}{\beta(g')}\right), \tag{5.144}$$

where

$$t = \ln\frac{Q^2}{\mu^2}. \tag{5.145}$$

We use the expansions as powers of g^2, defined above for $C_{i,n}$ and earlier for β, and also define the expansion:

$$\gamma_n = g^2\gamma_{n0} + g^4\gamma_{n1} + \cdots \tag{5.146}$$

(Note that to obtain γ_{n1} and β_1 one must evaluate Feynman diagrams involving two loops. This is know as a **next-to-leading order (NLO)** calculation.) Then Eq. (5.144) yields the renormalization group improved coefficients;

$$C_{i,n}(Q^2) = \left(c_{i,n}(0) + c_{i,n}(1)\bar{g}^2(t)\right) \exp\left(\int_{\bar{g}(0)}^{\bar{g}(t))} dg' \, \frac{g'^2\gamma_{n0} + g'^4\gamma_{n1}}{\beta_0 g'^3 + \beta_1 g'^5}\right), \tag{5.147}$$

where we have neglected terms of higher order in g^2. Separating the fraction in the numerator as

$$\frac{g'^2\gamma_{n0} + g'^4\gamma_{n1}}{\beta_0 g'^3 + \beta_1 g'^5} = \left(\frac{g'^2\gamma_{n0} + g'^4\gamma_{n1}}{\beta_0 g'^3 + \beta_1 g'^5} - \frac{\gamma_{n0} g'^2}{\beta_0 g'^3}\right) + \frac{\gamma_{n0} g'^2}{\beta_0 g'^3}, \tag{5.148}$$

we can perform the integrals yielding:

$$C_{i,n}(Q^2) = \left(c_{i,n}(0) + c_{i,n}(1)\bar{g}^2(t)\right) \left(\frac{\bar{g}(t)}{\bar{g}(0)}\right)^{\frac{\gamma_{n0}}{\beta_0}} \left(\frac{\beta_0 + \beta_1 \bar{g}^2(t)}{\beta_0 + \beta_1 \bar{g}^2(0)}\right)^{-\frac{1}{2}\left(\frac{\gamma_{n0}}{\beta_0} - \frac{\gamma_{n1}}{\beta_1}\right)} . \tag{5.149}$$

With these coefficients now evaluated within renormalization group improved QCD, we can return to the analysis of deep inelastic scattering data on any target. From Eqs. (5.131) and (5.132) we see that, if we know the structure function of, say, a proton at any scale, Q_0^2, then we can find the non-perturbative target matrix elements, $a_{n,i}$, by dividing the n'th moment of the measured structure function at Q_0^2 by $C_{i,n}(Q_0^2)$, for example:

$$a_{n,i} = \frac{1}{C_{i,n}^{(2)}(Q_0^2)} \int_0^1 dx \, x^{n-2} F_{2i}(x, Q_0^2). \tag{5.150}$$

The result, multiplied by the coefficient at any other Q^2, then yields the moment of the structure function at that new scale and the structure function itself can be reconstructed through a standard Mellin transform:

$$F_{2i}(x, Q^2) = \int_{K-i\infty}^{K+i\infty} dn \, x^{1-n} a_{n,i} C_{i,n}^{(2)}(Q^2), \tag{5.151}$$

where the constant K is a real positive constant to the right of any singularities of the integrand.

It is a straightforward property of the Mellin transform that the Mellin transform of a product of moments is a convolution of the Mellin transforms of the individual moments. This immediately leads us to an understanding of the Altarelli-Parisi (DGLAP) equations which were introduced in Chapter 4. The so-called splitting functions introduced there are just the Mellin transforms of the coefficient functions which we have just explained how to calculate. We note that the leading order scaling violation described in Chapter 4 – c.f. Eq. (4.29) – follows directly from Eq. (5.149) if we set the next-to-leading order terms, namely β_1 and γ_{n1} to zero and neglect the $\mathcal{O}(g^2)$ correction ($c_{i,n}(1)$) to the coefficient function. Clearly the present analysis is limited only by one's ability to evaluate Feynman integrals and can be carried through to any order in g.

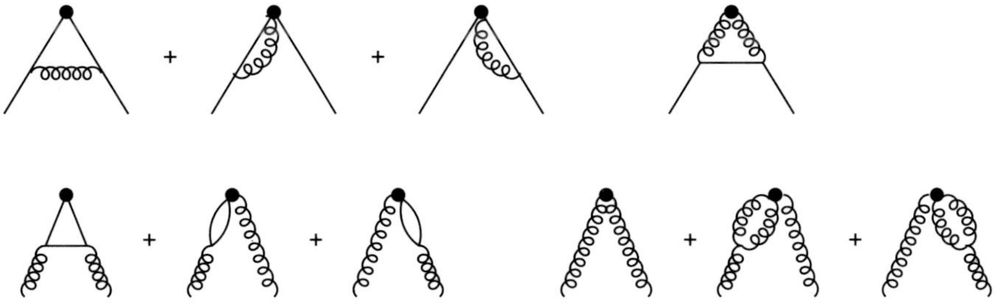

Figure 5.14: Diagrams involved in the renormalization of the flavor singlet operators to one loop. Notice that the renormalization of the quark operators necessarily involves the corresponding gluon operator.

5.6.4 Singlet Case

The diagrams which enter in the renormalization of the singlet quark operators are illustrated in Fig. 5.14. We see that in order to evaluate these diagrams one must also include the renormalization of the purely gluonic operator (5.137). These operators are said to mix under renormalization and the coefficient functions and the corresponding anomalous dimensions are then described by 2×2 matrices. As this analysis is slightly more complicated than that which we have already described, but introduces nothing conceptually new, we leave the details of this analysis to standard texts [Mut 98, BL 86, ESW 96]. For the explicit formulae for a NLO analysis we refer to some key articles dealing with the evolution of parton distributions within perturbative QCD [Bur 80, CTEQ 95, HSV 96, GRV 98, KS 98, MRS 98].

6 Aspects of Non-Perturbative QCD

At low energies and momenta, $Q < 1$ GeV, the running coupling strength $\alpha_s(Q)$ of QCD is of order one. An expansion in powers of α_s is no longer useful: we are entering the region commonly referred to as *non-perturbative QCD*.

Many important features of nucleon structure are determined by non-perturbative QCD for which no rigorous solution exists. Several different strategies have been developed to face this challenge, every one of them with its own specific limitations. Considerable progress has been made in attempts to solve QCD on a discretized, Euclidean space-time lattice using powerful computers. We will briefly review this approach inasmuch as it relates to questions of nucleon structure. Another method goes under the heading of QCD Sum Rules. It combines the operator product expansion techniques introduced in Chapter 5 with phenomenology based on dispersion relations. Last, but not least, one can make active use of basic QCD symmetries and the corresponding conserved currents. This latter framework has proved efficient as a guiding principle to construct effective Lagrangians which approximate QCD at low energy. It will be the basis for our discussion of chiral symmetry in Chapter 7. In the present chapter we prepare the ground for these developments.

6.1 Symmetries, Currents and Anomalies

As we explained in the preceding chapter, QCD is built upon local $SU(3)$ color gauge invariance. The interaction between quarks and gluons is completely determined by this gauge symmetry. It is independent of the various quark flavors $(u, d, s, \ldots)$, so that the distinction between quarks, as far as their strong interactions are concerned, comes only from their different masses. Apart from its local color gauge symmetry, QCD has a number of global unitary symmetries which govern the strong interactions of hadrons. The significance of these symmetries is that they impose a well defined order on the spectrum of eigenstates of the QCD Hamiltonian, and that they imply conserved currents which set important constraints on the dynamics of strongly interacting systems, irrespective of whether their relevant degrees of freedom are elementary quarks and gluons or composite hadrons.

In the following our starting point is QCD with $N_f = 3$ flavors, that is the light quarks u, d and s. They are represented by the three-component field $\psi(x)$, as in Eq. (1.6). Let Γ_a be the generators of a unitary group acting on ψ. For example, if the group is $SU(N_f)$, then the index a labels the $N_f^2 - 1$ generators of that group. Suppose now that the Lagrangian $\mathcal{L}(\psi, \partial_\mu \psi)$ is invariant under the global transformation

$$\psi(x) \rightarrow \exp[i\theta^a \Gamma_a]\, \psi(x), \tag{6.1}$$

where θ^a denotes a set of arbitrary "angles", which are independent of the space-time coordinate x. The gluon field is not affected by these transformations. Noether's theorem states that for each Γ_a there exists a conserved current

$$J_a^\mu(x) = \overline{\psi}(x)\,\gamma^\mu\,\Gamma_a\psi(x), \tag{6.2}$$

with

$$\partial_\mu J_a^\mu(x) = 0. \tag{6.3}$$

Each of these currents has a charge

$$Q_a = \int d^3x\, J_a^0(x), \tag{6.4}$$

which is also conserved:

$$\frac{dQ_a}{dt} = \int d^3x\, \frac{\partial J_a^0}{\partial t} = -\int d^3x\, \vec{\nabla}\cdot\vec{J}_a = 0. \tag{6.5}$$

(Note that we have assumed that there are no subtle topological effects which might give rise to a contribution from the surface at infinity.) Equivalently, these charges commute with the QCD Hamiltonian:

$$\dot{Q}_a = i[H, Q_a] = 0. \tag{6.6}$$

6.1.1 Baryon and Flavour Currents

The simplest unitary symmetry of the QCD Lagrangian is its invariance under a $U(1)$ transformation which multiplies the quark field by a global phase factor:

$$\psi(x) \to e^{i\theta}\,\psi(x). \tag{6.7}$$

The corresponding conserved current is the baryon current, $\frac{1}{3}\overline{\psi}\gamma_\mu\psi$, with its conserved charge, the baryon number

$$B = \frac{1}{3}\int d^3x\,\psi^\dagger\psi, \tag{6.8}$$

properly normalized to one third for each quark.

Next, consider three quarks of equal mass ($m_u = m_d = m_s$). In this limit the QCD Lagrangian has a global $SU(3)$ flavor symmetry. It is invariant under the transformation

$$\psi(x) \to \exp\left[i\theta^a\,\frac{\lambda_a}{2}\right]\psi(x), \tag{6.9}$$

where $\lambda_1, \ldots, \lambda_8$ are the standard Gell-Mann $SU(3)$ matrices (see Appendix A.2). The corresponding eight conserved vector currents are:

$$V_a^\mu(x) = \overline{\psi}(x)\,\gamma^\mu\,\frac{\lambda_a}{2}\,\psi(x). \tag{6.10}$$

Two of these currents, the ones related to the diagonal matrices λ_3 and λ_8, have already been encountered in the discussion of the electromagnetic quark currents: the isospin current

$$V_3^\mu = \frac{1}{2}\left[\overline{\psi}_u \gamma^\mu \psi_u - \overline{\psi}_d \gamma^\mu \psi_d\right], \tag{6.11}$$

and the hypercharge current

$$J_Y^\mu = \frac{2}{\sqrt{3}} V_8^\mu = \frac{1}{3}\left[\overline{\psi}_u \gamma^\mu \psi_u + \overline{\psi}_d \gamma^\mu \psi_d\right] - \frac{2}{3}\overline{\psi}_s \gamma^\mu \psi_s. \tag{6.12}$$

The conserved charges

$$Q_a = \int d^3x\, \psi^\dagger(x) \frac{\lambda_a}{2} \psi(x) \tag{6.13}$$

form a Lie algebra specified by the commutation relations

$$[Q_a, Q_b] = i f_{abc} Q_c, \tag{6.14}$$

where f_{abc} are the $SU(3)$ structure constants. In deriving this commutator, use has been made of the canonical anticommutation relations for the quark fields, together with $[\lambda_a, \lambda_b] = 2i f_{abc}\lambda_c$. A useful identity for practical purposes is

$$[A\lambda_a, B\lambda_b] = \frac{1}{2}\{A, B\}[\lambda_a, \lambda_b] + \frac{1}{2}[A, B]\{\lambda_a, \lambda_b\}, \tag{6.15}$$

where A, B are any of the Dirac matrices γ^μ or $\sigma^{\mu\nu}$. (N.B. $[A, B]$ and $\{A, B\}$ denote, respectively, the commutator and anti-commutator of A and B.)

The $SU(3)$ flavor symmetry is broken by the quark mass differences. Consider the mass term of the QCD Lagrangian,

$$\mathcal{L}_{\text{mass}} = -\overline{\psi}\, \mathcal{M}\, \psi \tag{6.16}$$

with the quark mass matrix $\mathcal{M} = \text{diag}(m_u, m_d, m_s)$. The divergence of the flavor current now becomes

$$\partial_\mu V_a^\mu = i\overline{\psi}\left[\mathcal{M}, \frac{\lambda_a}{2}\right]\psi. \tag{6.17}$$

Hence flavor changing vector currents have divergences proportional to quark mass differences. Take for example the combinations

$$V_\pm^\mu = V_1^\mu \pm i V_2^\mu, \tag{6.18}$$

which convert d into u quarks and vice versa. One finds

$$\partial_\mu V_+^\mu = i(m_u - m_d)\,\overline{\psi}_u \psi_d, \tag{6.19}$$

$$\partial_\mu V_-^\mu = -i(m_u - m_d)\,\overline{\psi}_d \psi_u. \tag{6.20}$$

6.1.2 Massless Quarks: Chiral Symmetry

Apart from a renormalization scale, the only remaining parameters in the QCD Lagrangian are the running quark masses, $\bar{m}_u, \bar{m}_d, \bar{m}_s, \ldots$ of Eq. (5.103). Their absolute values at a scale of order 1 GeV are [PP 00, Mal 99, PDG 98/00]:

$$m_u \equiv \bar{m}_u(1\ \text{GeV}) = (4 \pm 2)\ \text{MeV}, \tag{6.21}$$

$$m_d \equiv \bar{m}_d(1\ \text{GeV}) = (8 \pm 4)\ \text{MeV}, \tag{6.22}$$

$$m_s \equiv \bar{m}_s(1\ \text{GeV}) = (164 \pm 33)\ \text{MeV}. \tag{6.23}$$

These masses are small on characteristic hadronic scales such as the proton mass:

$$\frac{m_u}{M_p} \sim 5 \cdot 10^{-3}, \qquad \frac{m_d}{M_p} \sim 10^{-2}, \qquad \frac{m_s}{M_p} \sim 0.2. \tag{6.24}$$

It is therefore meaningful to examine first the limit $m_u = m_d = m_s = 0$ and then treat the light quark masses as perturbations. On the other hand, the c, b and t quarks are heavy. At low energy it is a good first approximation to treat them as infinitely heavy. In this limit they "freeze", and the only active degrees of freedom are those associated with the light u, d and s quarks.

In the limit of massless quarks, the QCD Lagrangian has an extra symmetry related to the conserved right- or left-handedness (chirality) of zero mass spin-1/2 particles. We introduce right- and left-handed quark fields,

$$\psi_{R,L} = \frac{1}{2}(1 \pm \gamma_5)\psi, \tag{6.25}$$

and observe that separate global unitary transformations of these fields,

$$\psi_R \to \exp\left[i\theta_R^a \frac{\lambda_a}{2}\right]\psi_R, \qquad \psi_L \to \exp\left[i\theta_L^a \frac{\lambda_a}{2}\right]\psi_L, \tag{6.26}$$

leave the QCD Lagrangian with $\mathcal{L}_{\text{mass}} = 0$ invariant. That is, the right- and left-handed components of the *massless* quark fields do not communicate. The group of these transformations is called $SU(3)_R \times SU(3)_L$, and the invariance of the Lagrangian under this group is referred to as the chiral symmetry of QCD. It implies sixteen conserved Noether currents,

$$J^\mu_{R,a} = \overline{\psi}_R \gamma^\mu \frac{\lambda_a}{2}\psi_R, \qquad J^\mu_{L,a} = \overline{\psi}_L \gamma^\mu \frac{\lambda_a}{2}\psi_L, \tag{6.27}$$

with $\partial_\mu J^\mu_{R,a} = \partial_\mu J^\mu_{L,a} = 0$. Instead of right- and left-handed currents, it is common to introduce the vector current (already encountered),

$$V^\mu_a(x) = J^\mu_{R,a} + J^\mu_{L,a} = \overline{\psi}(x)\gamma^\mu \frac{\lambda_a}{2}\psi(x), \tag{6.28}$$

and the axial current

$$A^\mu_a(x) = J^\mu_{R,a} - J^\mu_{L,a} = \overline{\psi}(x)\gamma^\mu\gamma_5 \frac{\lambda_a}{2}\psi(x). \tag{6.29}$$

The corresponding charges

$$Q_a^V = \int d^3x\, \psi^\dagger(x)\, \frac{\lambda_a}{2}\, \psi(x), \qquad Q_a^A = \int d^3x\, \psi^\dagger(x)\, \gamma_5\, \frac{\lambda_a}{2}\, \psi(x), \tag{6.30}$$

generate the algebra;

$$[Q_a^V, Q_b^V] = i f_{abc}\, Q_c^V, \tag{6.31}$$
$$[Q_a^V, Q_b^A] = i f_{abc}\, Q_c^A, \tag{6.32}$$
$$[Q_a^A, Q_b^A] = i f_{abc}\, Q_c^V. \tag{6.33}$$

The algebra involving the currents themselves, known as "current algebra", was applied extensively to low energy hadronic processes in the sixties and seventies [Alf 73, AD 68].

The chiral $SU(3)_R \times SU(3)_L$ symmetry is explicitly broken by the small but finite quark masses. The divergence of the axial current becomes:

$$\partial_\mu A_a^\mu = i\bar{\psi} \left\{ \mathcal{M}, \frac{\lambda_a}{2} \right\} \gamma_5 \psi. \tag{6.34}$$

As examples, consider the axial current components $A^\mu_{1,2,3}$ which operate in the sector of u and d quarks, with the combinations $A^\mu_\pm = A^\mu_1 \pm i A^\mu_2$ relevant for beta decay matrix elements. One finds

$$\partial_\mu A^\mu_+ = (m_u + m_d)\, \overline{\psi}_u i\gamma_5\, \psi_d, \tag{6.35}$$
$$\partial_\mu A^\mu_- = (m_u + m_d)\, \overline{\psi}_d i\gamma_5\, \psi_u, \tag{6.36}$$
$$\partial_\mu A^\mu_3 = m_u\, \overline{\psi}_u i\, \gamma_5\, \psi_u - m_d\, \overline{\psi}_d i\, \gamma_5\, \psi_d. \tag{6.37}$$

The divergence of the axial current thus produces sources of pseudoscalar quark-antiquark pairs. This is the microscopic basis of the PCAC (partially conserved axial current) hypothesis, which plays a key role in weak interactions of the nucleon (as we have already seen in Chapter 3) and low energy pion-nucleon dynamics (discussed in Chapter 7).

The physical meaning of the axial charge Q^A becomes obvious in the same context. Take for example the flavor changing axial current A^μ_+. The corresponding axial charge

$$Q_+^A = \int d^3x\, \psi_u^\dagger(x)\, \gamma_5\, \psi_d(x)$$

is the volume integral over the axial source density which converts a $d-$quark into a $u-$quark.

6.1.3 Spontaneous Breaking of Chiral Symmetry

In the following we first return to the exact chiral limit, with vanishing quark masses. There is evidence both from low-energy hadron phenomenology and from lattice QCD calculations that chiral symmetry is *spontaneously* broken: the ground state (vacuum) of QCD does not share the chiral symmetry of the Lagrangian. If the ground state were symmetric, both vector and axial charge operators would annihilate the vacuum:

$$Q_a^V |0\rangle = Q_a^A |0\rangle = 0. \tag{6.38}$$

This is the **Wigner-Weyl** realization of chiral symmetry with a "trivial" vacuum. Given the different parity assignments of Q^V and Q^A, one would expect that, in this case, the ground state property, Eq. (6.38), implies a close correspondence between the spectra of positive and negative parity excited states. For example, correlation functions of vector and axial vector currents should be identical: $\langle 0|V_a^\mu V_b^\nu|0\rangle = \langle 0|A_a^\mu A_b^\nu|0\rangle$. Consequently, the spectra of vector ($J^\pi = 1^-$) and axial vector ($J^\pi = 1^+$) mesonic excitations of the vacuum should also be identical, which they are not: the mass of the ρ meson ($m_\rho \simeq 0.77$ GeV) is far smaller than that of the axial a_1 meson ($m_{a_1} \simeq 1.23$ GeV). In general, for a vacuum state that satisfies Eq. (6.38), the complete symmetry between right- and left-handed quarks generates a parity doublet spectrum: for each positive parity state there should exist a negative parity state of equal mass. This is most strikingly **not** the case for the light pseudoscalar ($J^\pi = 0^-$) mesons, $(\pi, K, \ldots)$, which have masses much lower than the lightest scalar ($J^\pi = 0^+$) mesons.

The approximate validity of $SU(3)$ flavor symmetry, which is observed, is compatible with the assumption

$$Q_a^V|0\rangle = 0. \tag{6.39}$$

Therefore one must conclude that

$$Q_a^A|0\rangle \neq 0. \tag{6.40}$$

That is, the $SU(3)_R \times SU(3)_L$ symmetry is spontaneously broken down to the flavor group $SU(3)_V$. This is the **Nambu-Goldstone** realization of chiral symmetry.[1]

Goldstone bosons

Goldstone's theorem [Go 61, Go 62, Na 60] states that for any spontaneously broken global symmetry of a given Lagrangian, there exists a zero frequency (or massless) mode, the Goldstone boson. Applied to the spontaneously broken chiral symmetry of QCD this theorem reads as follows. If the axial charges Q_a^A do not leave the vacuum invariant ($Q_a^A|0\rangle \neq 0$), then there must be a physical state, $|\phi_a\rangle = Q_a^A|0\rangle$, which is energetically degenerate with the ground state $|0\rangle$. This is immediately obvious since the QCD Hamiltonian, H, commutes with the axial charge. Setting the ground state energy equal to zero for convenience, we find $H|\phi_a\rangle = Q_a^A H|0\rangle = 0$. Therefore, since Q_a^A is an *axial* charge, the state $|\phi_a\rangle$ represents eight massless **pseudoscalar mesons**.

The eight lightest hadrons observed in nature are indeed the pseudoscalar mesons $(\pi^\pm, \pi^0, K^\pm, K^0, \overline{K}^0, \eta)$. The fact that they are not exactly massless reflects the explicit breaking of $SU(3)_R \times SU(3)_L$ symmetry by the small but nonzero masses of the light quarks. If the quark mass parameters were varied, then the π, K and η would all become massless as m_u, m_d and m_s tend to zero.

[1] In the context of QCD, we observe that there is a very general result [Hoo 80] that if quarks are confined, then chiral symmetry **must** be spontaneously broken. The reason is simply that in this case positive parity states of (say) a $\bar{q}q$ system must have odd angular momentum and will not, in general, be degenerate with the negative parity state of the system. Of course, the converse is not true, that is the spontaneous breaking of chiral symmetry does not imply confinement.

This very special role of the pseudoscalar mesons is clearly visible in the hadronic mass spectrum. While all baryon states and almost all meson states are well separated from the ground state $|0\rangle$ by a characteristic gap

$$\Delta \sim m_{\rho-\text{meson}} \sim M_{\text{nucleon}} \sim 1 \text{ GeV}, \tag{6.41}$$

the pion mass lies well within that gap. For the K and η the large mass of the strange quark leads to a less impressive result, $m_{K,\eta}/\Delta \sim 0.5$. Thus one must exercise caution concerning the predictions of $SU(3)$ chiral symmetry.

Chiral condensate

Spontaneous chiral symmetry breaking is accompanied by a qualitative rearrangement of the ground state, an entirely non-perturbative phenomenon. The vacuum is now populated by scalar quark-antiquark pairs. Their density has a non-zero expectation value:

$$\langle 0|\bar{\psi}\psi|0\rangle \;=\; \langle 0|\bar{\psi}_u\psi_u|0\rangle + \langle 0|\bar{\psi}_d\psi_d|0\rangle + \langle 0|\bar{\psi}_s\psi_s|0\rangle. \tag{6.42}$$

(The precise definition of this quantity is given in terms of the full quark propagator, $S_F(x,y) = -i\langle 0|T\psi(x)\overline{\psi}(y)|0\rangle$, as $\langle 0|\overline{\psi}\psi|0\rangle = -i\,\mathrm{Tr}\lim_{y\to x_+} S_F(x,y)$.[2]) This so-called **chiral condensate** (often called the "quark condensate") represents the order parameter for spontaneous chiral symmetry breaking. For convenience we frequently use the notation

$$\langle 0|\bar{\psi}\psi|0\rangle \;\equiv\; \langle\bar{\psi}\psi\rangle \;=\; \langle\overline{u}u\rangle + \langle\overline{d}d\rangle + \langle\overline{s}s\rangle, \tag{6.43}$$

and sometimes introduce the generic expression $\langle\bar{q}q\rangle$ for any one of the quantities $\langle\overline{u}u\rangle$, $\langle\overline{d}d\rangle$ or $\langle\overline{s}s\rangle$. To see why these expectation values of scalar quark bilinears play the role of order parameters, recall that $\overline{\psi}\psi = \overline{\psi}_R\psi_L + \overline{\psi}_L\psi_R$ mixes right- and left-handed quarks and therefore breaks chiral symmetry. Consequently, a non-zero vacuum expectation value of $\overline{\psi}\psi$ signals the fact that the ground state is restructured in such a way that it has lost the chiral symmetry of the underlying Lagrangian. To establish this connection more formally, note first that the vacuum expectation value of $[Q_a^V, \overline{\psi}\lambda_b\psi] = i f_{abc}\overline{\psi}\lambda_c\psi$ vanishes because $Q_a^V|0\rangle = 0$. From this it follows that all non-diagonal condensates (such as $\langle\overline{u}d\rangle$) vanish and

$$\langle\overline{u}u\rangle \;=\; \langle\overline{d}d\rangle \;=\; \langle\overline{s}s\rangle.$$

Next, we can use the relation

$$[Q_1^A, \bar{\psi}\gamma_5\lambda_1\psi] = -(\bar{\psi}_u\psi_u + \bar{\psi}_d\psi_d). \tag{6.44}$$

Taking the vacuum expectation value of this expression, we see that

$$Q_a^A|0\rangle \;\neq 0 \tag{6.45}$$

is consistent with non-zero values for the quark condensates:

$$\langle\overline{q}q\rangle \;\neq 0. \tag{6.46}$$

[2] We emphasize that $\langle 0|\bar{\psi}_i\psi_i|0\rangle$ are strictly not observables in QCD, because under renormalization they become scale dependent – see Eq. (5.103). It should always be understood that such quantities will be multiplied by a mass so that the product is renormalization group invariant.

PCAC and the Gell-Mann-Oakes-Renner relation

Goldstone's theorem implies the existence of eight pseudoscalar Goldstone bosons. Let their state vectors be $|\pi_a(p)\rangle$, with four-momentum $p^\mu = (E_p, \vec{p})$ and the normalization $\langle \pi_a(p)|\pi_b(p')\rangle = 2E_p\delta_{ab}(2\pi)^3\delta^3(\vec{p}-\vec{p}\,')$. The theorem also implies non-vanishing transition matrix elements of the axial current which connect $|\pi_a(p)\rangle$ with the vacuum:

$$\langle 0|A_a^\mu(x)|\pi_b(p)\rangle = ip^\mu f_0 \delta_{ab} e^{-ip\cdot x}. \tag{6.47}$$

The constant f_0 is called the pion decay constant. Here, as indicated by the subscript, it is the decay constant in the chiral limit – i.e., for vanishing quark masses. It is related to the physical pion decay constant,

$$f_\pi = (92.4 \pm 0.3)\ \text{MeV}, \tag{6.48}$$

determined in the charged pion decay $\pi \to \mu\nu_\mu + \mu\nu_\mu\gamma$ [PDG 98/00], by

$$f_\pi = f_0(1 + \mathcal{O}(m_q)). \tag{6.49}$$

Note that the difference between f_π and f_0 is linear in the small quark mass, m_q.

As already mentioned, the non-zero quark masses shift the masses of the Goldstone bosons from zero to the observed values of the physical pseudoscalar mesons. In order to derive the relationship between the pion mass and the u and d quark masses, we begin with the PCAC relation, Eq. (6.34), which yields

$$\partial_\mu A_1^\mu = (m_u + m_d)\overline{\psi} i\gamma_5 \frac{\lambda_1}{2}\psi. \tag{6.50}$$

Combining this with Eq. (6.44) we find:

$$\langle 0|[Q_1^A, \partial_\mu A_1^\mu]|0\rangle = -\frac{i}{2}(m_u + m_d)\langle \overline{u}u + \overline{d}d\rangle. \tag{6.51}$$

We now insert a complete set of states on the left and assume, in the spirit of PCAC, that this is saturated by the pion, so that

$$\int \frac{d^3p}{2E_p(2\pi)^3}\,|\pi_a(p)\rangle\langle \pi_a(p)| = 1. \tag{6.52}$$

Using Eq. (6.47) and

$$\langle 0|Q_a^A(t=0)|\pi_b(p)\rangle = i\delta_{ab} f_0 E_p (2\pi)^3\delta^3(\vec{p}), \tag{6.53}$$

one finds the Gell-Mann-Oakes-Renner relation [GOR 68]

$$m_\pi^2 = -\frac{1}{2f_0^2}(m_u + m_d)\langle \overline{u}u + \overline{d}d\rangle + \mathcal{O}(m_{u,d}^2). \tag{6.54}$$

Neglecting terms of order $m_{u,d}^2$ in this equation, setting $f_0 = f_\pi$ and using $m_u + m_d \simeq$ 14 MeV (corresponding to a scale of order 1 GeV), one obtains

$$\langle \overline{u}u\rangle \simeq -(225\ \text{MeV})^3 \simeq -1.5\ \text{fm}^{-3} \tag{6.55}$$

at a scale of 1 GeV. An analogous relation holds for the kaon mass:

$$m_K^2 = -\frac{1}{2f_0^2}(m_u + m_s)\langle \overline{u}u + \overline{s}s\rangle + \mathcal{O}(m_s^2). \tag{6.56}$$

Here the corrections are not so well under control because of the much larger strange quark mass.

The GOR relation establishes a connection between QCD properties, expressed in terms of fundamental quark degrees of freedom, and the pion mass and decay constant. Such relations, based entirely on the symmetries and symmetry breaking pattern of QCD, are characteristic of low energy hadron physics. In the present context, the pion decay constant, f_π, is a measure of spontaneously broken chiral symmetry, just like the quark condensate $\langle \overline{q}q\rangle$.[3] In Chapter 7 it reappears in a variety of so-called low-energy theorems which govern pion-pion and pion-nucleon interactions, as well as the photon-induced production of pions from nucleons close to threshold.

6.1.4 Dynamical Symmetry Breaking and Fermion Mass Generation

In QCD the concept of mass is closely tied to strong interaction dynamics itself. Just like the running coupling strength α_s, of Eq. (5.90), the quark mass depends on the momentum scale at which it is probed. In this context, the spontaneous breakdown of chiral symmetry has profound consequences. The presence of a chiral condensate in the QCD ground state dramatically alters the perturbative behavior, Eq. (5.103), of the quark masses. Initially massless quarks can develop large dynamical effective masses when exposed to a condensate, a phenomenon frequently referred to as *dynamical symmetry breaking*.

Consider for example a massless quark travelling through the condensed vacuum with large, spacelike four-momentum $p^\mu = (p_0, \vec{p})$ – a situation which can be realized, for example, in the deep inelastic scattering of leptons on hadrons. The (renormalized) quark propagator has the general form

$$S_F(p) = \frac{1}{(1 - A(p^2))\gamma.p - B(p^2)} = \frac{Z(p^2)}{\gamma.p - M(p^2)}, \tag{6.57}$$

where $M(p^2) = B(p^2)/(1 - A(p^2))$ defines the mass function. The forces experienced by the quark are summarized in the self-energy, $\Sigma(p)$, such that the propagator of an originally massless quark subject to strong interactions can be written:

$$S_F(p) = \frac{1}{\gamma.p - \Sigma(p)}, \tag{6.58}$$

or

$$\Sigma(p) = A(p^2)\gamma.p + B(p^2). \tag{6.59}$$

Purely perturbative interactions of the quark, involving the emission and absorption of virtual gluons, contribute only to $A(p^2)$, the reason being that gluons are vector fields. Their coupling to quarks respects chiral symmetry: a massless quark will stay massless to all orders of perturbation theory.

[3] In fact the gap (6.41) is commonly identified as $\Delta = 4\pi f_\pi$.

This situation changes qualitatively in the presence of a non-perturbative scalar field provided by the condensed QCD vacuum. We are now interested in the self-energy resulting from the process shown in Fig. 6.1, in which the quark, while exchanging a gluon, experiences the background field produced by the quark condensate. Because of translational invariance, there is no exchange of momentum with the vacuum condensate, so that the intermediate gluon carries all of the large momentum p along with it.

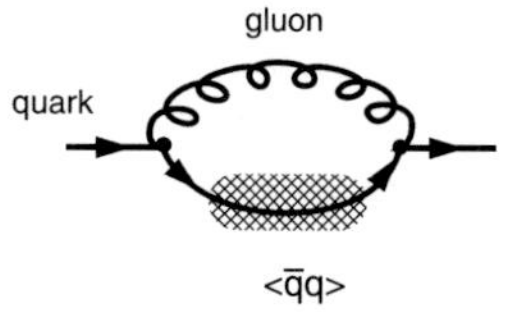

Figure 6.1: Self-energy of a quark in the presence of a quark condensate

The actual calculation of $\Sigma(p)$ from this process involves some subtleties of gauge fixing (c.f. Eq. (5.71) in the perturbative case) [Mut 98]. To leading order in the condensate one finds the following result (written in the Feynman gauge, setting $\lambda = 1$ in Eq. (5.48), for which the gluon propagator has the simple form $D_{ij}^{\mu\nu}(p) = -i\delta_{ij}g^{\mu\nu}(p^2 + i\epsilon)^{-1}$ with color indices i, j and Lorentz tensor indices μ, ν) [Po 76, La 97]:

$$\Sigma(p) = \frac{3\,\pi}{2}\,\frac{N_c^2 - 1}{N_c^2}\,\frac{\alpha_s(p)}{p^2}\,\langle \bar{q}q \rangle. \tag{6.60}$$

This expression holds in the limit of large spacelike quark momentum with $-p^2 \equiv Q^2 \gg p_0^2$, and we have omitted small perturbative corrections. The scalar nature of the condensate implies that one can identify this $\Sigma(p)$ with the mass function. With $N_c = 3$ one finds a dynamical quark mass

$$M(Q) = -\frac{4\,\pi}{3}\,\frac{\alpha_s(Q)}{Q^2}\,\langle \bar{q}q \rangle + \cdots, \tag{6.61}$$

which is plotted in Fig. 6.2 as a function of momentum Q, using α_s of Eq. (4.19). Of course, this result should only be taken for qualitative orientation. It cannot be used at small Q where $\alpha_s(Q)$ becomes large. In fact, $M(Q)$ is not even directly observable. Nevertheless Fig. 6.2 illustrates how effective quark masses of several hundreds of MeV can easily develop in the presence of a condensate.

The dynamical symmetry breaking scenario just outlined, which turns light QCD quarks into massive fermions, is a likely origin of the so-called **constituent quarks**. Constituent quarks are hypothetical, massive quasi-particles which have traditionally played an important part in phenomenological descriptions of nucleon structure (see also Section 8.2).

6.1.5 A Schematic Model: Nambu and Jona-Lasinio

In fermionic many-body systems such as atoms or nuclei, valuable insights are often gained, even at the level of the Hartree-Fock approximation, by using suitable effective interactions. A similar approach in hadron physics at the quark level is provided by the model of Nambu and

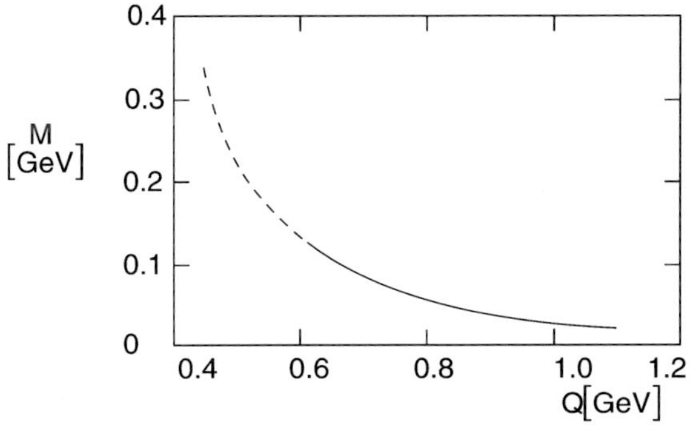

Figure 6.2: Running quark mass driven by a quark condensate.

Jona-Lasinio [NJL 61]. It offers a simple and practical illustration of the basic mechanisms that drive spontaneous chiral symmetry breaking. For simplicity we shall restrict ourselves here to the original flavor $SU(2)$ version of the model which was worked out in the early pioneering papers.

The NJL model is based on an effective Lagrangian of relativistic fermions (quarks) which interact through local current-current couplings, assuming that the gluonic degrees of freedom can be frozen into pointlike effective interactions between quarks. In QCD one can start from the non-local interaction between two color currents, $J_i^\mu = \bar{\psi}\gamma^\mu t_i \psi$, where t_i are the generators of the $SU(3)_c$ color gauge group. The contribution of such a current-current coupling to the action is (see Eq. (5.47)):

$$S_{\text{int}} \sim \int d^4x \int d^4y \, J_i^\mu(x) \, g_s^2 D_{\mu\nu}^{ij}(x,y) \, J_j^\nu(y), \tag{6.62}$$

where $D_{\mu\nu}^{ij}$ is the full gluon propagator and g_s is the QCD coupling constant. In perturbative QCD Eq. (6.62) generates the familiar one-gluon exchange interaction between quarks. At low-energy or long-distance scales, however, the propagating gluons experience strong screening effects which cannot be handled perturbatively. Lattice QCD results for the gluonic field strength correlation function indicate [DiG 92] that the space-time distance over which such colored gluon fields can propagate is small, of order 0.2 fm, corresponding to a momentum scale Λ of order 1 GeV (not to be confused with the QCD scale parameter Λ_{QCD}).

In order to provide an effective description of the physics at momenta smaller than Λ, we can replace the short-ranged, non-local interaction (6.62) by an approximate local one:

$$\mathcal{L}_{\text{int}}(x) \simeq G_s \, J_i^\mu(x) \, J_\mu^i(x). \tag{6.63}$$

The effective coupling strength, $G_s = \bar{g}_s^2 \Lambda^{-2}$, combines the averaged strong interaction coupling constant, $\bar{g}_s$, with the correlation length, Λ^{-1}, over which color can propagate. This interaction is invariant under chiral transformations and therefore strictly preserves chiral symmetry: the chiral transformation (6.26) leaves the vector currents J_i^μ invariant. In order to understand how it operates in color singlet channels, it is customary to make a Fierz rearrangement of the color octet current-current interaction (6.63) [VW 91]. In the two-flavor case, with $\psi^T = (\psi_u, \psi_d)$, this Fierz rearrangement produces, among other terms, the combination $(\bar{\psi}\psi)^2 + (\bar{\psi}i\gamma_5\boldsymbol{\tau}\psi)^2$, which is manifestly invariant under chiral $SU(2)_R \times SU(2)_L$ transformations.

This is the starting point for the schematic NJL model. In its simplest version, the NJL Lagrangian with massless u- and d-quarks is

$$\mathcal{L}_{\text{NJL}} = \bar{\psi}i\gamma_\mu\partial^\mu\psi + \frac{G}{2}\left[(\bar{\psi}\psi)^2 + (\bar{\psi}i\gamma_5\boldsymbol{\tau}\psi)^2\right], \tag{6.64}$$

where G is a coupling strength of dimension (length)^2. This Lagrangian generates a non-linear equation of motion for ψ. In the mean field or Hartree approximation, this equation is linearized by replacing $\bar{\psi}\psi$ by its vacuum expectation value $\langle\bar{\psi}\psi\rangle$. The resulting Dirac equation for a single quark becomes

$$(i\gamma_\mu\partial^\mu + G\langle\bar{\psi}\psi\rangle)\psi(x) = 0. \tag{6.65}$$

One sees that, provided the condensate $\langle\bar{\psi}\psi\rangle$ is non-zero, quarks which are massless at the level of the Lagrangian develop a dynamical mass determined by the **gap equation**:

$$M = -G\langle\bar{\psi}\psi\rangle. \tag{6.66}$$

The introduction of a small current quark mass, m, does not alter this discussion in any significant way, one just adds m on the right hand side of the gap equation.

The condensate, defined by

$$\langle\bar{\psi}\psi\rangle \;=\; -\operatorname{Tr}\lim_{x\to 0_+}\langle 0|\mathcal{T}\psi(0)\bar{\psi}(x)|0\rangle = -i\operatorname{Tr}\lim_{x\to 0_+} S_F(0,x), \tag{6.67}$$

becomes

$$\langle\bar{\psi}\psi\rangle = -4N_c \int\limits_{|\vec{p}|\le\Lambda} \frac{d^3p}{(2\,\pi)^3}\frac{M}{\sqrt{M^2+\vec{p}^{\,2}}}. \tag{6.68}$$

(In order to derive this result, insert the quark propagator

$$S_F(0,x) = \int \frac{d^4p}{(2\,\pi)^4}\exp(-ip\cdot x)(\gamma\cdot p + M)(p^2 - M^2 + i\epsilon)^{-1}$$

into Eq. (6.67), take the trace and perform the p_0-integration using the residue calculus.) The condensate in Eq. (6.68) represents the scalar density (including the color factor $N_c = 3$) of the negative energy Dirac sea, integrated up to the characteristic cutoff (Λ) in momentum space. (While we have chosen a non-covariant cutoff just for illustration, covariant cutoff schemes are usually employed in the actual calculations.)

In essence, the quark develops a large dynamical mass by polarizing the Dirac sea. There is an "active" part of that Dirac sea, with low momenta, $|\vec{p}| \le \Lambda$, which participates in the strong interactions, whereas the QCD forces are weak at high momenta (the phenomenon of asymptotic freedom). The NJL model draws a schematic picture of this by simply turning off the interaction for $|\vec{p}| \ge \Lambda$. The self-consistent solution of the gap equation gives an

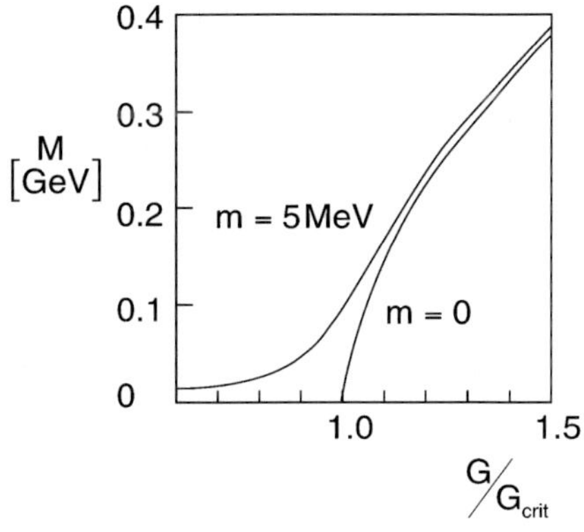

Figure 6.3: Dynamical quark mass M as a function of the coupling strength G, for different values of the "current" quark mass m (from [VW 91]).

effective quark mass as shown in Fig. 6.3. The empirical value of the quark condensate is reproduced with a cutoff Λ around 1 GeV. Together with a dimensionless scalar coupling constant $g^2 = G\Lambda^2$ of about 15, this generates a "constituent" quark mass $M \simeq 350$ MeV, just about the value commonly used in simple non-relativistic quark models.

The vacuum energy density derived from the NJL Lagrangian (6.64) has an effective potential which shows the characteristic features of spontaneous chiral symmetry breaking. A short digression to make this more explicit may be helpful. The reduced Lagrangian density obtained in the mean field approximation from Eq. (6.64) is actually

$$\mathcal{L}_{\text{MFA}} = \bar{\psi} i \gamma_\mu \partial^\mu \psi + G \langle \bar{\psi}\psi \rangle \bar{\psi}\psi - G \langle \bar{\psi}\psi \rangle^2, \tag{6.69}$$

where the last term must be included to avoid double counting of the interaction – a feature familiar from the Hartree-Fock method in many-body theory. It is useful to treat $\langle \bar{\psi}\psi \rangle$ as a background field variable, σ_0, so that (in terms of the dimensionless coupling constant, $g^2 = G\Lambda^2$):

$$M(\sigma_0) = g\sigma_0 = -G\langle \bar{\psi}\psi \rangle,$$

or equivalently, $\sigma_0 = -g\langle \bar{\psi}\psi \rangle / \Lambda^2$. The Hamiltonian density derived from $\mathcal{L}_{\text{MFA}}$ is then expressed as a function of σ_0 in the following form:

$$\mathcal{H}_{\text{MFA}}(\sigma_0) = \bar{\psi} \left[-i\vec{\gamma} \cdot \vec{\nabla} + M(\sigma_0) \right] \psi + \frac{1}{2} \Lambda^2 \sigma_0^2. \tag{6.70}$$

The "trivial" vacuum with $\sigma_0 \sim \langle \bar{\psi}\psi \rangle = 0$ has an energy density

$$\mathcal{E}(0) = \langle 0 | \mathcal{H}_{\text{MFA}}(0) | 0 \rangle, \tag{6.71}$$

whereas the ground state energy density with a vacuum that hosts a condensate ($\sigma_0 > 0$) is

$$\mathcal{E}(\sigma_0) = \langle \sigma_0 | \mathcal{H}_{\text{MFA}}(\sigma_0) | \sigma_0 \rangle. \tag{6.72}$$

The effective potential is defined as

$$\mathcal{U}(\sigma_0) = \mathcal{E}(\sigma_0) - \mathcal{E}(0) \tag{6.73}$$

and has the form shown in Fig. 6.4. Once the coupling strength exceeds a critical value, $G > G_{\text{crit}}$, $\mathcal{U}$ develops a minimum at a non-zero value of the order parameter $\langle \bar{\psi}\psi \rangle$. In fact, minimizing the vacuum energy density in the mean field approximation with respect to $\langle \bar{\psi}\psi \rangle$ just leads to the gap equation (6.66). In this picture, spontaneous chiral symmetry breaking and dynamical fermion mass generation evidently have the same origin. This mechanism works when the curvature of the effective potential, $\mathcal{U}''(\sigma_0) = \Lambda^2$, is sufficiently large that quantum fluctuations do not become overwhelmingly important.

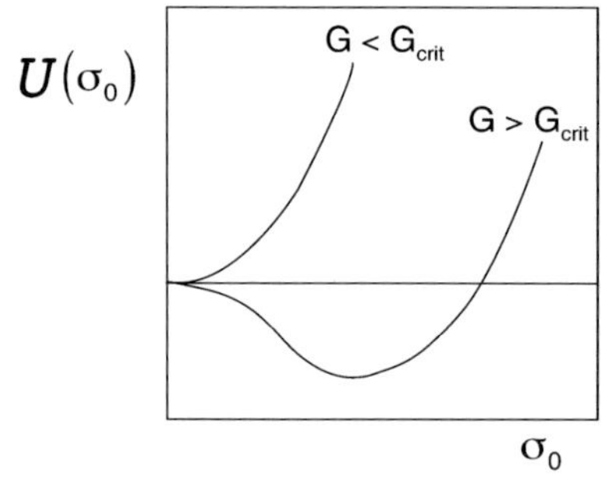

Figure 6.4: Illustration of the effective potential of the NJL model, showing spontaneous chiral symmetry breaking for $G > G_{\text{crit}}$.

The critical coupling strength, G_{crit}, depends on the number of flavors involved. For $N_f = 3$, the dimensionless constant g^2 has a critical value, $g^2_{\text{crit}} = 4\,\pi^2/3$, and the corresponding "fine structure constant" becomes $\alpha_{\text{crit}} = g^2_{\text{crit}}/4\pi \sim 1$.

By solving a Bethe-Salpeter equation with the NJL effective interaction, one finds that the pion emerges as a massless quark-antiquark bound state as required by Goldstone's theorem. When small quark masses m_u and m_d are introduced the Gell-Mann-Oakes-Renner relation, Eq. (6.54), comes as a natural output of this model. Detailed results can be found in several reviews [VW 91, Kl 92, HK 94, Bij 96].

The NJL model, which has been used here to illustrate the condensation of quark-antiquark pairs in the QCD ground state, actually dates back to the times when the BCS theory of superconductivity was developed, long before the foundation of QCD. This is not accidental; in fact the **gap equation** (6.66) indicates some analogy with a BCS type superconductor, the ground state of which is populated by a condensate of spin-singlet electron Cooper pairs, $\langle \psi_\uparrow \psi_\downarrow \rangle$.

The simple, local four-quark interaction given in Eq. (6.63) has been generalized to model the non-perturbative behavior of gluon mediated quark-quark interactions within QCD [PRC 87]. This has led to extensive investigations of hadronic properties based on the self-consistent solution of the Schwinger-Dyson equations [RW 94]. For a survey of the phenomenological achievements of this approach we refer to the appropriate reviews [PRC 89, RW 94, Tan 97, SWT 98]. One of the exciting aspects of this work is that it may be possible to connect the phenomenological input to these calculations to the corresponding quantities (e.g. gluon propagator, quark-gluon vertex etc.) calculated within lattice QCD (in a particular gauge) [Lei 99]. On the other hand, in terms of the insight into the mechanism of spontaneous symmetry breaking, such models are essentially equivalent to the simple NJL model.

As we have just hinted with regard to studies of QCD based upon Schwinger-Dyson equations, the spontaneous symmetry breaking scenario just outlined is of a very general nature, quite independent of the detailed form of the interaction. All that is required is that the effective coupling strength exceeds a certain critical value. This is a widespread phenomenon, and analogues exist in various other fields of physics. A standard textbook example is the ferromagnet. The Hamiltonian of this spin system is invariant under rotations in space. However, its low temperature phase is characterized by a non-vanishing magnetization. In the ground state, the magnetization points in a definite direction in space: rotational symmetry is spontaneously broken. This is a collective phenomenon; many spins cooperate to form the macroscopically magnetized material. Slow variations in the direction of the magnetization produce a collective low-frequency, long wavelength motion of the spins. This spin wave is the Goldstone boson of spontaneously broken rotational symmetry. In our case of spontaneous chiral symmetry breaking, the analogue of the non-vanishing magnetization is the chiral condensate; the analogue of the spin wave is the pion.

6.1.6 The Axial Anomaly

Apart from its chiral $SU(3)_R \times SU(3)_L$ flavor symmetry, the QCD Lagrangian is also invariant under a global *axial* $U(1)$ transformation,

$$\psi(x) \rightarrow e^{i\theta\gamma_5}\,\psi(x). \tag{6.74}$$

Application of Noether's theorem would then suggest that there should be a ninth classically conserved axial current, the flavor singlet current

$$A_0^\mu = \bar{\psi}\gamma^\mu\gamma_5\psi = \bar{\psi}_u\gamma^\mu\gamma_5\psi_u + \bar{\psi}_d\gamma^\mu\gamma_5\psi_d + \bar{\psi}_s\gamma^\mu\gamma_5\psi_s, \tag{6.75}$$

with $\partial_\mu A_0^\mu = 0$. However, this $U(1)_A$ symmetry turns out to be violated in the full **quantum** field theory. This phenomenon is called the **axial U(1) anomaly**.

Anomalies are symmetries of classical Lagrangians that are broken at the quantum level. The Noether current is no longer divergenceless, but receives a contribution from quantum corrections [Tr 85]. In carrying through the renormalization procedures for the singlet axial current one faces a dilemma. One can choose to renormalize in such a way that $\partial_\mu A_0^\mu = i2m\bar{\psi}\gamma_5\psi$, which vanishes (so A_0^μ is conserved) in the massless quark limit. However, the consequence of this choice is that this axial current is not gauge invariant (under the $SU(3)_c$ transformations of QCD). Since QCD is built upon gauge invariance this is not an acceptable option. Once one renormalizes in such a way that gauge invariance is preserved we find (for massless quarks):

$$\partial_\mu A_0^\mu = \frac{\alpha_s}{2\pi} N_f \ \mathrm{Tr}(G_{\mu\nu}\tilde{G}^{\mu\nu}), \tag{6.76}$$

where N_f is the number of quark flavors, $G^{\mu\nu} = G_i^{\mu\nu} t_i \equiv G_i^{\mu\nu}\frac{\lambda_i}{2}$ is the gluon field tensor contracted with the generators t_i $(i = 1, \ldots, 8)$ of the $SU(3)_{\text{color}}$ group, and $\tilde{G}_{\mu\nu} = \frac{1}{2}\epsilon_{\mu\nu\alpha\beta}G^{\alpha\beta}$

is the adjoint field tensor. The expression $\frac{1}{2}\,\mathrm{Tr}\, G_{\mu\nu}\tilde{G}^{\mu\nu}$ is actually the product of color electric and magnetic fields, $\boldsymbol{E}\cdot\boldsymbol{B}$.

The axial U(1) anomaly also occurs in QED (indeed it was discovered there), where the strong coupling α_s is replaced by $\alpha = \frac{e^2}{4\pi}$ and the gluon field tensor is replaced by that of the photon. The original discussion of Adler, concerning the impossibility of preserving both a conserved axial current **and** gauge invariance is still one of the clearest expositions available to the interested student [Ad 69]. This anomaly is responsible for the decay $\pi^0 \to 2\gamma$ of the neutral pion.

In QCD, the axial anomaly is held responsible for the fact that the mass of the η' meson ($m_{\eta'} = 958$ MeV) is considerably larger than all other pseudoscalar meson masses. Without the $U(1)_A$ anomaly, QCD with massless quarks would have a chiral $U(3)_R \times U(3)_L$ symmetry, and its spontaneous breakdown would lead to *nine* rather than eight pseudoscalar Goldstone bosons. The axial anomaly "removes" the $U(1)_A$ symmetry, keeping $SU(3)_R \times SU(3)_L \times U(1)_V$ intact which is then spontaneously broken down to $SU(3)_V \times U(1)_V$. With inclusion of quark masses, the divergence of the flavor singlet axial current within QCD becomes:

$$\partial_\mu A_0^\mu = \frac{\alpha_s}{2\pi} N_f\ \mathrm{Tr}(G_{\mu\nu}\tilde{G}^{\mu\nu}) + 2i(m_u\bar{\psi}_u\gamma_5\psi_u + m_d\bar{\psi}_d\gamma_5\psi_d + m_s\bar{\psi}_s\gamma_5\psi_s), \quad (6.77)$$

and we note that the quark masses do not modify the coefficient of the anomaly. The reason is that it is the ultraviolet divergences of the quantum corrections which give rise to the anomaly and these are insensitive to masses.

As first realized by Kodaira [Kod 80], the axial anomaly plays an important role in the discussion of the spin structure of the nucleon. Nucleon matrix elements of the singlet axial current are deduced from deep-inelastic scattering of polarized leptons on polarized protons in combination with $\beta-$decay processes. The close connection of the anomaly with gluon dynamics raises the interesting question how much of the nucleon spin is of gluonic origin (see Section 4.6).

6.1.7 Scale Invariance and the Trace Anomaly

In the limit of massless quarks, the QCD Lagrangian has no dimensional parameter. As a consequence, a scale transformation (dilatation) which replaces the quark field by

$$\psi(x) \to \lambda^{3/2}\,\psi(\lambda x), \quad (6.78)$$

and the gluon field by

$$A^\mu(x) \to \lambda A^\mu(\lambda x), \quad (6.79)$$

with an arbitrary constant λ, does not change the classical QCD action $\int d^4x\mathcal{L}_{\mathrm{QCD}}$. This property is called classical scale invariance.

Assume, more generally, that a given Lagrangian $\mathcal{L}(\phi_j, \partial^\mu\phi_j)$ is scale invariant, i.e. its classical action does not change under the transformation $\phi_j \to \lambda^d\phi_j(\lambda x)$, with $d = 3/2$ for spin 1/2 fields and $d = 1$ for boson fields. There exists a corresponding conserved current,

$$S^\mu = x_\nu\Theta^{\mu\nu}, \quad (6.80)$$

expressed in terms of the energy-momentum tensor

$$\Theta^{\mu\nu} = \sum_j \frac{\partial \mathcal{L}}{\partial(\partial_\mu \phi_j)} \partial^\nu \phi_j - g^{\mu\nu} \mathcal{L}. \tag{6.81}$$

As a consequence of space-time translational invariance one has $\partial_\mu \Theta^{\mu\nu} = 0$. The divergence of the scale current, S^μ, is the trace of the energy momentum tensor, and scale invariance implies that this trace must vanish:

$$\partial_\mu S^\mu = \Theta^\mu_\mu = 0. \tag{6.82}$$

Now consider a nucleon with four-momentum P^μ, denoted $|N(P)\rangle$. The nucleon matrix element of the energy momentum tensor operator is

$$\langle N(P)|\Theta^{\mu\nu}|N(P)\rangle = \frac{P^\mu P^\nu}{M_N} \bar{u}(P)\, u(P), \tag{6.83}$$

where $u(P)$ is the nucleon Dirac spinor normalized as $\bar{u}u = 2M_N$ (c.f. Appendix A) and M_N the nucleon mass. Clearly, Eq. (6.83) implies that:

$$\langle N(P)|\Theta^\mu_\mu|N(P)\rangle = 2M_N^2, \tag{6.84}$$

and $\partial_\mu S^\mu = 0$ would then imply that the nucleon mass should vanish in the chiral limit. This is not the case. Indeed, setting the up and down quark masses equal to zero would reduce the nucleon mass by approximately the value of the sigma commutator, which will be discussed in Section 7.2. Its phenomenological value seems to be around 45 MeV, which is far below the observed mass of the nucleon.

The resolution of this apparent contradiction is once again that QCD is a **quantum** field theory and the quantum corrections destroy scale invariance. In particular, renormalization introduces a scale. For example, the running coupling strength α_s needs to be determined at some scale (e.g. $\alpha_s(M_Z = 91.2\ \text{GeV}) \simeq 0.12$). This fixes a scale parameter, Λ (see Eq. (5.91)), which controls the variation of α_s with Q^2. As a quantum field theory, QCD requires that one not only write down the Lagrangian, but also specify Λ. This in turn violates scale invariance.

In full QCD the trace of the energy momentum tensor becomes

$$\Theta^\mu_\mu = \partial_\mu S^\mu = \frac{\beta(g)}{2g} \mathrm{Tr}(G_{\mu\nu}G^{\mu\nu}) + \sum_{\text{flavors}} m_i \bar{\psi}_i \psi_i, \tag{6.85}$$

where $\beta(g)$ is the QCD β-function (see Section 5.3.2). Apart from the quark mass term, which breaks scale invariance explicitly, the divergence of the scale current now has an extra term proportional to the squared gluon field tensor, which is referred to as the **trace anomaly**.

We ignore the heavy quarks (using $N_f = 3$) and work to leading order in α_s. Then the nucleon mass becomes

$$M_N = \langle N(P)| - \frac{9\,\alpha_s}{4\,\pi} \mathrm{Tr}(G_{\mu\nu}G^{\mu\nu}) + m_u \bar{\psi}_u \psi_u + m_d \bar{\psi}_d \psi_d + m_s \bar{\psi}_s \psi_s |N(P)\rangle, \tag{6.86}$$

which survives in the chiral limit. This fundamental expression emphasizes the important role played by gluons in building up the nucleon mass. (Note that $\mathrm{Tr}(G_{\mu\nu}G^{\mu\nu}) = \boldsymbol{B}^2 - \boldsymbol{E}^2$, in terms of the color magnetic and electric fields.) The contributions from the u and d quark masses are small. We use the nucleon sigma term, $\sigma_N \sim 45$ MeV, as an estimate, as explained earlier. Furthermore, we anticipate the (rough) estimate of the strange quark contribution from Section 7.2.3:

$$\Delta M_N^{s-\text{quarks}} = \frac{y m_s}{m_u + m_d}\,\sigma_N, \tag{6.87}$$

with $y = 0.2 \pm 0.2$. Inserting typical values for the quark masses,

$$\frac{m_s}{m_u + m_d} \sim 10,$$

this gives an impression of the large gluonic field strength inside the nucleon: the matrix element $\langle N|\frac{\alpha_s}{\pi}(\boldsymbol{E}^2 - \boldsymbol{B}^2)|N\rangle$, taken as a measure, can range between 0.35 and 0.4 times the nucleon mass [Ji 95].

6.2 Structure of the QCD Vacuum: brief survey

Even now, with our limited ability to solve non-perturbative QCD, the QCD vacuum is known to have a remarkably complex structure. This section gives a short exposition and summary of the properties of the QCD vacuum in so far as they are important in the context of nucleon structure. In the following the QCD ground state or vacuum is always denoted by $|0\rangle$.

Quark condensates

To summarize previous discussions, we recall that chiral $SU(3)_R \times SU(3)_L$ is broken to $SU(3)_V$ by the strong correlations present in the non-perturbative vacuum. The corresponding order parameters are the quark condensates $\langle \bar{q}q\rangle$. In the chiral limit, $m_u = m_d = m_s = 0$, their values are

$$\langle \bar{u}u\rangle = \langle \bar{d}d\rangle = \langle \bar{s}s\rangle = -(225 \pm 25\ \text{MeV})^3 \tag{6.88}$$

at a renormalization scale of about 1 GeV.

Gluon condensate

The energy density of the vacuum is lowered in the presence of electric and magnetic gluon fields. An estimate of the gluonic vacuum field strengths has been given by Shifman, Vainshtein and Zakharov [SVZ 79] who introduced the gluon condensate and derived its value using QCD sum rules applied to charmonium states. The result is:

$$\begin{aligned}\left\langle 0\left|\frac{\alpha_s}{\pi}\,G^i_{\mu\nu}\,G^{\mu\nu}_i\right|0\right\rangle &= \left\langle 0\left|\frac{2\alpha_s}{\pi}\,(\boldsymbol{B}^2 - \boldsymbol{E}^2)\right|0\right\rangle \\ &= (350 \pm 30\ \text{MeV})^4,\end{aligned} \tag{6.89}$$

or about $2\text{GeV}/\text{fm}^3$. From the discussion of the trace anomaly in Section 6.1.7 we observe the close relationship between this condensate and scale breaking in the QCD vacuum.

Instantons

The axial anomaly, discussed in Section 6.1.6, has an interesting connection with yet another important non-perturbative property of the QCD vacuum, the instanton. The non-Abelian (Yang-Mills) field equations of the pure gluon gauge theory admit solutions for which the classical, Euclidean action, $\int d^4x\, \boldsymbol{E} \cdot \boldsymbol{B}$, is finite, so-called instantons [Bel 75]. These instantons give rise to zero modes for massless quarks. It was suggested by 't Hooft [Hoo 76] that the resulting, instanton-induced effective low-energy interaction between quarks could be represented by a flavor determinant:

$$\delta\mathcal{L} \sim G_{\text{inst}} \left[\det J^{+}_{ij} \;+\; \det J^{-}_{ij}\right], \tag{6.90}$$

with $J^{\pm}_{ij} = \overline{\psi}_i(1 \pm \gamma_5)\psi_j$, where i and j run over the three quark flavors u, d and s. This interaction therefore involves up, down and strange quarks simultaneously, in combinations such as $(\bar{u}_R u_L)(\bar{d}_R d_L)(\bar{s}_R s_L)$. It explicitly breaks $U(1)_A$ symmetry at the quark level. For example, in the $SU(3)$ version of the NJL model, discussed briefly in the previous section, such an interaction when combined with a ($U(3)_R \times U(3)_L$ symmetric) local, color current-current interaction between quarks, can account well for the symmetry breaking pattern seen in the pseudoscalar meson nonet [VW 91]. It seems likely that spontaneous chiral symmetry breaking is at least partly driven by instantons [Dia 84, Schä 97/98].

6.3 QCD Sum Rules

QCD has two distinct limits with very different physical properties. First, at short-distance we have the asymptotically free limit of high energy and momentum transfers, as explored, for example, in deep-inelastic scattering processes. Here the appropriate degrees of freedom are undeniably the quarks and gluons. Secondly, at large distances we have the low-energy limit, where the relevant degrees of freedom are composite hadrons and the non-perturbative character of the vacuum governs much of the physics. The QCD sum rule approach, developed by Shifman, Vainshtein and Zakharov [SVZ 79], is designed to interpolate between the perturbative and non-perturbative sectors using dispersion relation methods in combination with the operator product expansion. While this approach cannot claim to be rigorously derived from QCD, its intuitive value is that it provides a bridge between the low and high energy properties of QCD. Here we summarize those aspects of QCD sum rules which are specifically relevant to the nucleon. Some preparation is needed for this purpose.

6.3.1 Correlation Functions

The basic quantities entering the QCD sum rules are current-current correlation functions. Let $\mathcal{J}(x)$ be a current, given as a composite operator of quark fields, and define the *correlator*

$$\Pi(q^2) = i \int d^4x\, e^{iq\cdot x} \langle 0|\mathcal{T}\mathcal{J}(x)\mathcal{J}^\dagger(0)|0\rangle, \tag{6.91}$$

where $\mathcal{T}$ denotes the time ordered product. The singularities of this correlator for $q^2 > 0$ correspond to excitations carrying the quantum numbers of the current $\mathcal{J}$. The singularity structure of $\Pi(q^2)$ in the complex q^2 plane, combined with Cauchy's theorem, leads to a dispersion relation

$$\Pi(q^2) = \int_{s_0}^{\infty} ds \, \frac{\rho(s)}{s - q^2 - i\epsilon}, \tag{6.92}$$

where the function

$$\theta(q_0)\rho(q^2) = \frac{1}{\pi} \operatorname{Im}\Pi(q^2) = (2\pi)^3 \sum_m \delta^4(p_m - q)\left|\langle 0|\mathcal{J}(0)|m\rangle\right|^2 \tag{6.93}$$

describes the spectrum of physical intermediate states and s_0 is the threshold for the excitation of these states. Subtraction constants are required if the spectral function, $\rho(s)$, does not fall fast enough as $s \to \infty$.

In practice, one often encounters cases in which the spectral function has a resonance of mass M dominating the low-mass part of the spectrum and a continuum, Π_c, extending to larger s. In the QCD sum rule analysis this is commonly parametrized in the simple form

$$\operatorname{Im}\Pi(s) = \lambda^2 \delta(s - M^2) + \theta(s - s_c) \operatorname{Im}\Pi_c(s), \tag{6.94}$$

with adjustable constants λ and s_c.

6.3.2 Operator Product Expansion

The correlator can also be written in terms of the operator product expansion, which was explained in Chapter 5 (see Eq. (5.106)):

$$i \int d^4x \, e^{iq\cdot x} \langle 0|\mathcal{T}\mathcal{J}(x)\mathcal{J}^\dagger(0)|0\rangle = \sum_n C_n(q^2)\langle 0|\hat{O}_n|0\rangle. \tag{6.95}$$

Here the $\hat{O}_n$ are *local* operators and the c-number functions, $C_n(q^2)$, are called Wilson coefficients. The operators $\hat{O}_n$ are constructed from quark and gluon fields and organized according to their dimensions. Apart from the trivial unit operator of dimension $d = 0$, the operators of lowest dimension, $d = 4$, are $m_q\bar{q}q$ and $G^i_{\mu\nu}G^{\mu\nu}_i$. The next higher ones, with $d = 6$, include four-quark operators and combinations of quark and gluon fields.

The point is now that for sufficiently large spacelike values of $Q^2 = -q^2 > 0$, one can expect that the operator product expansion (6.95) converges rapidly since the Wilson coefficients, $C_n(q^2)$, decrease with inverse powers of Q^2 for each higher dimension operator to which they are attached. One may then apply QCD perturbation theory to calculate the Wilson coefficients as a power series in the running coupling strength α_s,

$$C_n(q^2) = A_n(q^2) + B_n(q^2)\,\alpha_s + \cdots, \tag{6.96}$$

with the expansion terminated after few powers of α_s.

6.3.3 The Sum Rule Method

The essence of the QCD sum rule approach is to match the dispersion relation for the correlator $\Pi(q^2)$ with its operator product representation for sufficiently large $Q^2 = -q^2 > 0$:

$$\begin{aligned}\Pi(-Q^2) &= \frac{1}{\pi}\int_{s_0}^{\infty} ds\, \frac{\operatorname{Im}\Pi(s)}{s+Q^2} \quad (+\text{subtractions}) \\ &= C_0(-Q^2) + C_1(-Q^2)\langle m_q\bar{q}q\rangle + C_2(-Q^2)\langle G_{\mu\nu}G^{\mu\nu}\rangle + \cdots ,\end{aligned} \tag{6.97}$$

which now involves vacuum expectation values (condensates) of the operators $\hat{O}_n$ on the right hand side of the equation. Clearly, for this equality to be useful, the Q^2 range has to be restricted to an interval where the short-distance, large Q^2 expansion and the resonance-plus-continuum ansatz for $\operatorname{Im}\Pi$ are both valid. This is presumed to happen in the region around $Q^2 \sim 1\ \text{GeV}^2$, although accurate criteria for this choice cannot easily be given.

To carry out the sum rule analysis, it is useful to improve the convergence properties of both sides of Eq. (6.97) by performing a differential operation, called the **Borel transform**. This operation keeps the dispersion relation free of subtraction constants (at the expense of introducing a technical parameter, the **Borel mass** $\mathcal{M}$) and suppresses operators of high dimension on the right hand side of Eq. (6.97). The Borel transform of a function, $f(Q^2)$, is defined by

$$\mathcal{B}f(Q^2) = \lim_{n,Q^2\to\infty} \frac{(Q^2)^n}{(n-1)!}\left(-\frac{d}{dQ^2}\right)^n f(Q^2), \tag{6.98}$$

with $Q^2/n \equiv \mathcal{M}^2$. For example, the Borel transform replaces an inverse power of Q^2 by an exponentially decreasing function of the Borel parameter $\mathcal{M}$:

$$\mathcal{B}\frac{1}{Q^2+m^2} = \mathcal{M}^{-2}\exp\left(-\frac{m^2}{\mathcal{M}^2}\right). \tag{6.99}$$

Application of the Borel transform to $\Pi(-Q^2)$ thus improves the stability of the analysis, with $\mathcal{M}$ defining the scale at which short and long distance expansions meet.

6.3.4 Ioffe's Formula

The application of this method to the nucleon now proceeds as follows. First one introduces a current which carries the quantum numbers of the nucleon. A convenient choice for the proton with its uud valence structure is [Io 81]:

$$\mathcal{J}_N(x) = \epsilon_{ijk}(\psi_u^i)^T(x)\, C\gamma^\mu\, \psi_u^j(x)\, \gamma_5\, \gamma_\mu\, \psi_d^k(x), \tag{6.100}$$

where the quark fields ψ_u^i etc. are denoted by their color index i and flavor u or d, and C is the charge conjugation operator. The construction is such that $\mathcal{J}_N$ represents a spin $\frac{1}{2}$-particle. The general form of the nucleon correlator has two terms:

$$\begin{aligned}\Pi(q^2) &= i\int d^4x\, e^{iq\cdot x}\langle 0|\mathcal{T}\mathcal{J}_N(x)\overline{\mathcal{J}}_N(0)|0\rangle \\ &= \Pi_1(q^2) + \gamma_\mu q^\mu \Pi_2(q^2).\end{aligned} \tag{6.101}$$

It has a pole at the nucleon mass, $q^2 = M_N^2$:

$$\Pi(q^2 \simeq M_N^2) = \lambda_N^2 \frac{\gamma_\mu q^\mu + M_N}{q^2 - M_N^2 + i\epsilon}, \tag{6.102}$$

where λ_N is the amplitude to find all three quarks simultaneously at the origin.

The calculation of the Wilson coefficients then yields the following leading terms:

$$\Pi_1(-Q^2) \simeq \frac{Q^2}{4\,\pi^2} \ln \frac{Q^2}{\mu^2} \langle \overline{q}q \rangle + \cdots, \tag{6.103}$$

$$\Pi_2(-Q^2) \simeq \frac{Q^4}{64\,\pi^3} \ln \frac{Q^2}{\mu^2} + \cdots, \tag{6.104}$$

where μ^2 is the renormalization scale at which the condensates are determined. The gluon condensate, $\langle G^2 \rangle \equiv \langle 0 | \frac{\alpha_s}{\pi} G^i_{\mu\nu} G^{\mu\nu}_i | 0 \rangle$, adds a correction of order $\langle G^2 \rangle / Q^4$ to Π_2 which can safely be neglected. Performing the Borel transformation on $\Pi_{1,2}$ one finds two conditions:

$$\mathcal{M}^6 + \frac{4}{3} a^2 = c, \tag{6.105}$$

$$2a\mathcal{M}^4 = cM_N, \tag{6.106}$$

with $a = -4\,\pi^2 \langle \overline{q}q \rangle$ and $c = 32\,\pi^4 \lambda_N^2 \, \exp(-M_N^2/\mathcal{M}_B^2)$. Eliminating c one obtains in leading order:

$$M_N = -\frac{8\,\pi^2}{\mathcal{M}^2} \langle \overline{q}q \rangle + \cdots. \tag{6.107}$$

This is Ioffe's formula. For $\mathcal{M} \sim 1$ GeV one finds $M_N \sim 1$ GeV. (In his original paper [Io 81], Ioffe arbitrarily set $M_N = \mathcal{M}$, with the amusing accidental result $M_N = (-8\,\pi^2 \langle \overline{q}q \rangle)^{\frac{1}{3}} \sim$1 GeV.)

While the numerical agreement between Eq. (6.107) and the observed mass of the nucleon is impressive, it is somewhat accidental. Technically, this analysis is too simple. The quark condensate is more closely connected to the continuum states than the nucleon itself and one must include higher order condensates in order to obtain theoretically reliable results [Lei 97]. For example, the variation of the mass of the nucleon with the quark condensate predicted by Eq. (6.107) is not consistent with that found in lattice QCD. Nevertheless, the Ioffe formula does demonstrate the important truth that chirally odd operators, such as $\langle \overline{q}q \rangle$, play an essential role in giving the nucleon its mass.

6.4 Lattice QCD and the Nucleon

The genuinely non-perturbative nature of QCD at large distances, or low momentum scales, sets obvious limits to the applicability of analytic approaches. However, important conceptual and technical progress has been made toward solving QCD on a lattice (a discretized, finite space-time volume) using advanced numerical methods, and first steps have been undertaken to investigate the properties of the nucleon in this framework. These developments have been made possible by great theoretical efforts in understanding quantum field theory on a lattice, and by an enormous increase in the available computer power. Here we discuss results from lattice QCD to the extent that they relate to nucleon structure. One must be aware of the fact that this particular field of research is in rapid development, and it should be clear that this section can only present a snapshot taken at the time of writing this book.

6.4.1 Field Theory on a Lattice

Framework

Before focusing on specific applications of lattice QCD to the structure of the nucleon, it is useful to introduce the main theoretical ideas underpinning the lattice approach at a basic level. Detailed introductions can be found in [Cr 83, MM 94, Rot 97].

Let us start from a field theoretical Lagrangian density $\mathcal{L}(\Phi, \partial^\mu \Phi)$, given in terms of a suitable set of fields $\Phi(x)$, where $x^\mu = (x^0, \vec{x}\,) \equiv (t, \vec{x}\,)$ is a vector in Minkowski space-time. In QCD, the Φ's should be identified with the gluon and quark fields. In general, a quantum field theory (QFT) is determined by the vacuum expectation values of time ordered products of the fields, the *Green functions* of the theory:

$$\mathcal{G}^{(n)}(x_1, \ldots, x_n) = \mathcal{N} \langle 0 |\, \mathcal{T}[\Phi(x_1) \cdots \Phi(x_n)]\, | 0 \rangle, \tag{6.108}$$

where $\mathcal{N}$ is a normalization constant.

The basis of modern quantum field theory is the functional integral approach, which was outlined in Chapter 5 (see also Appendix E). Let

$$S[\Phi] = \int d^4x \; \mathcal{L}(\Phi(x), \partial^\mu \Phi(x)) \tag{6.109}$$

be the action. Then the Green functions, $\mathcal{G}^{(n)}$, can be formulated in terms of path integrals taken over all possible field configurations:

$$\mathcal{G}^{(n)}(x_1, \ldots, x_n) = \frac{\int [d\Phi] \; \Phi(x_1) \cdots \Phi(x_n) \; e^{iS[\Phi]}}{\int [d\Phi] \; e^{iS[\Phi]}}. \tag{6.110}$$

All physically relevant quantities (such as masses, form factors and scattering amplitudes) of a system described by the action S can be derived from these Green functions. In essence, solving the quantum field theory is equivalent to calculating the functional integrals in Eq. (6.110).

The Lattice Approach

The task of computing the Green functions (6.110) on a lattice involves the following steps:

i) Any renormalizable quantum field theory, such as QCD, requires an ultraviolet (short distance) regularization in order to eliminate infinities from calculated physical quantities. Here this cutoff procedure is introduced by defining the fields on a mesh of discrete lattice points which replaces the space-time continuum. Ultimately one must ensure that the dependence of any calculated observable on the lattice spacing satisfies the appropriate renormalization group equation. That is, the calculated values of physical observables should scale correctly with a, as $a \to 0$.

ii) An infrared (long-wavelength) cutoff is imposed by working on a finite lattice volume. This is a technical necessity in order to be able to perform the numerical calculation of the integrals in Eq. (6.110) which unavoidably requires a finite number of variables. Again, one must extrapolate the value of any observable to the infinite volume limit.

iii) In general, the functional integrals in Minkowski space are complex and oscillate rapidly, so that they cannot be treated numerically in this form. In order to provide a sound basis, both for formal mathematical reasons and for practical numerical purposes, it is normal to calculate the path integrals for imaginary times:

$$t \equiv x^0 = -ix_4 \tag{6.111}$$

This change of variables, together with the appropriate analytic continuation of the Green functions, defines a **Euclidean** quantum field theory: that is, the fields now live on a four-dimensional Euclidean space $\{x = (x_1, \ldots, x_4); x_\mu \in \boldsymbol{R}\}$. Lattice QFT is then represented in terms of well defined functional integrals taken over the Euclidean lattice hypercube

$$\boldsymbol{\Lambda} = \{x;\ x_\mu/a \in \boldsymbol{Z}, \mu = 1, \ldots, 4\}, \tag{6.112}$$

where a is the lattice spacing.

Ideally, the **continuum limit** ($a \to 0$) should be approached as closely as possible, and where necessary the theory should be analytically continued back into Minkowski space-time, once all the procedures have been completed. In practice these steps cannot be taken. The accuracy of this approach is therefore limited by technical constraints. One must work with a reasonable compromise between maximizing the total lattice volume and minimizing the lattice spacing.

The transformation (6.111) leads naturally to a Euclidean action, S_E, in which the fields and their derivatives are expressed in terms of $x \in \boldsymbol{\Lambda}$. For example, the kinetic term of the Lagrangian density that appears in S_E is $\frac{1}{2}\partial_\mu \Phi \partial_\mu \Phi$, with $\mu = 1, \ldots, 4$. The relation between S_E and the original action is

$$S = iS_E, \tag{6.113}$$

so that the Euclidean functional integral which replaces Eq. (6.110) now involves $\exp(-S_E)$ instead of $\exp(iS)$. Strong oscillations at large S are turned into exponential damping at large S_E, and the functional integral changes from complex to real. These features make the integrations numerically tractable.

Let us explore how a one-particle Green function, or two-point function,

$$\mathcal{G}_E(x, y) = \frac{\int [d\Phi]\ \Phi(x)\Phi(y)\ e^{-S_E[\Phi]}}{\int [d\Phi]\ e^{-S_E[\Phi]}} \tag{6.114}$$

behaves in Euclidean space-time. Assume that the mass spectrum $\rho(m^2)$ of states represented by Φ starts at mass m_0. The original Green function in Minkowski space-time has the spectral representation:

$$\mathcal{G}(x,y) = i \int_{m_0^2}^{\infty} dm^2 \rho(m^2)\, \Delta_F(x-y;m^2) \tag{6.115}$$

with

$$\Delta_F(x;m^2) = \int \frac{d^4k}{(2\,\pi)^4} \frac{e^{-ik\cdot x}}{k^2 - m^2 + i\epsilon} \tag{6.116}$$

and $k \cdot x = k_0 x^0 - \vec{k} \cdot \vec{x}$. Here the masses m are determined by the singularities of Δ_F. The Euclidean Green function, on the other hand, has the representation

$$\mathcal{G}_E(x,y) = \int_{m_0^2}^{\infty} dm^2 \rho(m^2)\, \Delta_E(x-y;m^2) \tag{6.117}$$

with

$$\Delta_E(x;m^2) = \int \frac{d^4k}{(2\,\pi)^4} \frac{e^{ik\cdot x}}{k^2 + m^2} \tag{6.118}$$

and $k \cdot x = \vec{k} \cdot \vec{x} + k_4 x_4$. Integrating Eq. (6.117) over three-space we obtain

$$\int d^3x\, \mathcal{G}_E(\vec{x}, x_4; \vec{y}, y_4) = \int_{m_0}^{\infty} dm\ \rho(m^2)\, e^{-m|x_4 - y_4|}. \tag{6.119}$$

Hence we see that for large Euclidean time separations the lowest mass, m_0, dominates and can be determined as:

$$m_0 = -\lim_{T\to\infty} \frac{1}{T} \ln \int d^3x\, \mathcal{G}_E(\vec{x}, \tau; \vec{y}, 0). \tag{6.120}$$

This is the basic principle for generating the low-mass hadron spectrum in lattice QCD. One has to keep in mind, of course, that the limiting procedure just mentioned requires that one take the continuum limit on an infinite lattice. Corrections arise from finite volume effects and the finite lattice spacing. The actual quantity determined on the lattice is

$$m_0\, a = -\frac{1}{N} \ln \int d^3x\, \mathcal{G}_E(\vec{x}, Na; \vec{y}, 0), \tag{6.121}$$

where N is the maximum number of lattice sites available along the time direction.

QCD on the Lattice

In the original pioneering work by Wilson [Wi 74] the main motivation for introducing a lattice was to permit a completely non-perturbative formulation of QCD, in order to see whether the phenomenon of confinement could be explained in terms of strong, long-distance color forces. QCD involves gauge fields (gluons) coupled to fermion fields (quarks), both of which require some careful explanation concerning their path integral treatment on a discretized lattice space.

i) **Gauge Fields** are intimately connected with the concept of **parallel transport** in differential geometry. Let $\mathcal{C}_{yx}$ be a curve leading from a point x to a point y in space-time. With $\mathcal{C}_{yx}$ we associate a matrix $U(\mathcal{C}_{yx})$. When U acts on any field at point x it "transports" this field from x to y along the curve $\mathcal{C}$. Now let $y = x + dx$ differ from x by an infinitesimal distance. The corresponding transporter

$$U(\mathcal{C}_{x+dx,x}) = 1 - A_\mu(x)\, dx^\mu + \mathcal{O}[(dx^\mu)^2], \tag{6.122}$$

defines the gauge field A_μ. In QCD with its color gauge group $SU(3)_c$ we have $A_\mu \equiv -ig\, A_\mu^b t_b$, where t_b are the generators of the gauge group and g is the bare coupling constant.[4] The covariant derivative is $D_\mu = \partial_\mu + A_\mu$. Taking a closed curve, which begins and ends at point x, along an infinitesimal parallelogram with sides dx and dy, one finds

$$U(\mathcal{C}_{x,x}) = 1 - F_{\mu\nu}(x)\, dx^\mu dy^\nu, \tag{6.123}$$

where

$$F_{\mu\nu} = [D_\mu, D_\nu] = \partial_\mu A_\nu - \partial_\nu A_\mu + [A_\mu, A_\nu] \tag{6.124}$$

is the field strength tensor.

On the Euclidean lattice, $\mathbf{\Lambda}$, the smallest distance over which parallel transport can take place is the distance between two neighbouring lattice points separated by the lattice spacing a. The pair $(x + dx, x)$ is replaced by the pair of neighbouring lattice points

$$c = (x + a\hat{e}_\mu, x), \tag{6.125}$$

where $\hat{e}_\mu$ $(\mu = 1, \ldots, 4)$ are the unit vectors which span the four-dimensional Euclidean space. The transporter associated with a link c,

$$U(c) \equiv U(x + a\hat{e}_\mu, x) \tag{6.126}$$

is called a **link variable**. (Note that $U(c)$ is usually called $U_\mu(a)$ in the lattice QCD literature and it is formally equal to $\mathcal{P}\exp(-\int_x^{x+a\hat{e}_\mu} dx'\, A_\mu(x'))$, with $\mathcal{P}$ indicating path ordering.) For an arbitrary path $\mathcal{C}$ on the lattice, given as a superposition of links, the parallel transporter is a product of link variables:

$$U(\mathcal{C}) = U(c_n)\ \cdots\ U(c_1). \tag{6.127}$$

The set $\{U(c)\}$, of all possible link variables, is the **lattice gauge field**, generically denoted U. The smallest closed loops on a lattice, called **plaquettes**, consist of four links forming a parallelogram. The corresponding transporter, $U(p)$, is called a **plaquette variable**.

[4] Note that A_μ here differs by a factor of $-ig$ from the definition in Chapter 5. We choose the different convention here because it matches the convention usually adopted in the lattice QCD literature.

The essential step is now to construct the lattice counterpart of the gluonic action, $-\frac{1}{2g^2}\operatorname{Tr}\int d^4x F_{\mu\nu}F^{\mu\nu}$. Wilson [Wil 69] and Wegner [Weg 71] have shown how to build up the lattice action for a gauge field in terms of a sum over plaquette variables. The key is to observe that

$$\operatorname{Tr}(U(p) + U(p)^{-1}) = \text{const.} + a^4 \operatorname{Tr}(F_{\mu\nu}(x))^2. \tag{6.128}$$

Apart from an irrelevant additive constant, one ends up with Wilson's action

$$S_g = -\frac{1}{2g^2}\sum_x a^4 \operatorname{Tr}(F_{\mu\nu}(x))^2, \tag{6.129}$$

expressed in terms of plaquette variables, where the sum is taken over all possible plaquettes on the lattice.

In concluding this discussion it is vital to appreciate that the choice to work with the lattice gauge field, U, rather than A_μ itself, is not a trivial matter. The reason for this choice is that in this way gauge invariance is preserved **exactly** for **all** choices of lattice spacing. We have already emphasized the fundamental role played by local gauge invariance in QCD and we saw, for example, that the axial anomaly was a consequence of enforcing gauge invariance over chiral symmetry. In the lattice context it is crucial that in attempting to extrapolate to the continuum limit ($a \to 0$) we are not also extrapolating quantities which violate gauge invariance at some level. Of course, the price one pays for this is that one effectively sums terms of order a to infinite orders and in order to improve the convergence of the lattice calculations (as $a \to 0$) it has been found desirable to incorporate certain mean-field corrections. Such improvements go under the heading of "improved actions" [LM 93] and these will be vital in practical calculations in the foreseeable future.

ii) **Fermion Fields** require a special treatment, for two reasons. First, the functional integral formalism must be prepared to accommodate anticommuting fields. Second, naive lattice discretization of a Dirac Hamiltonian leads to an unwanted replication of fermion states which needs to be repaired.

As explained in Chapter 5 (see also Appendix C), the first part of the problem is solved by describing the fermion fields using Grassmann variables: c-number quantities with the interesting algebraic property that any pair of them *anti*-commutes. The integration rules implied by this algebra are used to perform functional integrals involving fermion fields. In particular, the quark fields in QCD enter the Lagrangian density quadratically and the fermion action has the generic form:

$$S_q = \int d^4x\, \overline{\psi}(x)\mathcal{Q}\psi(x), \tag{6.130}$$

where $\mathcal{Q}$ is referred to as the **quark matrix**. Following the procedures outlined in Chapter 5, one can carry out the functional integral over $\overline{\psi}$ and ψ exactly:

$$\int [d\overline{\psi}d\psi]\, e^{-S_q} = \det \mathcal{Q}. \tag{6.131}$$

The result is called the **fermion determinant**.

The lattice discretization procedure seems, on first sight, a straightforward step, simply replacing the continuous Euclidean x by the lattice points $x \in \mathbf{\Lambda}$. However, the following complication arises. Given periodic boundary conditions on the lattice[5], the energy spectrum of Dirac particles is now discrete, and the allowed quark momenta are points in the Brillouin zone of the lattice. The periodicity leads to a multiplication in the number of fermion species at each given energy. In four dimensions, there are $2^4 = 16$ corners in a given Brillouin zone which all correspond to the same energy. In order to avoid this "Fermion doubling" in each dimension, two alternative methods are commonly used. The first one (**Wilson fermions**) introduces an additional term in the action such that 15 out of the 16 quark species become infinitely heavy in the continuum limit. The second method (**Kogut-Susskind** or **staggered fermions**) puts only a single fermion field component on each lattice site.

The lattice action of QCD has the following form:

$$S[U,\psi,\overline{\psi}] = S_g[U] + S_q[U,\psi,\overline{\psi}], \tag{6.132}$$

where S_g is the pure gluonic action and S_q is the action of quarks coupled to gluons. The task is then to compute, for a given physical quantity, $F[U,\psi,\overline{\psi}]$, the expectation value

$$\langle F\rangle = \frac{\int [dU\, d\overline{\psi}\, d\psi]\; e^{-S}\; F}{\int [dU\, d\overline{\psi}\, d\psi]\; e^{-S}}. \tag{6.133}$$

The next formal step is the integration over the quark fields using Eq. (6.131). Under this formal manipulation the original action S now turns into an effective action, $S_{\text{eff}}[U]$, and the same is true for F which turns into $F_{\text{eff}}[U]$. The result

$$\langle F\rangle = \frac{\int [dU]\; e^{-S_{\text{eff}}}\; F_{\text{eff}}}{\int [dU]\; e^{-S_{\text{eff}}}}, \tag{6.134}$$

does not lead any further because the formally eliminated quark degrees of freedom are still present implicitly in the effective action

$$S_{\text{eff}}[U] \equiv S_g[U] - \ln\det \mathcal{Q}[U], \tag{6.135}$$

and in F_{eff}, which depends on the inverse of the quark matrix $\mathcal{Q}$ representing the quark propagator.

The main problem with the calculation of the fermion determinant, $\det \mathcal{Q}[U]$, is that it includes rapid fluctuations and involves the computational expense of inverting a huge matrix. On the other hand, an enormous simplification occurs if one makes the so-called **quenched approximation**, which consists of replacing the fermion determinant, $\det \mathcal{Q}[U]$, by a constant which is independent of the gauge fields. In this limit the effective action, Eq. (6.135), reduces to the pure gauge field part, S_g. A perturbative analysis in powers of the gauge coupling reveals that the quark determinant generates quark loops – i.e., quantum fluctuations of the quark fields, such as the virtual creation and annihilation of quark-antiquark pairs. The quenched approximation suppresses these quark loops – indeed, the quenched approximation

[5] Note that for fermions the boundary conditions must be anti-periodic in the Euclidean time direction [Cr 83, MM 94].

is exact in the limit of infinitely massive quarks. Physical quantities computed in quenched lattice QCD therefore emphasize the role of valence quarks, while the chiral aspects of hadron structure are inconsistent with QCD [LS 96]. Nevertheless, the majority of lattice calculations have made use of this approximation.

6.4.2 Nucleon Mass

After these preparations we are ready to discuss the properties of the nucleon as they emerge from lattice QCD. The most basic quantity is, of course, the nucleon mass. As outlined in Eq. (6.120), it can be deduced from the asymptotic behavior of the lattice Green function of the nucleon for large Euclidean time intervals. In order to construct this Green function one needs a suitable interpolating field, $N(x)$, expressed in terms of the quark fields $\psi(x)$, that represents a three-quark system with the quantum numbers of the nucleon. One possible choice is to use the Ioffe current, $\mathcal{J}_N$, introduced in Eq. (6.100), or a simpler variant with the $\gamma^\mu \cdots \gamma_\mu \gamma_5$ combination in $\mathcal{J}_N$ replaced by γ_5. All that is required, in principle, is that the choice of current should have a reasonable overlap with the wave function of the physical nucleon. In practice, the greater this overlap to start with, the more rapidly the asymptotic behavior will be dominated by the mass of the nucleon.

Given $N(x)$ one computes the nucleon Green function

$$\mathcal{G}_N(\vec{x}, \tau; 0) = \langle 0|N(\vec{x}, \tau)\overline{N}(0)|0\rangle = \frac{\int [dU\, d\overline{\psi}\, d\psi]\; e^{-S}\; N(x)\overline{N}(0)}{\int [dU\, d\overline{\psi}\, d\psi]\; e^{-S}}. \tag{6.136}$$

Its asymptotic behavior for large τ determines the nucleon mass,

$$M_N \;= -\frac{1}{\tau} \ln \int d^3x\, \mathcal{G}_N(\vec{x}, \tau; 0), \tag{6.137}$$

where τ is a large multiple of the lattice spacing, a. Evidently, a mass on the lattice cannot be determined on an absolute scale, but only in units of a^{-1}. The most satisfactory way to convert dimensionless quantities, such as $M_N a$, to physical units is to determine a by demanding that the string tension be correctly reproduced.

The string tension, $\sigma \sim 800$ MeV/fm, is the energy per unit length required to move two static quarks apart. It is traditionally determined from the slope of Regge trajectories or more accurately from the potential needed to reproduce the energy levels of heavy quarkonia [CP-P 99, UKQ 99]. The systematic uncertainty in setting the absolute value of the lattice spacing through σ is of order 10 %. Another common alternative is to compute the ratio of various hadron masses, for example, the ratio $\frac{m_\rho}{M_N}$. The major reason for preferring the string tension to mass ratios is that at present the lattice calculations are performed either at large quark masses, for full QCD with dynamical fermions, or in quenched approximation. In either case the effects of chiral symmetry, which on quite general grounds are expected to lead to rapid variation of hadron masses at small quark masses, will be incorrect. These errors will not, in general, cancel in mass ratios. In order to appreciate this issue better we consider next the role of quark masses.

The only scale that enters the QCD Lagrangian is the quark mass, m_q. In lattice computations this mass can be dialed (again in lattice units), and the nucleon mass and other hadron

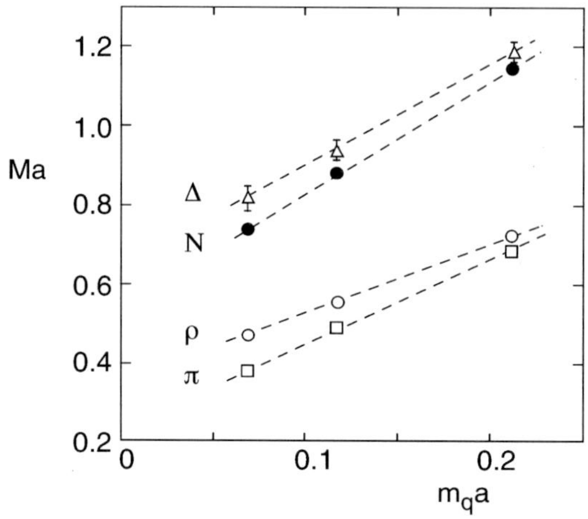

Figure 6.5: Masses of the nucleon and other hadrons (in lattice units) as a function of the quark mass $m_q a$ from lattice QCD in the quenched approximation [Liu 99].

masses can then be discussed as a function of $m_q a$ (see Fig. 6.5). Within full QCD, that is retaining the fermion determinant, the calculation time rises like $1/m_q^p$, with $p \sim 2-3$, as $m_q \rightarrow 0$. For this reason calculations are commonly restricted to quark masses above 100 MeV, rather than the 5-10 MeV required for a realistic calculation. The difficulty that this raises is that it is possible to show, using the fact that chiral symmetry is spontaneously broken in QCD, that the mass of the nucleon is not an analytic function of m_q. In particular, the self-energy correction involving a Goldstone pion loop gives rise to a **model independent** term in the mass of the nucleon:

$$\delta^{\mathrm{LNA}} M_N = -\frac{3\,g_A^2}{32\,\pi\,f_\pi^2}\,m_\pi^3, \tag{6.138}$$

where g_A is the axial charge of the nucleon.[6] The superscript LNA indicates that this is the "leading non-analytic" component of the nucleon mass. It is non-analytic in m_q because $m_\pi^2 \propto m_q$ and hence this term involves $m_q^{3/2}$, which has a branch point at $m_q = 0$. It is the leading non-analytic piece because it is the lowest odd power of m_π in a series expansion of M_N.

Numerically $\delta^{\mathrm{LNA}} M_N$ does not seem very important, it is only -17 MeV at the physical pion mass. On the other hand, at just three times the physical pion mass it is -460 MeV, half the mass of the nucleon and growing like m_π^3. Moreover all the lattice data with dynamical fermions (i.e. not quenched) is at larger pion masses than this. Clearly in extrapolating lattice data to the physical pion mass one *must* deal with the issue of chiral corrections. Figure 6.6 shows lattice data in full QCD from CP-PACS [CP-P 99] and UKQCD [UKQ 99] for the nucleon and the Δ, as a function of the mass of the pion. The physical scale has been set

[6] Strictly both g_A and f_π should be evaluated in the chiral limit, but this is not a big correction and we shall use the physical values in our numerical estimates.

using the string tension, as explained above. We also show a fit to the data which for the nucleon is based upon the form [Lei 00]:

$$M_N(m_\pi^2) = \alpha + \beta m_\pi^2 + \sigma_{NN}(m_\pi^2, \Lambda) + \sigma_{N\Delta}(m_\pi^2, \Lambda). \tag{6.139}$$

Here α, β and Λ are parameters fitted to the data and σ_{Nj} are the self-energy diagrams involving a pion loop with an intermediate baryon j, with ultra-violet cut-off Λ. Eq. (6.139) guarantees the correct leading and next-to-leading chiral behavior outlined above, but suppresses the chiral corrections at large m_π in a very natural way. It is not possible to simply fit the data with a cubic polynomial in m_π while preserving the correct LNA coefficient. We note that the extrapolated values of the nucleon and Δ masses are in quite good agreement with the observed masses, within the 10 % systematic uncertainty noted earlier.

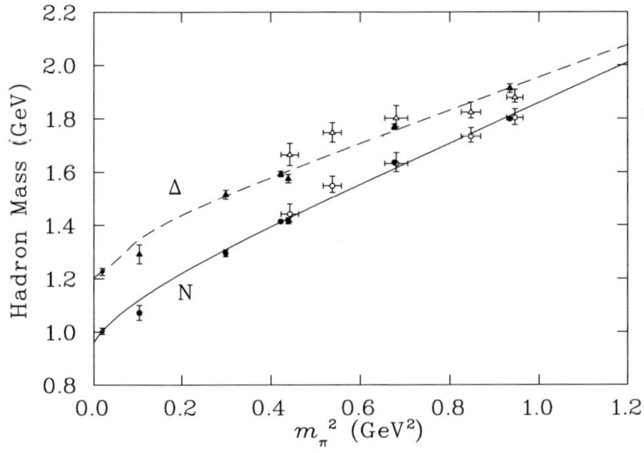

Figure 6.6: Lattice results from UKQCD (open points) and CP-PACS (solid points) for the nucleon and Δ masses in full QCD – with 5 % adjustments in the scale parameters to ensure consistency between the two data sets. The solid (dashed) curves illustrate fits to the combined data sets which preserve the correct leading and next-to-leading chiral behavior as $m_\pi \to 0$ – from Ref. [Lei 00].

6.4.3 Nucleon Form Factors

The next level in complexity is the computation of current matrix elements of the nucleon. They have the form

$$\begin{aligned} \mathcal{G}_{\mathcal{J}}(x,y) &= \langle 0|N(x)\mathcal{J}(y)\overline{N}(0)|0\rangle \\ &= \frac{\int [dU\, d\overline{\psi}\, d\psi]\, e^{-S}\, N(x)\mathcal{J}(y)\overline{N}(0)}{\int [dU\, d\overline{\psi}\, d\psi]\, e^{-S}}. \end{aligned} \tag{6.140}$$

Here $\mathcal{J}(y)$ stands generically for any quark current, such as the electromagnetic current

$$J^\mu(y) = \overline{\psi}(y) Q \gamma^\mu \psi(y), \tag{6.141}$$

where Q is the quark charge matrix, or the axial current

$$A_b^\mu(y) = \overline{\psi}(y)\gamma^\mu\gamma_5\frac{\lambda_b}{2}\psi(y). \tag{6.142}$$

In order to generate form factors on the lattice one carries out a three-dimensional discrete Fourier transform,

$$G_{\mathcal{J}}(\tau_1,\tau_2;\vec{p},\vec{q}) = \sum_{\vec{x},\vec{y}} e^{i\vec{q}\cdot\vec{y}}e^{-i\vec{p}\cdot\vec{x}}\mathcal{G}_{\mathcal{J}}(\tau_1,\vec{x};\tau_2,\vec{y}), \tag{6.143}$$

inserts complete sets of states in Eq. (6.140) and takes $\tau_1 \gg \tau_2 \gg a$, the lattice spacing. This procedure is used to compute, for example, the electromagnetic form factors $G_E(q^2), G_M(q^2)$ and the axial form factor, $G_A(q^2)$, of the nucleon for a set of values $\vec{q}^{\,2}$, corresponding to spacelike momentum transfers $q^2 = -\vec{q}^{\,2}$ in Minkowski space-time.

In Fig. 6.7 we show examples of results from such a lattice computation [Liu 99]. The numerical effort to produce these form factors is formidable: those shown in the figure were generated on a lattice with $16^3 \times 24$ sites and took several thousands of hours on a supercomputer – even though this was still done in the quenched approximation. Together, the quenched approximation and the difficult extrapolation to small quark masses, which was carried out using a naive linear extrapolation in quark mass, are the major uncertainties in this work. (The scale for the momentum transfer q was set by the nucleon mass, which was computed simultaneously.)

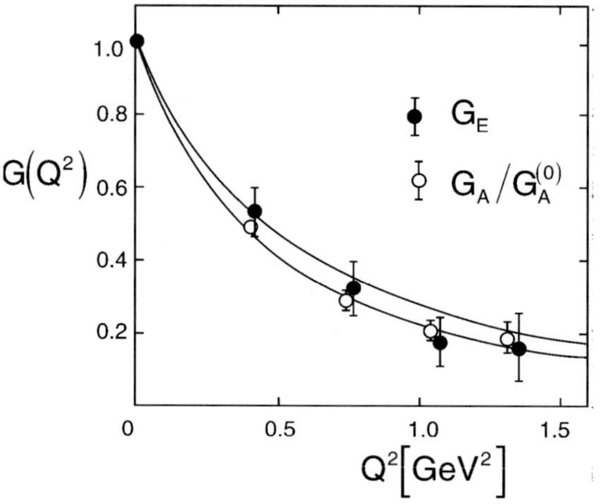

Figure 6.7: Lattice QCD results for the electric (G_E) and axial (G_A) form factors of the proton in the quenched approximation [Liu 99]. The lines show empirical dipole fits to the form factors.

Considering the limitations inherent in the approach, the results shown in Fig. 6.7 compare remarkably well with empirical dipole fits to those form factors. The axial vector coupling constant obtained in these simulations is $g_A = 1.18 \pm 0.11$, somewhat low compared to the experimental $g_A = 1.26$, but certainly consistent within the statistical errors. In the computation of the electromagnetic form factor, current conservation has been checked to be accurately fulfilled at the level of 10^{-4}.

One should also note that important aspects of vector meson dominance in G_E are already included in these calculations, even in the quenched approximation. The point is that a quark-antiquark pair, with inclusion of all possible strong interactions via gluons, can be formed and attached to the external photon starting from a single connected valence quark line in the nucleon, without the need to introduce quark loops. One feature which is lost in the quenched approximation, however, is the width of the vector meson associated with its decay into pions. The other key feature that is not correctly incorporated is chiral symmetry. For the magnetic moments has it been possible to improve this aspect of the calculation and we briefly consider that next.

Magnetic Moments

Just as for the mass of the nucleon, one can show that the magnetic moments of the proton and neutron must exhibit certain model-independent non-analytic behavior as a function of quark mass. Indeed, the leading non-analytic (LNA) behavior begins at order $m_q^{1/2}$ in this case – i.e. at $\mathcal{O}(m_\pi)$. For small masses chiral symmetry demands that the proton and neutron magnetic moments satisfy:

$$\mu^{p(n)} = \mu_0^{p(n)} \mp \alpha m_\pi + \mathcal{O}(m_\pi^2), \tag{6.144}$$

where $\alpha = \frac{M_N g_A^2}{8\pi f_\pi^2}$. Using the one loop chiral perturbation theory value for g_A, we find $\alpha = 4.41$ if μ is in nuclear magnetons and m_π in GeV. This is a big coefficient. It means that at the physical pion mass the LNA contribution to the neutron magnetic moment is +0.62 nuclear magnetons, or roughly 30 % of the measured value. Because this is non-analytic in the quark mass it cannot be included in any simple valence quark model that does not respect chiral symmetry (c.f. Chapter 8). In a quenched lattice calculation it will be incorrect.

One proposal to incorporate the correct LNA behavior in an extrapolation of lattice data at large quark masses has produced a very promising result. The idea is to make the simplest extrapolation formula that incorporates both the correct LNA behavior as $m_\pi \to 0$ and the correct large mass limit. For the latter, heavy quark effective theory tells us that the magnetic moment should behave as $1/m_q$ (or $1/m_\pi^2$). Thus Leinweber *et al.* [LLT 99] proposed the form:

$$\mu^{p(n)} = \frac{\mu_0^{p(n)}}{1 \pm \frac{\alpha}{\mu_0^{p(n)}} + \beta^{p(n)} m_\pi^2}, \tag{6.145}$$

where $\mu_0^{p(n)}$ (the magnetic moment in the chiral limit) and $\beta^{p(n)}$ are parameters to be adjusted to fit the lattice data. While the existing lattice calculations of $\mu^{p(n)}$ are all quenched, model calculations suggest that at the large quark masses where these calculations were performed the errors should not be large. The effects of unquenching grow dramatically as m_π approaches zero. Figure 6.8 shows the extrapolation of the proton and neutron magnetic moment data as a function of m_π, using the simple phenomenological form given in Eq. (6.145).

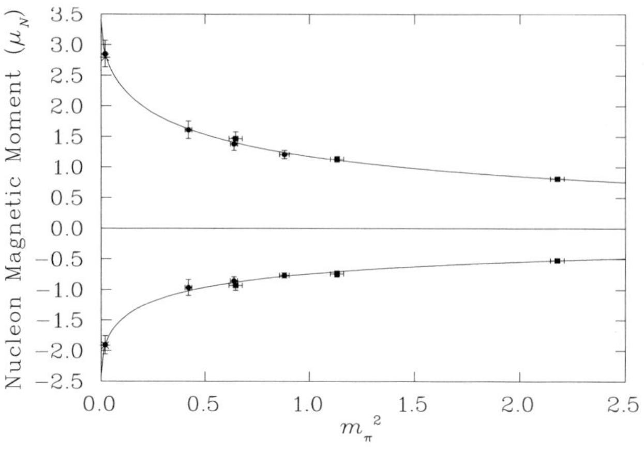

Figure 6.8: Extrapolation to the chiral limit (from Ref.[LLT 99]) of lattice QCD magnetic moments calculated in Refs. [LWD 91] and [WDL 92] for the proton (upper) and neutron (lower). The curves illustrate a two parameter fit to the lattice data using Eq. (6.145), in which the LNA chiral coefficient (i.e. the coefficient of m_π) is taken from chiral perturbation theory. The experimentally measured moments are indicated by asterisks.

The nucleon magnetic moments at the physical pion mass obtained from this extrapolation were

$$\mu_p = 2.85(22)\,\mu_N \quad \text{and} \quad \mu_n = -1.90(15)\,\mu_N, \tag{6.146}$$

which agree surprisingly well with the experimental values, 2.793 and -1.913 μ_N, respectively. It is essential to follow up this result with calculations with dynamical fermions at light quark masses, but this is computationally extremely demanding. For the present, the values given in Eq. (6.146) are the most accurate theoretical estimates of the nucleon magnetic moments within QCD.

6.4.4 Moments of Structure Functions

Advanced developments in lattice QCD are now aiming at the computation of at least the lower moments of the nucleon structure functions measured in deep inelastic lepton scattering. We recall from Chapters 4 and 5 that the calculation of these structure functions is basically a non-perturbative problem. Perturbative QCD can only describe how a given structure function evolves from one Q^2 scale to another one, provided Q^2 is large enough.

Structure functions are related to nucleon matrix elements of the correlation function of two electromagnetic currents J^μ, so that the principal problem is to compute the Euclidean functional integral

$$\begin{aligned} \mathcal{G}^{\mu\nu}_{JJ}(x,y,z) &= \langle 0|N(x)J^\mu(y)J^\nu(z)\overline{N}(0)|0\rangle \\ &= \frac{\int [dU\, d\overline{\psi}\, d\psi]\, e^{-S}\, N(x)J^\mu(y)J^\nu(z)\overline{N}(0)}{\int [dU\, d\overline{\psi}\, d\psi]\, e^{-S}}. \end{aligned} \tag{6.147}$$

Such highly non-local objects are extremely difficult to handle. Instead, the strategy is to use the operator product expansion, discussed in Chapter 5, which expresses the non-local product of currents as a series of *local* operators, $\mathcal{O}_\alpha$. For example, the type of quark operators appearing in the moments of the unpolarized structure functions F_1, F_2 are (see the discussion around Eq. (5.123)):

$$\mathcal{O}_{\mu_1\cdots\mu_n} = \overline{\psi}\gamma_{\mu_1}\overleftrightarrow{D}_{\mu_2}\cdots\overleftrightarrow{D}_{\mu_n}\psi, \tag{6.148}$$

where D_μ is the gauge covariant derivative. Similarly, for the polarized structure functions g_1 and g_2 we are dealing with local operators of the type

$$\mathcal{O}^5_{\mu_1\cdots\mu_n} = \overline{\psi}\gamma_5\gamma_{\mu_1}\overleftrightarrow{D}_{\mu_2}\cdots\overleftrightarrow{D}_{\mu_n}\psi. \tag{6.149}$$

The problem is therefore reduced to the computation of a set of three-point correlation functions. One can fix the operators at a space point on the lattice and write schematically, displaying only the Euclidean time variable:

$$\begin{aligned}\langle 0|N(\tau_1)\mathcal{O}_\alpha(\tau_2)\overline{N}(0)|0\rangle &= \\ &\sum_{n,m}\langle 0|N|n\rangle e^{-E_n(\tau_1-\tau_2)}\langle n|\mathcal{O}_\alpha|m\rangle e^{-E_m\tau_2}\langle m|\overline{N}|0\rangle,\end{aligned} \tag{6.150}$$

where complete sets of states have been inserted. Taking the limit, $\tau_1 \gg \tau_2 \gg a$, along the Euclidean time axis, only the lowest mass (nucleon) states survive. Dividing by $\langle 0|N(\tau_1)\overline{N}(0)|0\rangle$ one then focuses directly on the relevant nucleon matrix elements $\langle N|\mathcal{O}_\alpha|N\rangle$.

Consider first the unpolarized structure functions (Eq. (4.7)) $F_1(x, Q^2)$ and $F_2(x, Q^2)$. Henceforth x refers to the Bjorken variable, $x = Q^2/2M\nu$. Using the operator product expansion, moments of these structure function are given by

$$\langle x^{n-1}\rangle_1 = \int_0^1 dx\, x^{n-1}F_1(x, Q^2) = \frac{1}{2}\sum_\alpha C^{(\alpha)}_{1,n}(Q^2)\, v^{(\alpha)}_n, \tag{6.151}$$

$$\langle x^{n-2}\rangle_2 = \int_0^1 dx\, x^{n-2}F_2(x, Q^2) = \sum_\alpha C^{(\alpha)}_{2,n}(Q^2)\, v^{(\alpha)}_n, \tag{6.152}$$

where the Q^2 dependent Wilson coefficients, $C^{(\alpha)}_n$, are determined by perturbative QCD at some renormalization scale which must be properly matched to the scale introduced by the lattice spacing, a. The task is then to compute the matrix elements $v^{(\alpha)}_n$ of the appropriate local operators $\mathcal{O}_\alpha$. Their interpretation in parton language is as follows. We have

$$v^{(q)}_n = \langle x^{n-1}\rangle^{(q)} = \int_0^1 dx\, x^{n-1}[q_\uparrow(x) + q_\downarrow(x)] \tag{6.153}$$

for the quark distributions of flavor $q = u, d$ and spin "up" or "down", where the lattice simulation is performed in the quenched approximation. The gluonic contribution is

$$v^{(g)}_n = \langle x^{n-1}\rangle^{(g)} = \int_0^1 dx\, x^{n-1}g(x), \tag{6.154}$$

where $g(x)$ is the gluon distribution function.

The moments of the polarized structure function g_1 are analogous to $v_n^{(q)}$, except that the sum of spin up and spin down distribution functions is now replaced by their difference. The moments of g_2 involve additional "twist-3" terms which, as we saw in Section 4.6.4, do not have an interpretation within the parton model.

The actual calculations have been done on large (up to $24^3 \times 32$) lattices, with a spacing $a \simeq 0.1$ fm. Several quark masses in the range between 30 and 200 MeV were used, and the results shown in Tables 6.1, 6.2 and 6.3 were obtained by naive, linear extrapolation down to zero quark mass.

Table 6.1: Moments of unpolarized structure functions from lattice QCD extrapolated to the chiral limit [Gö 97]. The lattice values are compared with results from best fits to empirical valence quark distributions functions [Lai 95].

Moment	Lattice (quenched, $\mu^2 \approx 5$ GeV^2)	Experiment ($\mu^2 = 4$GeV^2)
$\langle x\rangle^{(u)}$	0.410(34)	0.284
$\langle x\rangle^{(d)}$	0.180(16)	0.102
$\langle x^2\rangle^{(u)}$	0.108(16)	0.083
$\langle x^2\rangle^{(d)}$	0.036(8)	0.025
$\langle x^3\rangle^{(u)}$	0.02(1)	0.032
$\langle x^3\rangle^{(d)}$	0.000(6)	0.008
$\langle x\rangle^{(g)}$	0.53(23)	0.441

Let us have a first look at the moments of the unpolarized structure functions (Table 6.1). Since they were obtained in the quenched approximation, one would expect the results to be most meaningful at the level of the valence quark distributions. One observes that the first moments on the lattice are too large as compared with the valence distributions extracted from fits to data. This is not surprising, however. In the quenched approximation one expects the structure of the proton to be dominated by valence quarks and gluons, with the sea being underestimated. The calculated momentum sum rule:

$$\langle x\rangle^{(u)} + \langle x\rangle^{(d)} + \langle x\rangle^{(g)} = 1.12 \pm 0.23 \tag{6.155}$$

is consistent with the general requirement that the quark and gluon contributions must add up to one. In the real proton there will be additional sea quarks which contribute to the momentum sum. One might therefore expect that the valence quark contribution would be smaller, and this is indeed seen in the table. The non-singlet moments, $\langle x\rangle^{(u)} - \langle x\rangle^{(d)}$, should be less sensitive to sea quark effects. Here the lattice and empirical values are closer, as expected.

The lattice moments of the polarized structure functions (Table 6.2) agree surprisingly well with experiment. The spin fractions Δu and Δd are not far from the empirical valence quark values, although the axial vector coupling constant, $g_A = \Delta u - \Delta d$, comes out too low. The second moments compare reasonably well, within errors, with a determination from the SMC measurement of the charge asymmetry of fast pions in polarized muon-nucleon collisions.

Table 6.2: Moments of polarized proton structure functions from lattice QCD extrapolated to the chiral limit [Gö 97]. The lattice values are compared with experiment [Ab 96] and with fits to the valence distributions [GS 96].

Moment	Lattice (quenched, $\mu^2 \approx 5\text{GeV}^2$)	Experiment ($\mu^2 = 4\text{GeV}^2$)
Δu	$0.841(52)$	0.823 (val.)
Δd	$-0.245(15)$	-0.303 (val.)
g_A	$1.086(67)$	$1.267(4)$
$\int_0^1 dx(g_1^p - g_1^n)$	$0.176(14)$	$0.163(17)$
$\int_0^1 dx x^2 g_1^p$	$0.0150(32)$	$0.012(1)$
$\int_0^1 dx x^2 g_1^n$	$-0.0012(20)$	$-0.004(3)$
$\int_0^1 dx x^2 g_2^p$	$-0.0261(38)$	$-0.006(2)$
$\int_0^1 dx x^2 g_2^n$	$-0.0004(22)$	$0.005(8)$

Higher moments are small and should not be very much influenced by sea quarks which have their effects concentrated at small Bjorken x.

6.4.5 Nucleon Spin Structure

Even within the quenched approximation it has been possible to estimate the effect of a quark loop when the external current couples directly to it, rather than to the valence quarks. This is particularly interesting in the context of QCD investigations of the spin structure of the nucleon, where we already know from our discussion of the axial anomaly that quark loops (i.e., vacuum polarization effects in the presence of gluon background fields) play an essential role in the flavor singlet case.

The quantities of interest in this context are the nucleon matrix elements of the diagonal flavor-$SU(3)$ axial currents

$$A_\mu^a = \overline{\psi}\gamma_\mu\gamma_5\frac{\lambda^a}{2}\psi \tag{6.156}$$

with $a = 3, 8$, together with the flavor singlet current

$$A_\mu^0 = \overline{\psi}\gamma_\mu\gamma_5\psi. \tag{6.157}$$

We recall that the individual flavor components of these currents are related to the "spin fractions" carried by quarks in the nucleon by

$$\langle N_s|\overline{q}\gamma_\mu\gamma_5 q|N_s\rangle = s_\mu \Delta q \qquad (q = u, d, s). \tag{6.158}$$

The axial vector coupling constants are

$$g_A^0 = \Delta u + \Delta d + \Delta s, \tag{6.159}$$
$$g_A^3 = \Delta u - \Delta d \equiv g_A, \tag{6.160}$$
$$g_A^8 = \Delta u + \Delta d - 2\Delta s. \tag{6.161}$$

Lattice explorations of these quantities have been performed by computing the matrix elements

$$G_A(\tau_1, \tau_2) = \sum_{\vec{x},\vec{y}} \langle 0|N(\tau_1, \vec{x}) A_\mu(\tau_2, \vec{y}) \overline{N}(0)|0\rangle, \tag{6.162}$$

and taking again the limit $\tau_1 \gg \tau_2 \gg a$. In the quenched approximation the physics content of such a matrix element is sketched in Fig. 6.9a, which shows the axial current attached to a valence quark line moving in the gluonic background field. We denote the results obtained in the quenched limit by Δu_0 and Δd_0. Obviously, $\Delta s_0 = 0$ since there are no strange sea quark contributions in that limit.

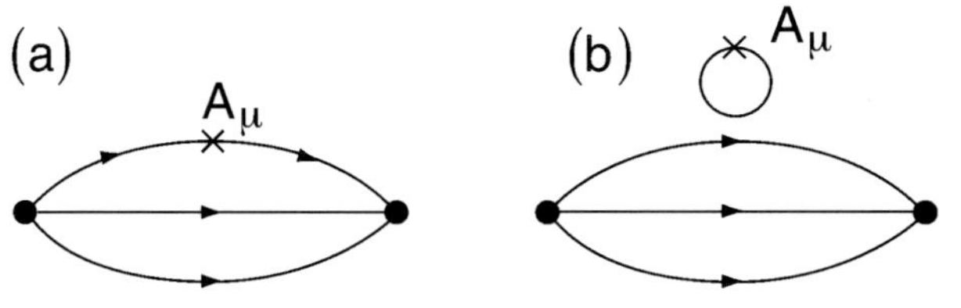

Figure 6.9: Connected (a) and disconnected (b) diagrams contributing to the axial current matrix elements. Diagram (a) is computed in the quenched approximation on the lattice. Diagram (b) represents additional vacuum polarization contributions involving a sea quark loop.

The actual computations were performed on a $16^3 \times 24$ lattice, with the scale $a \simeq 0.1$ fm set by the nucleon mass. Several values of the quark mass m_q were used as input, and the results extrapolated to the chiral limit, again using a naive linear extrapolation.

For heavy quarks one expects to approach the results of the non-relativistic quark model (see Section 8.2.2), $g_A^0 = 1$ and $g_A = 5/3$. Fig. 6.10 shows that the expected ratio $g_A^0/g_A = 0.6$ is indeed reached on the lattice for large quark mass. It tends to decrease toward smaller quark mass, showing the increasing influence of relativistic effects from the lower components of the quark spinors. The lattice result for the flavor singlet axial constant obtained in the quenched approximation is

$$(g_A^0)_{\text{quenched}} = 0.61 \pm 0.13. \tag{6.163}$$

At the same time $(g_A^3)_{\text{quenched}}$ is reduced from $5/3$ to about 1.2.

The next step is to go beyond the quenched approximation and incorporate vacuum polarization loops, as illustrated in the disconnected diagram of Fig. 6.9 (b)[7]. Such vacuum polarization effects, in the presence of the strongly interacting gluonic background, cause

[7] Of course, this sort of skeleton diagram is not disconnected in the sense of quantum field theory, in that infinitely many non-perturbative gluon exchanges are included in the lattice calculation.

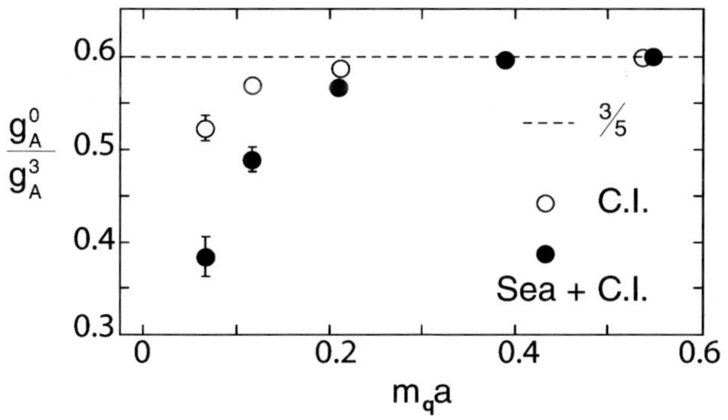

Figure 6.10: Ratio of flavor singlet and isovector axial constants on the lattice [Liu 99], plotted as a function of the quark mass given in lattice units. Here "C.I." corresponds to Fig. 6.9 (a) only, whereas "Sea + C.I." includes Fig. 6.9 (b) as well.

screening and depolarization corrections in the axial constants, especially in g_A^0. In the lattice simulation of [Liu 99], it turns out that the contributions from quark loops (pol) are flavor independent, with the result

$$\Delta u_{\text{pol}} = \Delta d_{\text{pol}} \simeq \Delta s_{\text{pol}} = -0.12 \pm 0.01. \tag{6.164}$$

While these contributions do not affect flavor non-singlet quantities like g_A^3 and g_A^8, the singlet axial charge decreases to

$$g_A^0 = \Delta u_{\text{C.I.}} + \Delta d_{\text{C.I.}} + 3\,\Delta u_{\text{pol}} = 0.25 \pm 0.12. \tag{6.165}$$

This is remarkably consistent with the empirical value deduced from polarized deep-inelastic scattering data (see Section 4.6).

Table 6.3: Lattice QCD results [Liu 99] for axial coupling constants and quark spin content of the proton, in comparison with experimental data [SMC 97] and with the non-relativistic quark model.

	Lattice	Experiments	quark model
$g_A^0 = \Delta u + \Delta d + \Delta s$	$0.25(12)$	$0.28(16)$	1
$g_A^3 = \Delta u - \Delta d$	$1.20(10)$	$1.267(4)$	$5/3$
$g_A^8 = \Delta u + \Delta d - 2\,\Delta s$	$0.61(13)$	$0.58(3)$	1
Δu	$0.79(11)$	$0.82(5)$	$4/3$
Δd	$-0.42(11)$	$-0.44(5)$	$-1/3$
Δs	$-0.12(1)$	$-0.10(5)$	0

The complete set of lattice results on this issue is summarized in Table 6.3. It demonstrates the importance of effects beyond the quenched approximation in the analysis of some, but not all aspects of nucleon spin structure. The present state of investigations is, however, still incomplete in the sense that quark loops which do not couple to the external current are being omitted and the treatment of the extrapolation in quark mass is not consistent with the chiral properties of QCD.

In closing, one should note that the results presented in this and the preceding subsections are to be seen as first major steps in applying lattice QCD to compute properties of the nucleon. This field is expected to make further substantial progress as the technical developments in supercomputer capacity and further advances in improved lattice actions take place.

7 Chiral Symmetry and Nucleon Structure

At low energies the relevant, effective degrees of freedom in QCD are no longer the elementary quarks and gluons, but composite hadrons. The task is then to identify the active hadronic degrees of freedom and to set up the rules for their dynamics. This is done by constructing an effective Lagrangian which replaces the original QCD Lagrangian in the low-energy, long wavelength limit, but which systematically incorporates the symmetries and symmetry breaking patterns of QCD. An important symmetry for low energy QCD is chiral symmetry. In this Chapter we investigate its impact on the long distance structure of the nucleon.

Spontaneous chiral symmetry breaking leads to a characteristic gap, $\Delta \sim 4\pi f_\pi \sim 1$ GeV, in the hadron spectrum – as discussed in Section 6.1.3. The strong interaction physics at energies small compared to this gap should be governed by the softest excitations of the QCD ground state, namely by the pseudoscalar Goldstone bosons. In fact the energy scale introduced by Δ offers a natural separation between "light" and "heavy" (or, correspondingly, "fast" and "slow") degrees of freedom. The active, light ones are primarily the pions, but for some purposes one may also need to include the kaons and the η meson. The passive, heavier ones are the baryons and the mesons with masses comparable to the gap Δ or larger.

7.1 Chiral Effective Field Theory

The basic idea of an effective field theory is to treat the active, light particles as collective degrees of freedom, while the heavy particles are frozen and reduced to (almost) static sources. The dynamics are described by an effective Lagrangian, $\mathcal{L}_{\text{eff}}$, which incorporates all relevant symmetries of the underlying fundamental theory.

We split the Lagrangian into a symmetric part, $\mathcal{L}_0$, and a symmetry breaking part, $\mathcal{L}_{\text{sb}}$. In QCD with u, d and s quarks, the conditions to be imposed on

$$\mathcal{L}_{\text{eff}} = \mathcal{L}_0 + \mathcal{L}_{\text{sb}} \tag{7.1}$$

are the following:

(a) $\mathcal{L}_0$ is symmetric under the chiral flavor group $SU(3)_R \times SU(3)_L$;

(b) chiral symmetry is spontaneously broken: the ground state is symmetric only under $SU(3)_V$;

(c) the massless particles of the theory are the eight Goldstone bosons $\pi_a (a = 1, \ldots, 8)$ associated with (b);

(d) $\mathcal{L}_0$ has eight conserved axial currents with

$$\langle 0|A_a^\mu(0)|\pi_b(p)\rangle = ip^\mu f\delta_{ab},$$

where f is the pseudoscalar decay constant;

(e) the symmetry breaking term $\mathcal{L}_{\text{sb}}$ has the same matrix elements as the quark mass term in the original QCD Lagrangian;

A Lagrangian, $\mathcal{L}_{\text{eff}}$, with these properties, which makes use of the relevant degrees of freedom offered by the spectrum of QCD, by construction should produce the same "low-energy" expansion as QCD itself [We 79, GL 94, Ec 95]. A low-energy expansion is an ordering in powers of energies and momenta (generically denoted by p) of the interacting particles, such that any matrix element or scattering amplitude derived from $\mathcal{L}_{\text{eff}}$ is organized as a Taylor series, $\Sigma c_n p^n$. The systematic framework for this is called chiral perturbation theory. The usefulness of such an expansion derives from two features:

(i) The Goldstone bosons interact weakly. In fact, if $|\phi\rangle = Q^A|0\rangle$ describes a massless Goldstone boson satisfying the equation $H|\phi\rangle = 0$, then a state with n Goldstone bosons, $|\phi^n\rangle = (Q^A)^n|0\rangle$, still represents a massless state, since the axial charges Q^A (see Sections 6.1.2, 6.1.3) all commute with the full Hamiltonian H, including interactions. Therefore, interactions involving any number of Goldstone bosons must vanish in the chiral limit.

(ii) The symmetry breaking mass term is small, so that it can be handled perturbatively, together with the power series in momentum.

On the negative side, the convergence radius of chiral perturbation theory is often quite limited. Nevertheless, within that limit, rigorous statements can be made. Note that, as a consequence of confinement in QCD, the quarks and gluons do not show up as explicit degrees of freedom in $\mathcal{L}_{\text{eff}}$. They become directly visible only at energies and momenta much larger than the gap Δ, as in deep inelastic lepton scattering.

We can now proceed to construct $\mathcal{L}_{\text{eff}}$. The presentation will be in the form of a short summary. More detailed developments are given, for example, in the textbooks [We 96, Do 92]. We first introduce the pure meson sector (with baryon number $B = 0$). Then we include baryons (the $B = 1$ sector) and discuss applications to nucleon properties. In particular, we develop some key low energy theorems based on chiral symmetry as they relate to nucleon structure. Examples of chiral perturbation theory applied to low energy processes will be given, such as π^0 photoproduction and Compton scattering on protons close to threshold. In most applications we restrict ourselves to the flavor $SU(2)$ sector, but the preparation of the effective Lagrangian will actually be done in $SU(3)$, including strange quarks for completeness. Once strange quarks are included, however, one must be aware of the fact that their mass, m_s, is an order of magnitude larger than $m_u + m_d$, and this can lead to a lack of convergence in the chiral expansion.

7.1.1 Chiral Effective Lagrangian: Meson Sector

In order to construct $\mathcal{L}_{\text{eff}}$ in the meson $(B = 0)$ sector with $N_f = 3$, it is convenient to introduce a representation of the pseudoscalar Goldstone bosons by a 3×3 matrix field

$U(x) \in SU(3)$, with $U^\dagger U = 1$ and $\det U = 1$. This matrix transforms linearly under the chiral rotations $\psi_R \to R\psi_R$, $\psi_L \to L\psi_L$ of the right and lefthanded quarks (see Eq. (6.26)):

$$U \to RUL^\dagger \tag{7.2}$$

The explicit representation of U in terms of the octet of pseudoscalar Goldstone boson fields, $\pi_a(x)$, is as follows:

$$U(x) = \exp\,[i\Phi(x)/f], \tag{7.3}$$

with

$$\Phi \equiv \lambda_a \pi_a = \sqrt{2}\begin{pmatrix} \frac{\pi^0}{\sqrt{2}} + \frac{\eta}{\sqrt{6}} & \pi^+ & K^+ \\ \pi^- & -\frac{\pi^0}{\sqrt{2}} + \frac{\eta}{\sqrt{6}} & K^0 \\ K^- & \bar{K}^0 & -\frac{2\,\eta}{\sqrt{6}} \end{pmatrix}. \tag{7.4}$$

Here the eta meson field, η, is the octet (η_8) component of the η-field, not the observed η meson.

The QCD Lagrangian is now replaced by an effective Lagrangian which involves the field $U(x)$ and its derivatives:

$$\mathcal{L}_{\text{QCD}} \to \mathcal{L}_{\text{eff}}(U, \partial^\mu U, \ldots). \tag{7.5}$$

At low energy, an expansion in powers of the meson momenta is equivalent to an expansion of $\mathcal{L}_{\text{eff}}$ in powers of the derivatives $\partial^\mu U$. Lorentz invariance restricts this series to terms with even numbers of derivatives. We write

$$\mathcal{L}_{\text{eff}} = \mathcal{L}^{(2)}_{\text{eff}} + \mathcal{L}^{(4)}_{\text{eff}} + \cdots \quad , \tag{7.6}$$

where the leading term, the so-called non-linear sigma model, involves two derivatives:

$$\mathcal{L}^{(2)}_{\text{eff}} = \frac{f^2}{4}\,\text{Tr}[\partial^\mu U^\dagger \partial_\mu U]. \tag{7.7}$$

It is completely determined by the pion decay constant f (evaluated at the chiral limit).

At this point, let us make a short digression and construct vector and axial vector currents as symmetry currents from $\mathcal{L}^{(2)}_{\text{eff}}$, using Noether's theorem. Recall the infinitesimal transformations which generate $U \to RUL^\dagger$, with $R = exp[i\theta^a_R\lambda_a/2] \simeq 1 + i\theta^a_R\lambda_a/2$, and L accordingly, replacing θ_R by θ_L. Vector and axial transformations are then generated by $\theta_{V,A} = (\theta_R \pm \theta_L)/2$. Under an infinitesimal vector transformation, $U \to U + i\theta^a_V[\lambda_a, U]$, while an infinitesimal axial vector transformation yields $U \to U + i\theta^a_A\{\lambda_a, U\}$, and we can summarize both operations formally by writing $U \to U + i\theta^a\mathcal{F}_a(U)$. Now, Noether's theorem says that the symmetry currents associated with these global transformations are constructed according to $J^\mu_a = -i[\partial\mathcal{L}_{\text{eff}}/\partial(\partial_\mu U)]\mathcal{F}_a(U)$, where this relation is generically understood to be applied for each component of the matrix field U. After a few algebraic steps, the resulting vector and axial vector currents are:

$$\begin{aligned} V^\mu_a &= -\frac{if^2}{4}\,\text{Tr}\left(\partial^\mu U^\dagger[\lambda_a, U]\right) \\ A^\mu_a &= -\frac{if^2}{4}\,\text{Tr}\left(\partial^\mu U^\dagger\{\lambda_a, U\}\right). \end{aligned} \tag{7.8}$$

At order p^4, symmetries permit three independent terms:

$$\mathcal{L}_{\text{eff}}^{(4)} = L_1 \left(\text{Tr}[\partial^\mu U^\dagger \partial_\mu U]\right)^2 + L_2 \,\text{Tr}[\partial^\mu U^\dagger \partial^\nu U]\,\text{Tr}[\partial_\mu U^\dagger \partial_\nu U] + L_3 \,\text{Tr}(\partial^\mu U^\dagger \partial_\mu U \partial^\nu U^\dagger \partial_\nu U), \tag{7.9}$$

which introduce three additional, effective coupling constants, L_1, L_2 and L_3.

We see that the low-energy expansion of $\mathcal{L}_{\text{eff}}$ has the form of a generalized Taylor series, with coefficients $f^2, L_1, L_2, \ldots$ that need to be determined by experiment. This expansion represents QCD, in the sense that it includes all possible terms permitted by symmetries [1].

Explicit chiral symmetry breaking introduces additional terms proportional to powers of the quark mass matrix $\mathcal{M}$ of Eq. (6.16). We already know from the discussion of the GOR relation, Eq. (6.54), that the leading (linear) order in $\mathcal{M}$ is related to the square of the pseudoscalar meson masses. The leading symmetry breaking term, linear in the quark masses, is therefore to be added to $\mathcal{L}_{\text{eff}}^{(2)}$ in such a way that it is still even in the Goldstone boson fields:

$$\mathcal{L}_{\text{eff}}^{(2)} = \frac{f^2}{4}\,\text{Tr}[\partial^\mu U^\dagger \partial_\mu U] + \frac{B}{2} f^2 \,\text{Tr}[\mathcal{M}(U + U^\dagger)]. \tag{7.10}$$

The fourth order term, $\mathcal{L}_{\text{eff}}^{(4)}$, also receives symmetry breaking contributions with additional effective coupling constants $L_4 \cdots L_8$.

For the purpose of illustration, let us briefly restrict ourselves to the $SU(2)_R \times SU(2)_L$ case, with

$$U(x) = \exp[i\boldsymbol{\tau} \cdot \boldsymbol{\pi}(x)/f], \tag{7.11}$$

where $\boldsymbol{\pi}$ is the isotriplet pion field. Expanding $\mathcal{L}_{\text{eff}}^{(2)}$ to fourth order in $\boldsymbol{\pi}$ one finds:

$$\begin{aligned}\mathcal{L}_{\text{eff}}^{(2)} &= \frac{1}{2}\partial_\mu \boldsymbol{\pi} \cdot \partial^\mu \boldsymbol{\pi} \\ &+ \frac{1}{6f^2}[(\boldsymbol{\pi} \cdot \partial_\mu \boldsymbol{\pi})(\boldsymbol{\pi} \cdot \partial^\mu \boldsymbol{\pi}) - (\boldsymbol{\pi} \cdot \boldsymbol{\pi})(\partial_\mu \boldsymbol{\pi} \cdot \partial^\mu \boldsymbol{\pi})] \\ &+ (m_u + m_d) B \left[f^2 - \frac{1}{2}\boldsymbol{\pi} \cdot \boldsymbol{\pi} + \frac{1}{24 f^2}(\boldsymbol{\pi} \cdot \boldsymbol{\pi})^2\right] + \mathcal{O}(\boldsymbol{\pi}^6).\end{aligned} \tag{7.12}$$

In the symmetry breaking part, proportional to the sum of quark masses, we have a constant vacuum contribution (with $U = U^\dagger = 1$). This should be compared with the vacuum expectation value of the mass term in the original QCD Lagrangian, $-(m_u\langle \bar{u}u\rangle + m_d\langle \bar{d}d\rangle)$, which gives

$$Bf^2 = -\langle \bar{q}q\rangle, \tag{7.13}$$

where we have introduced $\langle \bar{q}q\rangle = \langle \bar{u}u\rangle = \langle \bar{d}d\rangle$. Furthermore we can identify the pion mass term $(-m_\pi^2/2)\boldsymbol{\pi} \cdot \boldsymbol{\pi}$ and find

$$B = \frac{m_\pi^2}{m_u + m_d} = -\frac{\langle \bar{q}q\rangle}{f^2}, \tag{7.14}$$

[1] One caution, the QCD axial anomaly (discussed in Section 6.1.6) gives rise to an extra fourth order contribution, the Wess-Zumino term.

which is just the GOR relation (6.54). Finally the pionic vector and axial vector currents to leading order are:

$$\begin{aligned} V_a^\mu &= \epsilon_{abc}\pi_b\partial^\mu\pi_c + \cdots \quad , \\ A_a^\mu &= -f\partial^\mu\pi_a + \cdots \quad , \end{aligned} \tag{7.15}$$

with $\partial_\mu V_a^\mu = 0$ and $\partial_\mu A_a^\mu = f m_\pi^2 \pi_a$.

The effective Lagrangian (7.6) describes the low energy interactions of pseudoscalar mesons. It has been applied, in particular, to s-wave $\pi\pi$ and πK scattering. In fact, one of the classic test cases for chiral perturbation theory was the successful prediction of the $\pi\pi$ scattering lengths by Weinberg [We 66].

7.1.2 Chiral Effective Lagrangian including Baryons

We now move to the sector with baryon number $B = 1$ and incorporate, for the moment, only the ground state $SU(3)$ baryon octet. The baryons are represented by a Dirac spinor field $\Psi_B(x)$ which is again organized in the form of a 3×3 matrix:

$$\Psi_B \equiv \frac{1}{\sqrt{2}}\lambda_a\Psi_B^a = \begin{pmatrix} \frac{\Lambda}{\sqrt{6}} + \frac{\Sigma^0}{\sqrt{2}} & \Sigma^+ & p \\ \Sigma^- & \frac{\Lambda}{\sqrt{6}} - \frac{\Sigma^0}{\sqrt{2}} & n \\ \Xi^- & \Xi^0 & -\frac{2}{\sqrt{6}}\Lambda \end{pmatrix} . \tag{7.16}$$

The free field Lagrangian

$$\mathcal{L}_B = \mathrm{Tr}[\bar{\Psi}_B(i\gamma_\mu\partial^\mu - M_0)\Psi_B] \tag{7.17}$$

includes the baryon mass in the chiral $SU(3)$ limit, M_0.

Note that baryons, like their constituent quarks but unlike the pseudoscalar mesons, are supposed to have at least part of their large masses because of the strong scalar mean field provided by the quark condensate $\langle\bar{q}q\rangle$. In the Ioffe formula, Eq. (6.107), derived from a QCD sum rule, such a relationship between the nucleon mass and chiral symmetry breaking is explicit.

We can now construct the effective Lagrangian for interacting systems of pseudoscalar mesons and baryons. The previous pure meson Lagrangian $\mathcal{L}_{\text{eff}}$ will be replaced by $\mathcal{L}_{\text{eff}}(U, \partial^\mu U, \Psi_B, \ldots)$ which also includes the baryon fields. The additional term, denoted by $\mathcal{L}_{\text{eff}}^B$, is expanded again in powers of derivatives (external momenta) and quark masses:

$$\mathcal{L}_{\text{eff}}^B = \mathcal{L}_{\Phi B}^{(1)} + \mathcal{L}_{\Phi B}^{(2)} + \cdots . \tag{7.18}$$

Let us first discuss the leading term $\mathcal{L}_{\Phi B}^{(1)}$. Starting from the free baryon Lagrangian (7.17), we show that the primary modification is the replacement of the ∂^μ term by a chiral covariant derivative which introduces vector current interactions between the baryons and mesons. In addition we expect axial vector couplings.

Basic steps: $SU(2) \times SU(2)$ linear sigma model

In order to motivate and derive those interaction terms it is helpful to make a brief detour to the so-called *linear* sigma model. We do this for the simpler case with two flavors only.

In the linear sigma model, massless fermions have Yukawa couplings with a scalar field σ and a pseudoscalar, isovector field $\boldsymbol{\phi}$. The fermion part of this Lagrangian is

$$\mathcal{L}_F = \bar{\psi}[i\gamma_\mu \partial^\mu - g(\sigma - i\gamma_5\, \boldsymbol{\tau} \cdot \boldsymbol{\phi})]\psi \tag{7.19}$$

where g is a dimensionless coupling constant. Introducing left- and right-handed fermion fields we see that the Yukawa coupling term can be rewritten:

$$g\bar{\psi}(\sigma - i\gamma_5\, \boldsymbol{\tau} \cdot \boldsymbol{\phi})\psi = g[\bar{\psi}_L \Sigma^\dagger \psi_R + \bar{\psi}_R \Sigma \psi_L], \tag{7.20}$$

where

$$\Sigma \equiv \sigma + i\, \boldsymbol{\tau} \cdot \boldsymbol{\phi} \tag{7.21}$$

transforms like $\Sigma \to R \Sigma L^\dagger$ under chiral $SU(2) \times SU(2)$.

We decompose the σ field into its vacuum expectation value $\langle \sigma \rangle$ and a fluctuating part. The fermion mass thus becomes $M_0 = g\langle \sigma \rangle$. Later, when we identify this fermion with the nucleon, we find that $\langle \sigma \rangle = f$, where f is the pseudoscalar decay constant. Let us now make contact with the representation in terms of the chiral field $U = \exp(i\boldsymbol{\tau} \cdot \boldsymbol{\pi}/f)$, the one familiar from the non-linear sigma model discussed previously. We set $\Sigma = \langle \sigma \rangle U$, so that $\sigma = \langle \sigma \rangle \cos(\boldsymbol{\tau} \cdot \boldsymbol{\pi}/f)$ and $\boldsymbol{\tau} \cdot \boldsymbol{\phi} = \langle \sigma \rangle \sin(\boldsymbol{\tau} \cdot \boldsymbol{\pi}/f)$. Then $\mathcal{L}_F$ becomes

$$\mathcal{L}_F = \bar{\psi}_L i\gamma_\mu \partial^\mu \psi_L + \bar{\psi}_R i\gamma_\mu \partial^\mu \psi_R - M_0(\bar{\psi}_L U^\dagger \psi_R + \bar{\psi}_R U \psi_L). \tag{7.22}$$

Let us now change from "bare" to "dressed" baryons – i.e., baryon quasi-particles surrounded with an interacting meson cloud – by a redefinition of their fields. A useful choice turns out to be

$$\Psi_L \equiv \xi \psi_L, \qquad \Psi_R \equiv \xi^\dagger \psi_R, \qquad U = \xi\xi, \tag{7.23}$$

with $\xi = \exp(i\boldsymbol{\tau} \cdot \boldsymbol{\pi}/2f)$. Rewriting $\mathcal{L}_F$ in terms of the dressed fermion fields, $\Psi = \Psi_R + \Psi_L$, one finds

$$\mathcal{L}_F = \bar{\Psi}(i\gamma_\mu \mathcal{D}^\mu - M_0 + \gamma_\mu \gamma_5 a^\mu)\Psi, \tag{7.24}$$

with the chiral covariant derivative

$$\mathcal{D}^\mu = \partial^\mu - i v^\mu \tag{7.25}$$

and the vector and axial vector quantities

$$v^\mu = \frac{i}{2}(\xi^\dagger \partial^\mu \xi + \xi \partial^\mu \xi^\dagger), \tag{7.26}$$

$$a^\mu = \frac{i}{2}(\xi^\dagger \partial^\mu \xi - \xi \partial^\mu \xi^\dagger). \tag{7.27}$$

When expanded to second order in the pion field, these expressions reduce to

$$v^\mu = -\frac{1}{4f^2}\boldsymbol{\tau}\cdot(\boldsymbol{\pi}\times\partial^\mu\boldsymbol{\pi}), \tag{7.28}$$

$$a^\mu = -\frac{1}{2f}\boldsymbol{\tau}\cdot\partial^\mu\boldsymbol{\pi}. \tag{7.29}$$

Note that in the procedure of dressing the fermions by clouds of Goldstone pions, the Yukawa couplings of the "bare" fermions have now turned into vector and pseudovector couplings of the dressed fermions.

Generalization to $SU(3) \times SU(3)$

The generalization to $SU(3)$ is straightforward. We identify the dressed fermion fields Ψ with the baryon matrix field Ψ_B of Eq. (7.16). The only new feature is the appearance of two independent types of axial vector couplings. Altogether, the leading order term of the chiral $SU(3) \times SU(3)$ meson-baryon effective Lagrangian is given by

$$\begin{aligned}\mathcal{L}^{(1)}_{\Phi B} &= \mathrm{Tr}[\bar{\Psi}_B(i\gamma_\mu\mathcal{D}^\mu - M_0)\Psi_B] \\ &\quad + F\,\mathrm{Tr}\left(\bar{\Psi}_B\gamma_\mu\gamma_5[a^\mu,\Psi_B]\right) + D\,\mathrm{Tr}\left(\bar{\Psi}_B\gamma_\mu\gamma_5\{a^\mu,\Psi_B\}\right).\end{aligned} \tag{7.30}$$

(Note that the axial vector quantity a^μ, defined in Eq. (7.33) below, is now a 3×3 matrix, as is the baryon field Ψ_B, so the brackets, [ ] and { } denote commutators and anti-commutators, respectively.) The chiral covariant derivative is

$$\mathcal{D}^\mu\Psi_B = \partial^\mu\Psi_B - i[v^\mu,\Psi_B] \tag{7.31}$$

with

$$v^\mu = \frac{i}{8f^2}[\Phi,\partial^\mu\Phi] + \mathcal{O}(\Phi^4). \tag{7.32}$$

The axial vector terms involve the matrix field

$$a^\mu = -\frac{1}{2f}\,\partial^\mu\Phi + \mathcal{O}(\Phi^3). \tag{7.33}$$

We have given only the leading terms in the expansion of the vector and axial vector currents, v^μ and a^μ, which are the ones relevant for our applications. The $SU(3)$ axial vector coupling constants are determined by neutron and hyperon beta decay. One finds

$$F \simeq 0.51, \qquad D \simeq 0.76, \tag{7.34}$$

and the axial vector coupling constant of the nucleon is

$$g_A = F + D \simeq 1.27. \tag{7.35}$$

Actually, these are the physical constants, whereas the Lagrangian (7.30) involves those constants in the chiral limit. For both g_A and the baryon mass, the differences are of the order of the quark mass: $M = M_0(1 + \mathcal{O}(m_q))$ and $g_A = g_A^0(1 + \mathcal{O}(m_q))$.

Electromagnetic Interactions

The inclusion of electromagnetic couplings is straightforward: wherever $\partial^\mu \Phi$ or $\partial^\mu \Psi_B$ appears, it is replaced by the gauge covariant derivative which involves the photon field, $\mathcal{A}^\mu$. One finds

$$\mathcal{D}^\mu \Psi_B = \partial^\mu \Psi_B - ie\mathcal{A}^\mu [Q, \Psi_B] + \frac{1}{8f^2}\left[[\Phi, \partial^\mu \Phi], \Psi_B\right] + \cdots \tag{7.36}$$

and

$$a^\mu = -\frac{1}{2f}\partial^\mu \Phi + \frac{ie}{2f}\mathcal{A}^\mu [Q, \Phi] + \cdots , \tag{7.37}$$

with the quark charge operator

$$Q = \frac{1}{3}\begin{pmatrix} 2 & 0 & 0 \\ 0 & -1 & 0 \\ 0 & 0 & -1 \end{pmatrix}. \tag{7.38}$$

Next-to-Leading Order

At next-to-leading order, the quark mass terms enter. They are conveniently written using the mass matrix

$$\chi = \begin{pmatrix} m_\pi^2 & 0 & 0 \\ 0 & m_\pi^2 & 0 \\ 0 & 0 & 2m_K^2 - m_\pi^2 \end{pmatrix} = -\frac{2\langle \bar{q}q \rangle}{f^2}\begin{pmatrix} \bar{m} & 0 & 0 \\ 0 & \bar{m} & 0 \\ 0 & 0 & m_s \end{pmatrix}. \tag{7.39}$$

Here we set $m_u = m_d = \bar{m}$ for simplicity. Two symmetry breaking effects occur as a consequence of introducing χ. First, the baryon masses are shifted and split. Secondly, the meson-baryon interactions receive corrections from the non-zero quark masses. Both effects are generated by the symmetry breaking contributions to $\mathcal{L}^{(2)}_{\Phi B}$:

$$\Delta \mathcal{L}^{(2)}_{\Phi B} = b_0 \operatorname{Tr}(\bar{\Psi}_B \Psi_B) \operatorname{Tr} \hat{\chi} + b_F \operatorname{Tr}(\bar{\Psi}_B [\hat{\chi}, \Psi_B]) + b_D \operatorname{Tr}(\bar{\Psi}_B \{\hat{\chi}, \Psi_B\}) \tag{7.40}$$

with

$$\hat{\chi} = 2\chi - \frac{1}{4f^2}\{\Phi, \{\Phi, \chi\}\} + \mathcal{O}(\Phi^4). \tag{7.41}$$

The complete $\mathcal{L}^{(2)}_{\Phi B}$ includes all possible terms of order $(\partial\Phi)^2$ in addition. They can be found in the literature [Kr 90, Mu 97] and we do not list them here. For a consistent power counting in the chiral low energy expansion, it is useful at this order to shift the baryon mass scale to "remove" M_0, so that the chiral power counting begins with the kinetic energy term, $\vec{p}^{\,2}/2M_0$. This is systematically done in the so-called heavy baryon formalism, an elegant Lorentz covariant way to describe non-relativistic reductions (otherwise known as the Foldy-Wouthuysen transformation). In essence one writes $\Psi_B(x) = e^{-iM_0 t} B(x)$ and works with the reduced baryon fields $B(x)$. Detailed presentations of this scheme are given in [JM 92].

7.2 Baryon Masses and Sigma Terms

At tree level the parameters of the symmetry breaking part, $\Delta\mathcal{L}^{(2)}_{\Phi B}$, of the meson-baryon effective Lagrangian are fixed by using relations for mass splittings in the baryon octet. For example, to first order in the symmetry breaking terms, one finds:

$$M_\Sigma - M_\Lambda = \frac{16}{3}\, b_D(m_K^2 - m_\pi^2), \tag{7.42}$$

$$M_\Xi - M_N = 8\, b_F(m_\pi^2 - m_K^2), \tag{7.43}$$

and using the empirical values, $M_\Sigma - M_\Lambda = 77.5$ MeV and $M_\Xi - M_N = 379.2$ MeV (averaged over isospin multiplets), this yields:

$$\begin{aligned} b_D &\simeq 0.07\ \mathrm{GeV}^{-1}, \\ b_F &\simeq -0.21\ \mathrm{GeV}^{-1}. \end{aligned} \tag{7.44}$$

Since the $SU(3)$ symmetry breaking pattern generated by the effective Lagrangian is exactly that of QCD, it is not surprising that the Gell-Mann-Okubo formula

$$\begin{aligned} M_\Sigma - M_N &= \frac{1}{2}\,(M_\Xi - M_N) + \frac{3}{4}\,(M_\Sigma - M_\Lambda) \\ &= 4\,(b_D - b_F)\,(m_K^2 - m_\pi^2), \end{aligned} \tag{7.45}$$

is a natural outcome of this approach. This relation is remarkably accurate, agreeing with the empirical masses at the 3 % level. Unfortunately, very few relations based on first order $SU(3)$ symmetry breaking work this well.

7.2.1 Pion-Nucleon Sigma Term

Let us now return to our object of primary interest, the nucleon itself, and define a quantity σ_N, the so-called pion-nucleon sigma term:

$$\sigma_N = M_N - M_0 = \frac{\langle N|\bar{m}(\bar{u}u + \bar{d}d)|N\rangle}{2M_N} = \bar{m}\,\frac{\partial M_N}{\partial \bar{m}}. \tag{7.46}$$

It measures the contribution of the non-vanishing quark masses, $\bar{m} = (m_u + m_d)/2$, to the nucleon mass M_N.

The sigma term, σ_N, can be deduced from an extrapolation of low-energy pion-nucleon scattering data, as described in the next subsection. Its empirical value is [GL 91]:

$$\sigma_N = (45 \pm 8)\ \mathrm{MeV}. \tag{7.47}$$

In terms of the tree level, chiral effective Lagrangian, σ_N is determined from $\Delta\mathcal{L}^{(2)}_{\Phi B}$,Eq. (7.40), to be:

$$\sigma_N = -2m_\pi^2\,(2b_0 + b_F + b_D), \tag{7.48}$$

so that one obtains a constraint on b_0. Using b_F and b_D from Eqs. (7.44) we find:

$$b_0 = -(0.5 \pm 0.1)\ \mathrm{GeV}^{-1}. \tag{7.49}$$

A few words of interpretation may be useful at this point. At low energy, the nucleon can be viewed as a system of three constituent quarks surrounded by an interacting cloud of pions. This system represents a localized distribution of charge and baryon number embedded in the QCD vacuum. It also has a localized scalar density with a corresponding form factor $\sigma_N(t)$ given by

$$\bar{u}_N(P')u_N(P)\sigma_N(t) = \langle N(P')|\bar{m}(\bar{u}u + \bar{d}d)|N(P)\rangle, \tag{7.50}$$

where $t = (P' - P)^2$ and u_N is the Dirac spinor of the nucleon. The sigma term is identified with $\sigma_N(t = 0)$. In the Breit frame, defined by $\vec{P}\,' = -\vec{P}$, the Fourier transform of $\sigma_N(t = -4\vec{P}^{\,2})$ can be interpreted as a scalar density distribution in $\boldsymbol{r}$-space, apart from a factor $\bar{m}$. The sigma term, σ_N, is then just the volume integral of this scalar density (relative to the scalar density or quark condensate of the non-perturbative QCD vacuum), multiplied by the quark mass $\bar{m}$.

7.2.2 Empirical Sigma Term and Scalar Form Factor of the Nucleon

Here we briefly summarize the steps that lead to the empirical value of the nucleon sigma term, given in Eq. (7.47), from an extrapolation of low energy pion-nucleon scattering data. The on-shell scattering amplitude for $\pi_a(q) + N(P) \to \pi_b(q') + N'(P')$, where q, q' and P, P' are the pion and nucleon four-momenta, has the general form [Hö 83]:

$$\begin{aligned} T_{ba} &= \delta_{ba}\bar{u}_{N'}(P')[A^+(s,t) + \gamma^\mu q_\mu B^+(s,t)]u_N(P) \\ &\quad + i\epsilon_{bac}\bar{u}_{N'}(P')\tau_c[A^-(s,t) + \gamma^\mu q_\mu B^-(s,t)]u_N(P), \end{aligned} \tag{7.51}$$

expressed in terms of the isospin-even/-odd invariant amplitudes, $A^\pm$ and $B^\pm$, with the Mandelstam variables $s = (P + q)^2$ and $t = (q - q')^2$. The T-matrix (7.51) can also be expanded as

$$\begin{aligned} T_{ba} &= T_{N,ba} \\ &\quad + \frac{\delta_{ba}}{f_\pi^2}\langle N'(P')|\bar{m}(\bar{u}u + \bar{d}d)|N(P)\rangle\delta_{ab} \\ &\quad + \frac{q_\mu + q'_\mu}{2f_\pi^2} i\epsilon_{bac}\langle N'(P')|\bar{\psi}\,\gamma^\mu \frac{\tau_c}{2}\,\psi|N(P)\rangle + \cdots . \end{aligned} \tag{7.52}$$

This involves matrix elements of the isoscalar-scalar and isovector-vector quark currents, as well as the pseudovector Born (nucleon pole) term, T_N, which arises in second order perturbation theory from $\mathcal{L}^{(1)}$ of Eq. (7.30). The remaining terms are of higher order in the pion momenta. Further analysis reveals that the isospin-even amplitude, taken at certain unphysical values of the kinematical variables, the so-called *Cheng-Dashen* point with $s = M_N^2$ and $t = 2m_\pi^2$, isolates the scalar form factor σ_N of Eq. (7.50) as follows:

$$\sigma_N(t = 2m_\pi^2) = f_\pi^2\, A^+(s = M_N^2, t = 2m_\pi^2) - \frac{g_A^2}{M_N}, \tag{7.53}$$

where the contribution from the pseudovector Born term has been subtracted. Small corrections, which are of order m_π^4 or higher, have been omitted.

The Cheng-Dashen point is not directly accessible in πN scattering, but it can be reached by analytic continuation, using dispersion relation techniques, with the result [Koc 82]:

$$\sigma_N(2m_\pi^2) = (60 \pm 8)\ \text{MeV}. \tag{7.54}$$

The difference, $\sigma_N(t = 2m_\pi^2) - \sigma_N(0)$, reflects the relatively large size of the scalar density of the nucleon. One can write a dispersion relation for σ_N:

$$\sigma_N(t) = \sigma_N(0) + \frac{t}{\pi} \int_{4m_\pi^2}^{\infty} dt' \, \frac{\operatorname{Im} \sigma_N(t')}{t'(t' - t - i\epsilon)}. \tag{7.55}$$

The spectrum represented by $\operatorname{Im} \sigma_N$ turns out to be strongly dominated by the low-mass $\pi\pi$ continuum and reflects the soft, isoscalar pionic cloud surrounding the nucleon. The mean squared scalar radius

$$\langle r_S^2 \rangle = \frac{6}{\pi \sigma_N(0)} \int_{4m_\pi^2}^{\infty} dt \, \frac{\operatorname{Im} \sigma_N(t)}{t^2} \tag{7.56}$$

is large. With the spectrum shown in Fig. 7.1, which uses information about s-wave $\pi\pi$ interactions [GL 91], one finds

$$\langle r_S^2 \rangle^{1/2} \simeq 1.3\ \text{fm}, \tag{7.57}$$

which is much larger than the proton charge radius, $\langle r_E^2 \rangle_p^{1/2} \simeq 0.86$ fm. This implies that $\sigma_N(t)$ has a large slope around $t = 0$. Consequently the difference $\sigma_N(t = 2m_\pi^2) - \sigma_N(0)$ is large, about 15 MeV. This leads to the result, $\sigma_N(0) = (45 \pm 8)$ MeV, quoted earlier.

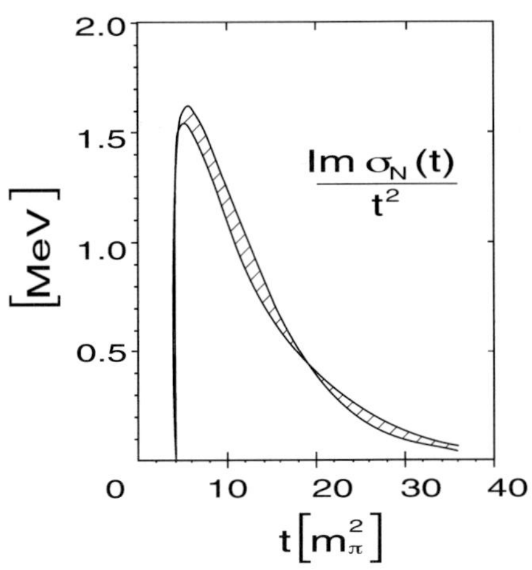

Figure 7.1: Spectrum $\operatorname{Im} \sigma_N(t)/t^2$ (with t in units of m_π^2) of the scalar nucleon form factor (adapted from [GL 91]).

7.2.3 Strange Quarks in the Nucleon

The simplest quark model pictures of the nucleon would suggest that it has no strange quark content. In particular, the expectation value of the scalar density $\bar{s}s$ in the proton would naively be zero. The experimental value of the sigma term has been interpreted as suggesting that this expectation may be incorrect.

The sigma term (7.46) can be rewritten exactly as:

$$\sigma_N = \frac{\langle N|\bar{m}(\bar{u}u + \bar{d}d - 2\,\bar{s}s)|N\rangle}{1-y}, \tag{7.58}$$

where the quantity

$$y = \frac{2\langle N|\bar{s}s|N\rangle}{\langle N|\bar{u}u + \bar{d}d|N\rangle} \tag{7.59}$$

is a measure of the (scalar) strange quark content of the nucleon. Under the assumption that $SU(3)$ symmetry breaking is small, one can use Eq. (7.58) and the octet baryon mass relations to express σ_N, to leading order in the quark masses, as:

$$(1-y)\sigma_N \simeq \frac{\bar{m}}{m_s - \bar{m}}(M_\Xi + M_\Sigma - 2M_N). \tag{7.60}$$

One should note, however, that Eq. (7.60) is based on a first order expansion in m_s about the limit of exact $SU(3)$ symmetry, which is not really reliable for this purpose, so that the conclusions can at most be qualitative. With $m_s \simeq 25\bar{m}$, one finds $(1-y)\sigma_N \simeq 26$ MeV. This value goes up to about (36 ± 7) MeV when one-loop corrections in chiral perturbation theory, involving the octet of pseudoscalar mesons, are included [JRT 94, Bor 97]. Together with the empirical $\sigma_N = (45 \pm 8)$ MeV this implies $y = 0.2 \pm 0.2$. The uncertainties of such estimates are obviously large. While they show a tendency for a non-vanishing admixture of strange quark pairs in the proton wave function, the result is also compatible with zero. To reinforce this remark, we note that a recent calculation of the sigma term from lattice QCD, using the chiral extrapolation described in Section 6.4.2, gave a value between 45 and 55 MeV [LTW 00]. This agrees rather well with the experimental value, even though the lattice data was based on QCD with dynamical u and d quarks (i.e., including the fermion determinant), but no strange quarks at all.

7.3 Chiral Low Energy Theorems

To leading order in the low energy expansion, the chiral effective Lagrangian implies a number of well established current algebra relations, so-called low-energy theorems, for pion-nucleon scattering and pion photoproduction amplitudes. We give here a brief summary of these relations. For simplicity we restrict ourselves to the reduced flavor $SU(2)$ version of the effective Lagrangian which describes interactions of nucleons ($\Psi_N = (p, n)$) and pions ($\boldsymbol{\pi}$). When expanded up to and including terms of order $\boldsymbol{\pi}^2$, and incorporating electromagnetic couplings, this effective Lagrangian becomes

$$\begin{aligned}\mathcal{L}_{\text{eff}}^{N} &= \bar{\Psi}_N(i\gamma_\mu\partial^\mu - M_N)\Psi_N - \frac{g_A}{2f_\pi}\bar{\Psi}_N\gamma_\mu\gamma_5\boldsymbol{\tau}\Psi_N\cdot\partial^\mu\boldsymbol{\pi} \\ &- \frac{1}{4f_\pi^2}\bar{\Psi}_N\gamma_\mu\boldsymbol{\tau}\Psi_N\cdot\boldsymbol{\pi}\times\partial^\mu\boldsymbol{\pi} + \frac{\sigma_N}{2f_\pi^2}\bar{\Psi}_N\Psi_N\boldsymbol{\pi}^2 \\ &+ \frac{e}{2}\bar{\Psi}_N\gamma_\mu(1+\tau_3)\Psi_N A^\mu + \frac{ieg_A}{4f_\pi}\bar{\Psi}_N\gamma_\mu\gamma_5[\tau_3,\boldsymbol{\tau}\cdot\boldsymbol{\pi}]\Psi_N A^\mu + \cdots\end{aligned} \tag{7.61}$$

with $f = f_\pi$ and $M_N = M_0 + \sigma_N$.

7.3.1 Goldberger-Treiman Relation

Computing the single pion emission amplitude $N \to \pi N$ from Eq. (7.61) one immediately obtains the Goldberger-Treiman relation [GT 66]

$$g_{\pi N} = g_A\frac{M_N}{f_\pi} = 12.9, \tag{7.62}$$

which connects the πN pseudoscalar (Yukawa) coupling constant with the axial constants of nucleon and pion. The empirical value, $g_{\pi N} = 13.2 \pm 0.1$ [Ar 94] is within less than 3 % of the righthand side of Eq. (7.62). The agreement is even better if one realizes that the empirical πN coupling constant is obtained at the nucleon pole, $q^2 = m_\pi^2$. Taking the empirical axial form factor as a guide (c.f. Chapter 3), this would reduce the coupling at $q^2 = 0$ (which is what should be compared with Eq. (7.62)) by about 4 %, giving very good agreement.

7.3.2 S-Wave Pion-Nucleon Scattering: Weinberg-Tomozawa Theorem

The T-matrix for pion-nucleon scattering, $\pi_a(q) + N(P) \to \pi_b(q') + N(P')$, with pion four-momenta $q^\mu = (\omega, \vec{q})$ and $q'^\mu = (\omega', \vec{q}\,')$, is decomposed into isospin-even and -odd parts as in Eq. (7.51):

$$T_{ba} = \delta_{ba}\,T^+ + \frac{1}{2}\left[\tau_b, \tau_a\right]T^-. \tag{7.63}$$

The corresponding s-wave scattering lengths are

$$a^\pm = \frac{1}{4\pi}\left(1 + \frac{m_\pi}{M_N}\right)^{-1} T^\pm\big|_{\text{threshold}}. \tag{7.64}$$

Consider now the forward scattering ($q = q'$) amplitude of a virtual pion with energy $\omega = P\cdot q/M_N$ in the laboratory frame and squared (virtual) mass $q^2 = \omega^2 - \vec{q}^{\,2}$. (On mass shell we have $q^2 = m_\pi^2$.) Systematic evaluation of this amplitude using the complete $\mathcal{L}_{\Phi B}^{(1)}$ and $\mathcal{L}_{\Phi B}^{(2)}$ (with proper inclusion of pseudoscalar source terms which also establish PCAC for virtual pions) gives for the isospin-even amplitude:

$$T^+ = T_N^+ + \frac{\sigma_N}{f_\pi^2}\left(\frac{2q^2}{m_\pi^2} - 1\right) + \frac{d}{f_\pi^2}\omega^2 + \frac{d'}{f_\pi^2}\vec{q}^{\,2}. \tag{7.65}$$

Here the terms involving the constants d and d' arise from combinations of order $(\partial\pi)^2$ in the Lagrangian, and

$$T_N^+ = \frac{g_A^2}{f_\pi^2} M_N \frac{q^4}{q^4 - 4M_N^2\omega^2} \tag{7.66}$$

is the contribution from the pseudovector nucleon Born term. The isospin-odd amplitude becomes

$$T^- = \frac{\omega}{2f_\pi^2} \tag{7.67}$$

to this order. This relation, combined with the statement that $T^+ = 0$ to $\mathcal{O}(q)$, is called the Weinberg-Tomozawa low energy theorem [Tom 66, We 66]. At threshold ($\omega = m_\pi, \vec{q} = 0$) this theorem implies that the isovector scattering length

$$\left(1 + \frac{m_\pi}{M_N}\right) a^- = \frac{m_\pi}{8\pi f_\pi^2} \simeq 0.13 \text{ fm} \tag{7.68}$$

vanishes in the chiral (soft-pion) limit $m_\pi \to 0$. On the other hand, the isoscalar scattering length

$$\left(1 + \frac{m_\pi}{M_N}\right) a^+ = \frac{m_\pi^2}{4\pi f_\pi^2}\left(\frac{\sigma_N}{m_\pi^2} + d - \frac{g_A^2}{4M_N}\right) \tag{7.69}$$

is suppressed in comparison with a^- by an additional power of m_π and involves a subtle cancellation between the sigma term σ_N and the parameter d, which must be fixed by comparison with experiment.

The empirical scattering lengths deduced from πN scattering experiments [Koc 86] together with accurate measurements of the shift of the 1s level in the $\pi^- p$ and $\pi^- d$ atoms [Schr 99], follow the expected pattern. We give here the results of the updated analysis of Ref. [ELT 00]:

$$\begin{aligned} a^+ &= (-0.003 \pm 0.002) \text{ fm}, \\ a^- &= (0.128 \pm 0.002) \text{ fm}. \end{aligned} \tag{7.70}$$

Corrections of order m_π^3 to the leading order Tomozawa-Weinberg result (7.63, 7.64) for a^- have been computed using chiral perturbation theory [BKM 95]. They increase the leading term by approximately 15 %, just enough to close the gap between the empirical a^- and the first order prediction.

7.3.3 Adler-Weisberger Sum Rule

When combined with dispersion relations, the chiral low energy theorems yield useful sum rules (see for example [Tr 85]). A famous example, first discussed by Adler [Ad 65] and Weisberger [Wei 65], connects the relevant low-energy parameters of the pion-nucleon system (the pion decay constant f_π and the nucleon axial vector coupling constant g_A) with an integral over the difference of the $\pi^+ p$ and $\pi^- p$ total cross sections, as follows.

Consider the isospin-odd pion-nucleon T-matrix for forward scattering, $T^-(\omega)$. We work in the laboratory frame and denote the pion energy by ω. An examination of the high energy behavior of $T^- = \frac{1}{2}(T_{\pi^- p} - T_{\pi^+ p})$ suggests that this amplitude should behave asymptotically as $T^- \sim \sqrt{\omega}$. In this case $T^-(\omega)/\omega$ satisfies an unsubtracted dispersion relation:

$$\frac{T^-(\omega)}{\omega} = \frac{2g_A^2}{f_\pi^2} \frac{m_\pi^2 M_N^2}{4M_N^2\omega^2 - m_\pi^4} + \frac{2}{\pi}\int_{m_\pi}^{\infty} d\omega' \frac{\mathrm{Im}\, T^-(\omega')}{\omega'^2 - \omega^2 - i\epsilon}, \tag{7.71}$$

where the first term on the right-hand side is the contribution from the nucleon pole piece of the πN amplitude.

Now consider the amplitude at threshold ($\omega = m_\pi$), use the Weinberg-Tomozawa low energy theorem (7.67), and neglect the small correction of order $m_\pi^2/4M_N^2 \ll 1$ in the denominator of the nucleon pole term. This gives

$$\frac{1}{2f_\pi^2} = \frac{g_A^2}{2f_\pi^2} + \frac{2}{\pi}\int_{m_\pi}^{\infty} d\omega \frac{\mathrm{Im}\, T^-(\omega)}{\omega^2 - m_\pi^2}. \tag{7.72}$$

Using the optical theorem to connect $\mathrm{Im}\, T^-$ with the $\pi^\pm p$ total cross sections, the result is the Adler-Weisberger sum rule:

$$g_A^2 = 1 + \frac{2f_\pi^2}{\pi}\int_0^{\infty} \frac{dq}{\omega}\left[\sigma_{\pi^+ p}(\omega) - \sigma_{\pi^- p}(\omega)\right] + \mathcal{O}(m_\pi^2), \tag{7.73}$$

where the integral is now taken over the pion laboratory momentum, $q = \sqrt{\omega^2 - m_\pi^2}$. This remarkable relation is often viewed as an illustration of how the surplus of the axial vector coupling constant beyond its "trivial" value $g_A = 1$ (for a pointlike, structureless nucleon) is tied to pion-nucleon dynamics and spontaneous chiral symmetry breaking. From a more fundamental point of view, the Adler-Weisberger sum rule is a consistency requirement on the internal structure of the nucleon, which not only yields $g_A \neq 1$ but also gives rise to the resonances seen in pion-nucleon scattering. The left and right hand sides of the sum rule turn out to be consistent at the percent level: using an accurate parametrization of the measured πp cross sections [ELT 00] and $f_\pi = 92.4$ MeV, the r.h.s. of Eq. (7.73) becomes 1.60 ± 0.02, to be compared with $g_A^2 = 1.61 \pm 0.01$, as determined from neutron β-decay (see Section 3.3.1).

7.3.4 Charged Pion Photoproduction: Kroll-Ruderman Theorem

The leading electromagnetic couplings of the πN system are displayed in the third line of Eq. (7.61). They lead to detailed predictions for the low energy behavior of pion photoproduction amplitudes.

The low energy $\gamma N \to \pi N$ processes involve electric and magnetic dipole amplitudes. As explained in Section 2.3.2, these are denoted by E_{0+} and $M_{1\pm}$ (the lower index refers to the πN partial wave l, and $\pm$ indicates the πN total angular momentum, $j = l \pm 1/2$). The cross section is

$$\sigma(\gamma N \to \pi N) = \frac{4\pi|\vec{q}|}{\omega}\left(|E_{0+}|^2 + 2|M_{1+}|^2 + |M_{1-}|^2 + \cdots\right), \tag{7.74}$$

where ω is the photon energy and $|\vec{q}\,|$ the momentum of the produced pion in the c.m. frame. For charged pion photoproduction, $\gamma p \to \pi^+ n$, the leading order prediction for the threshold amplitude, based on the chiral Lagrangian, is

$$E_{0+}(\gamma p \to \pi^+ n) = \frac{e g_A \sqrt{2}}{8\pi f_\pi}. \tag{7.75}$$

This is the Kroll-Ruderman theorem [KR 54]. In comparison with data this theorem is verified to within a few percent when including next-to-leading order corrections.

7.3.5 Charged Pion Electroproduction and the Axial Radius of the Nucleon

In Section 3.2 we have already encountered the axial form factor of the nucleon, $G_A(q^2)$. The electroproduction of charged pions from the proton, together with neutrino scattering, is the prime source of information about this form factor. It measures the combined distribution of spin and isospin, closely related to the Gamow-Teller operator $\vec{\sigma}\boldsymbol{\tau}$, in the nucleon. Neutrino scattering has been discussed in Section 3.2.1. When represented by a dipole form,

$$G_A(q^2) = g_A \left(1 - \frac{q^2}{M_A^2}\right)^{-2}, \tag{7.76}$$

the average value of the mass parameter determined from neutrino experiments is [Lie 99]

$$M_A = (1.03 \pm 0.02)\ \text{GeV}, \tag{7.77}$$

(see also Table 3.1 in Section 3.2.1) corresponding to an axial r.m.s. radius

$$\langle r_A^2 \rangle^{1/2} = \frac{2\sqrt{3}}{M_A} \simeq 0.66\ \text{fm}. \tag{7.78}$$

The axial radius also appears in the electric dipole amplitude for $ep \to e'\pi^+ n$ close to threshold. In the strict soft-pion limit ($m_\pi = 0$), this amplitude is subject to a low energy theorem first derived by Nambu, Lurie and Shrauner [Na 62]:

$$E_{0+}(\pi^+ n) = \frac{e g_A \sqrt{2}}{8\pi f_\pi}\left[1 + \frac{q^2}{6}\langle r_A^2 \rangle - \frac{q^2}{2M_N^2}\left(\kappa_n + \frac{1}{4}\right)\right] + \mathcal{O}(q^3), \tag{7.79}$$

where $\kappa_n = -1.913$ is the neutron magnetic moment and q^2 is the squared four-momentum transferred to the nucleon when it undergoes the transition $p \to n + \pi^+$. A detailed measurement of E_{0+} for this process, as a function of q^2, thus permits one to determine the axial radius.

Chiral perturbation theory gives a rigorous account of $\mathcal{O}(m_\pi^2)$ corrections to the soft-pion relation (7.79). When evaluating the corresponding pion loop diagrams, the term involving the axial radius receives an additional contribution [BKM 92] which effectively replaces $\langle r_A^2 \rangle$ by:

$$\langle \tilde{r}_A^2 \rangle = \langle r_A^2 \rangle + \frac{3}{64 f_\pi^2}\left(1 - \frac{12}{\pi^2}\right). \tag{7.80}$$

In summary, one therefore expects that the mean squared axial radius determined in charged pion electroproduction should be slightly smaller (by about 10 %) than that determined in neutrino scattering. The difference is related to pion cloud effects specific to electroproduction. The best fit value of the axial dipole mass parameter deduced from $ep \to e'\pi^+ n$ data, $M_A = (1.07 \pm 0.02)$ GeV [Lie 99], does indeed show such a tendency.

7.3.6 Neutral Pion Photoproduction

The low energy theorem for neutral pion photoproduction, $\gamma p \to \pi^0 p$, has an interesting history. To leading order this process is suppressed by a factor $m_\pi / M_N \simeq 1/7$, the ratio of the electric dipole moments of the $\pi^0 p$ and $\pi^+ n$ systems. Systematic chiral calculations performed to one-loop order lead to the following low-energy theorem [BKM 92a] for the threshold $\gamma p \to \pi^0 p$ amplitude:

$$E_{0+}(\gamma p \to \pi^0 p) = \frac{e g_A}{8\pi f_\pi}\left[-\frac{m_\pi}{M_N} + \frac{m_\pi^2}{2M_N^2}(3+\kappa_p) + \frac{m_\pi^2}{16 f_\pi^2}\right] + \mathcal{O}\left(\frac{m_\pi^3}{M_N^3}\right), \quad (7.81)$$

with $\kappa_p = 1.793$, the anomalous magnetic moment of the proton. The large term proportional to m_π^2/f_π^2 arises from the pion loop corrections, in which the incoming photon is absorbed on the pion cloud of the proton. This term was missed in the early days, before the more systematic development of chiral perturbation theory.

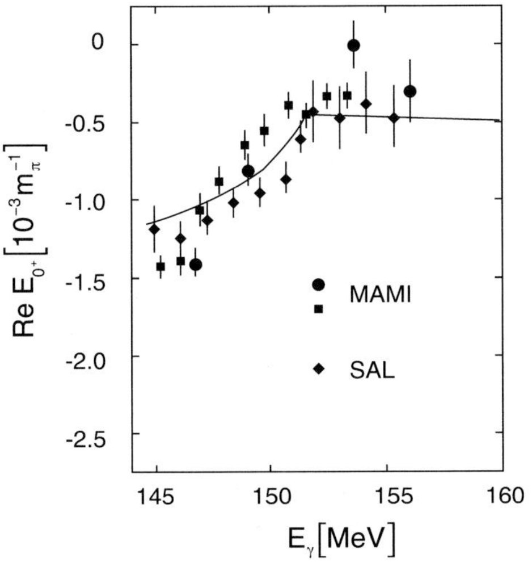

Figure 7.2: The electric dipole amplitude, E_{0+}, for $\gamma p \to \pi^0 p$, as a function of photon energy close to threshold. Data from Mainz (MAMI) [Be 90, Fu 96] and Saskatoon (SAL) [Ber 96] are compared with the result of a calculation based on chiral perturbation theory (solid curve [BKM 92a]).

The detailed quantitative analysis of the energy dependence of E_{0+} has been made possible by a new generation of precision experiments using the MAMI electron accelerator at

Mainz. It confirms the prediction (7.81), but only with inclusion of large $\mathcal{O}(m_\pi^3)$ corrections. The convergence of the chiral expansion is quite slow in this case. Figure 7.2 shows the E_{0+} amplitude deduced from the measured $\gamma p \rightarrow \pi^0 p$ cross section close to threshold, in comparison with the theoretical calculations of Ref. [BKM 92a].

7.3.7 Electromagnetic Polarizabilities of the Nucleon

The electric and magnetic polarizabilities, α_E and β_M, measure the response of the nucleon to low frequency, long wavelength electromagnetic fields as they are realized, for example, in low energy Compton scattering. We refer to Section 2.4 for definitions and phenomenology.

Because of the chiral structure of QCD, the charged pion cloud plays an active role in the electromagnetic response of the proton and neutron. The chiral effective Lagrangian provides a reliable framework to treat these "soft" parts of the structure of the nucleon. The fact that the pion mass is low enters crucially in the polarizabilities. Consider for illustration the Compton scattering process, Fig. 7.3, in which a low energy photon is absorbed and the nucleon responds by virtual excitation of πN intermediate states prior to radiating the photon again. If the pion were a strictly massless Goldstone boson, it would cost no energy to excite it with a low frequency electromagnetic field. One therefore expects that the leading contribution to both α_E and β_M should be proportional to the fine structure constant times $1/m_\pi$; the polarizabilities should diverge as $m_\pi \rightarrow 0$.

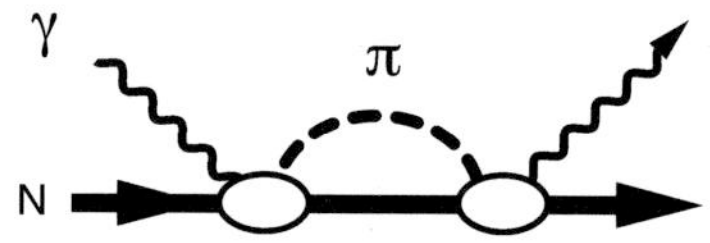

Figure 7.3: Illustration of low energy Compton scattering on the nucleon with virtual πN intermediate states.

This is indeed confirmed by a detailed calculation. The dependence of the polarizabilities on the pion mass has the generic form

$$\alpha_E\,,\beta_M \sim \frac{c_1}{m_\pi} + c_2 \log m_\pi + c_3, \tag{7.82}$$

where the constant c_1 is uniquely determined by leading order chiral perturbation theory. The second and third terms collect contributions from loop diagrams of order q^4 in powers of the momentum of the participating pion.

To leading order, the result for the electric polarizability of both the proton and neutron is [BKM 91]:

$$\alpha_E = \frac{5}{96\,\pi}\left(\frac{g_A}{f_\pi}\right)^2 \frac{\alpha}{m_\pi} \simeq 12.6 \cdot 10^{-4}\,\mathrm{fm}^3, \tag{7.83}$$

with the fine structure constant $\alpha = e^2/4\pi = 1/137.04$. The magnetic polarizability, to this order, turns out to be exactly ten times smaller:

$$\beta_M = \frac{1}{192\,\pi}\left(\frac{g_A}{f_\pi}\right)^2 \frac{\alpha}{m_\pi} \simeq 1.3 \cdot 10^{-4}\,\mathrm{fm}^3. \tag{7.84}$$

While these numbers compare quite well with the empirical values given in Section 2.4.3, it would be misleading to stop at this point since important physics is still missing. This becomes particularly evident when examining the magnetic polarizability. The prominent role of the Δ resonance in the pion photoproduction cross sections leads one to expect a large paramagnetic contribution to β_M from the strong $N \to \Delta$ magnetic dipole transition. Translating the corresponding cross sections into a polarizability gives

$$\delta\beta_\Delta \sim 7 \cdot 10^{-4}\ \mathrm{fm}^3, \tag{7.85}$$

an order of magnitude larger than the leading order chiral result (7.84). The question therefore arises as to what mechanism is responsible for the small net value of β_M found empirically.

Chiral perturbation theory gives a hint in next-to-leading order. The "chiral loop" contributions at that order, representing effects from the electromagnetic couplings of the charged pion cloud surrounding the nucleon, produce large negative terms, with a logarithmic dependence on the pion mass, in both α_E and β_M. In the proton magnetic polarizability one finds a diamagnetic term

$$\delta\beta_{\mathrm{dia}} \simeq -5 \cdot 10^{-4}\ \mathrm{fm}^3, \tag{7.86}$$

which corresponds to the negative induced magnetic dipole moment, generated by the pionic currents at the nucleon surface, as they respond to the external magnetic field. One concludes that the small net magnetic polarizability mainly results from the cancellation of individually large paramagnetic (Δ) and diamagnetic (pionic) contributions. Chiral perturbation theory has obvious limitations here, but it certainly points to the importance of the pion cloud in determining the low energy electromagnetic properties of the nucleon. When a realistic estimate of theoretical uncertainties is performed, the results from chiral perturbation theory, combined with Δ resonance contributions and smaller corrections from neutral pion and kaons, are as follows [BKM 94]:

$$\alpha_E^p = (10.5 \pm 2.0) \cdot 10^{-4}\ \mathrm{fm}^3, \tag{7.87}$$

$$\beta_M^p = (3.5 \pm 3.6) \cdot 10^{-4}\ \mathrm{fm}^3, \tag{7.88}$$

for the proton, and

$$\alpha_E^n = (13.4 \pm 1.5) \cdot 10^{-4}\ \mathrm{fm}^3, \tag{7.89}$$

$$\beta_M^n = (7.8 \pm 3.6) \cdot 10^{-4}\ \mathrm{fm}^3, \tag{7.90}$$

for the neutron. Within errors, these numbers are compatible with the empirical polarizabilities.

7.4 Summary: the Pion Cloud of the Nucleon

As the Goldstone boson of spontaneously broken chiral symmetry, the pion plays a very special and prominent role in low-energy QCD. Decades of research in nuclear physics have established it as the mediator of the long-range force between nucleons. In this Chapter we have accumulated some of the evidence that pions are also an important feature of nucleon structure itself. When probing the nucleon with long-wavelength, electroweak fields, a substantial part of the response comes from the pion cloud, the "soft" surface of the nucleon.

Complementary to the high-energy frontier in deep-inelastic lepton scattering which, together with theoretical QCD analysis, has resolved quarks and gluons as basic constituents of the nucleon, a second, high precision frontier at intermediate energies has opened which reveals the pionic structure of the nucleon surface. A well developed theoretical tool to deal with these aspects exists in the form of a low-energy effective Lagrangian based on the chiral symmetry of QCD, with the light pseudoscalar mesons acting as collective degrees of freedom. The efficient use of symmetries, their conserved currents and resulting low-energy theorems can, in this context, be considered as a modern extension of the powerful and successful current algebra approach of former times [Alf 73]. The parameters in this approach reflect the internal structure of the pion and the nucleon.

While details of the internal structure are not resolved in the low-energy, long-wavelength limit, the link to the high-energy, short-wavelength domain, discussed in Chapter 4, is nevertheless apparent – at least qualitatively. Pions are quark-antiquark excitations of the QCD vacuum. The existence of the pion cloud therefore implies a non-perturbative sea of $q\bar{q}$ pairs for the nucleon, as has indeed been established (see Section 4.4.4) in deep-inelastic lepton scattering and Drell-Yan processes.

8 Models of the Nucleon

The subtlety and complexity of Quantum Chromodynamics has so far prohibited analytic solutions which could be used in discussions of the non-perturbative aspects of nucleon structure at any practical level. The nucleon as a strongly interacting many-body system of quarks and gluons offers such a rich phenomenology that, despite massive numerical efforts in lattice QCD, for physical insights into its many facets, models are still helpful.

8.1 Survey

This chapter focuses on the three classes of models that have been frequently used to analyze and interpret the vast amounts of experimental data dealing with nucleon structure:

i) **Non-relativistic quark models** are based on the assumption that baryons are composed of three massive (non-relativistic) constituent quarks bound in a confining potential. One of the traditional successes of this model is that it gives a simple description of the anomalous magnetic moments of the proton and neutron together with those of the hyperons. With suitable residual interactions between quarks it can also explain most of the observed baryon spectrum.

ii) **Bag models** start from a picture of hadrons as color singlet cavities, or bags, of perturbative vacuum occupied by relativistic quarks and gluons. The bag is imbedded in the condensed medium which represents the QCD ground state. **Chiral bag models** incorporate important aspects of chiral symmetry into this framework by introducing interactions of quarks with the light pseudoscalar mesons, the Goldstone bosons of spontaneously broken chiral symmetry.

iii) **Soliton models** view the nucleon as a localized lump of energy density formed out of mesons and quarks. **Chiral solitons**, in particular, are non-linear interacting systems of quarks and pions. A conceptually interesting version is the topological soliton (or Skyrme soliton) based on the effective Lagrangian of the chiral non-linear sigma model. In this approach baryon number is identified with a topological winding number associated with the meson field configuration. Various hybrid forms of non-topological solitons with quarks bound to chiral meson fields also exist.

These models are constructed to mimic certain selected aspects of QCD. None of them *is* QCD. Each of them has its specific merits and limitations. Non-relativistic quark models emphasize the (hypothetical) role of massive constituent quarks as quasi-particles. These models have features reminiscent of phenomenological shell-models in nuclear physics. Soliton models are typical representatives of collective models: they operate with the low-mass mesons as collective degrees of freedom. Hybrid models such as the chiral bag attempt to unify quark and meson descriptions.

Despite its inherent limitations, the modeling program has been quite successful and contributed substantially to our understanding of baryon structure and spectroscopy. In this chapter we describe the characteristic features of each class of models and summarize their results, with special focus on the nucleon and its excited states.

8.2 Non-Relativistic Quarks Models

Historically, the first basic understanding of nucleon structure dates back to the early quark model, a very simple non-relativistic potential model pioneered by Zweig [Zw 64] and Gell-Mann [G-M 64], and further developed by Dalitz, Greenberg and many others (see historical recollections by O. W. Greenberg and H. J. Lipkin in [BAR 92]). In spite of its simplicity, the successes of this model and of its more recent variants in the analysis of the baryon spectrum are quite remarkable. Although the non-relativistic quark model lacks a field theoretic basis, its phenomenological value is beyond doubt, and it is still widely used.

8.2.1 Constituent Quarks

The basic hypothesis of non-relativistic quark models is the existence of constituent quarks as quasi-particle degrees of freedom inside hadrons. Constituent quarks share with the fundamental QCD quarks their conserved charges and quantum numbers: electric charge, baryon number, color and flavor. They have angular momentum $1/2$ and therefore behave as fermions, just like the fundamental quarks (although the interpretation of the "spin" of a constituent quark is much more involved). On the other hand, through their strong interactions, constituent quarks are supposed to become "heavy fermions" with effective masses exceeding the current quark masses of perturbative QCD by large factors.

Qualitatively, a constituent quark may be viewed as an object in which a "bare" valence quark is dressed by clouds of quark-antiquark pairs and gluons. Our earlier discussion of dynamical symmetry breaking (c.f. Section 6.1.4) certainly suggests that one should expect to find massive quasi-particles with the quantum numbers of quarks, coupled to Goldstone bosons. However, the Goldstone bosons have traditionally been omitted in the constituent quark models. The actual experimental observation of such composite quasi-particle structures inside nucleons is still a challenging problem, but there are many indications, as we shall see, that this concept is of some phenomenological significance.

In the following we assign a creation operator $b^\dagger_{\alpha i}$ to each constituent quark. Assuming that it moves in a confining potential, its single particle state is specified by the set $\alpha = \{n, \nu, q\}$ where n summarizes the orbital quantum numbers, $\nu = \pm\frac{1}{2}$ denotes the spin projection and $q = u, d, s$ the quark flavor (we restrict ourselves here to the flavor $SU(3)$ sector of the light quarks). The index $i = 1, 2, 3$ refers to the quark color. The single particle state of a constituent quark is defined by acting with $b^\dagger$ on the vacuum:

$$|Q_{\alpha i}\rangle = b^\dagger_{\alpha i}|\,0\rangle. \tag{8.1}$$

The quark operators satisfy the usual fermion anticommutation relations:

$$\begin{aligned} \{b_{\alpha i}, b^\dagger_{\beta j}\} &= \delta_{\alpha\beta}\delta_{ij}, \\ \{b^\dagger_{\alpha i}, b^\dagger_{\beta j}\} &= \{b_{\alpha i}, b_{\beta j}\} = 0. \end{aligned} \tag{8.2}$$

When considering ground state configurations, in which the constituent quarks are in their lowest orbits, we can suppress the orbital quantum numbers and reserve the index α for the remaining spin-flavor content:

$$\{\alpha\} = \{u\uparrow, u\downarrow, d\uparrow, d\downarrow, s\uparrow, s\downarrow\}. \tag{8.3}$$

The corresponding creation operators are conveniently denoted by $u^\dagger_{\nu i}$, $d^\dagger_{\nu i}$ and $s^\dagger_{\nu i}$, with spin indices $\nu = \uparrow, \downarrow$ and color indices i.

In the non-relativistic quark model, baryons are color singlet composites of three constituent quarks. For ground state baryons with the three quarks in their lowest (s-wave) orbitals, each given baryon state

$$|B\rangle = \sum_{\alpha\beta\gamma}\sum_{ijk} B_{\alpha\beta\gamma}\, \varepsilon_{ijk}\, b^\dagger_{\alpha i}\, b^\dagger_{\beta j}\, b^\dagger_{\gamma k}|\, 0\rangle \tag{8.4}$$

is antisymmetric in the color indices i, j, k. For $N_f = 3$, the set of spin-flavor coefficients $B_{\alpha\beta\gamma}$ forms a totally symmetric tensor with 56 independent components. These represent the observed ground state multiplets: the octet with total spin $J = \frac{1}{2}$ and the decuplet with $J = \frac{3}{2}$.

The simplest configuration is the Δ^{++} baryon with doubly positive electric charge, spin $J = \frac{3}{2}$ and spin projection $M_J = +\frac{3}{2}$. It has three $u-$quarks with their spins all aligned:

$$|\Delta^{++}_{M_J=+\frac{3}{2}}\rangle = \frac{1}{\sqrt{6}} \sum_{ijk} \varepsilon_{ijk}\, u^\dagger_{\uparrow i}\, u^\dagger_{\uparrow j}\, u^\dagger_{\uparrow k}|0\rangle. \tag{8.5}$$

Protons and neutrons with "spin up" have the following decomposition in terms of constituent quarks:

$$|p_\uparrow\rangle = \frac{1}{\sqrt{18}} \sum_{ijk} \varepsilon_{ijk}\, [u^\dagger_{\downarrow i}\, d^\dagger_{\uparrow j} - u^\dagger_{\uparrow i}\, d^\dagger_{\downarrow j}]\, u^\dagger_{\uparrow k}|0\rangle, \tag{8.6}$$

$$|n_\uparrow\rangle = \frac{1}{\sqrt{18}} \sum_{ijk} \varepsilon_{ijk}\, [d^\dagger_{\uparrow i}\, u^\dagger_{\downarrow j} - d^\dagger_{\downarrow i}\, u^\dagger_{\uparrow j}]\, d^\dagger_{\uparrow k}|0\rangle. \tag{8.7}$$

The state vectors of the remaining members of the baryon octet are:

$$
\begin{aligned}
|\Lambda_\uparrow\rangle &= \frac{1}{\sqrt{12}}\sum_{ijk} \varepsilon_{ijk}\,[u^\dagger_{\uparrow i}\,d^\dagger_{\downarrow j} - u^\dagger_{\downarrow i}\,d^\dagger_{\uparrow j}]\,s^\dagger_{\uparrow k}|0\rangle,\\
|\Sigma^0_\uparrow\rangle &= \frac{1}{6}\sum_{ijk} \varepsilon_{ijk}\,[s^\dagger_{\uparrow i}\,d^\dagger_{\downarrow j}\,u^\dagger_{\uparrow k} + s^\dagger_{\uparrow i}\,d^\dagger_{\uparrow j}\,u^\dagger_{\downarrow k} - 2s^\dagger_{\downarrow i}\,d^\dagger_{\uparrow j}\,u^\dagger_{\uparrow k}]|0\rangle,\\
|\Sigma^+_\uparrow\rangle &= \frac{1}{\sqrt{18}}\sum_{ijk} \varepsilon_{ijk}\,[s^\dagger_{\downarrow i}\,u^\dagger_{\uparrow j} - s^\dagger_{\uparrow i}\,u^\dagger_{\downarrow j}]\,u^\dagger_{\uparrow k}|0\rangle,\\
|\Sigma^-_\uparrow\rangle &= \frac{1}{\sqrt{18}}\sum_{ijk} \varepsilon_{ijk}\,[s^\dagger_{\uparrow i}\,d^\dagger_{\downarrow j} - s^\dagger_{\downarrow i}\,d^\dagger_{\uparrow j}]\,d^\dagger_{\uparrow k}|0\rangle,\\
|\Xi^0_\uparrow\rangle &= \frac{1}{\sqrt{18}}\sum_{ijk} \varepsilon_{ijk}\,[s^\dagger_{\downarrow i}\,u^\dagger_{\uparrow j} - s^\dagger_{\uparrow i}\,u^\dagger_{\downarrow j}]\,s^\dagger_{\uparrow k}|0\rangle,\\
|\Xi^-_\uparrow\rangle &= \frac{1}{\sqrt{18}}\sum_{ijk} \varepsilon_{ijk}\,[s^\dagger_{\uparrow i}\,d^\dagger_{\downarrow j} - s^\dagger_{\downarrow i}\,d^\dagger_{\uparrow j}]\,s^\dagger_{\uparrow k}|0\rangle.
\end{aligned}
\tag{8.8}
$$

The normalization constants are fixed by the condition that the baryon states form an orthonormal set, $\langle B_m|B_n\rangle = \delta_{mn}$.

8.2.2 Magnetic Moments and Axial Currents

Constituent quarks and baryon magnetic moments

Assigning phenomenological effective masses m_q to the constituent quarks, let us assume that the quark magnetic moments are given by their standard Dirac values

$$\mu_q = \frac{e_q}{2m_q} \tag{8.9}$$

with the quark electric charges $e_u = \frac{2e}{3}$, $e_d = e_s = -\frac{e}{3}$. Baryon magnetic moments are easily evaluated as matrix elements

$$\mu_B = \langle B\uparrow|\,\boldsymbol{m}_z\,|\,B\uparrow\rangle \tag{8.10}$$

of the non-relativistic magnetic moment operator

$$\vec{\boldsymbol{m}} = \sum_{q\ldots} \mu_q\,\langle\nu'|\,\vec{\sigma}\,|\nu\rangle\,b^\dagger_{q\nu',i}\,b_{q\nu,i}, \tag{8.11}$$

where the summation is understood to run over flavor (q), spin (ν,ν') and color (i). The results for the proton and neutron are:

$$
\begin{aligned}
\mu_p &= \frac{4}{3}\,\mu_u - \frac{1}{3}\,\mu_d,\\
\mu_n &= \frac{4}{3}\,\mu_d - \frac{1}{3}\,\mu_u.
\end{aligned}
\tag{8.12}
$$

In the limit of exact isospin symmetry with $m_u = m_d$ one finds

$$\frac{\mu_p}{\mu_n} = -\frac{3}{2}, \tag{8.13}$$

which is close to the observed ratio $\frac{\mu_p}{\mu_n} \simeq 1.46$. Fixing the (approximately degenerate) $u, d-$constituent quark mass by the proton magnetic moment,

$$\mu_p = \frac{3}{2}\,\mu_u = \frac{e}{2m_u} = 2.79\,\frac{e}{2M_p} \tag{8.14}$$

gives

$$m_{u,d} \simeq 0.34\ \text{GeV}. \tag{8.15}$$

The Λ hyperon magnetic moment equals μ_s since the $u, d-$pair in the Λ is coupled to spin zero:

$$\mu_\Lambda = \mu_s = -\frac{e}{6m_s} = -0.61\,\frac{e}{2M_p}, \tag{8.16}$$

which implies

$$m_s \simeq 0.51\ \text{GeV}. \tag{8.17}$$

The full listing of baryon magnetic moments is given in Table 8.1. A comparison is made between the measured moments and the quark model values for two cases. In fit A the $SU(3)$ limit of degenerate quark masses ($m_u = m_d = m_s$) is taken, with $\mu_d = \mu_s = -\frac{1}{2}\mu_u$. In fit B the individual quark magnetic moments are treated as free parameters adjusted to the proton, neutron and Λ magnetic moments, yielding $\mu_u = 1.852$, $\mu_d = -0.972$ and $\mu_s = -0.613$ in units of $\frac{e}{2M_p}$.

Considering the simplicity of the model, the general agreement is quite impressive. One should note, of course, that the magnetic moments are primarily determined by symmetry rather than by detailed dynamics, so that the success of the quark model at this point is mainly a measure of the accuracy to which the underlying $SU(6)$ spin-flavor symmetry works.

The nucleon $\rightarrow$ $\boldsymbol{\Delta(1232)}$ transition magnetic moment

The quark model also establishes a simple relationship between the proton magnetic moment and the matrix element for the magnetic dipole ($M1$) transition between the nucleon and the $\Delta(1232)$. For example, the matrix element:

$$\mu_{p\Delta} = \langle p\uparrow | \boldsymbol{m}_z | \Delta^+_{M_J=+1/2}\rangle, \tag{8.18}$$

enters the photoproduction amplitude for $\gamma p \rightarrow \Delta^+$. It is easily evaluated using the proton and delta isobar wave functions from Section 8.2.1, with the result

$$\mu_{p\Delta} = \frac{2}{3}\sqrt{2}\,\mu_p \tag{8.19}$$

in terms of the proton magnetic moment. This result strictly holds in the static limit when the energy and momentum transfer involved in the $N \rightarrow \Delta$ transition is small compared to the nucleon or Δ mass. In practice the photoexcitation of the Δ requires an energy transfer of about 300 MeV so that recoil corrections are non-negligible.

Table 8.1: Magnetic moments of the baryon octet. Experimental values [PDG 98/00] are compared with the quark model in the $SU(3)$ limit (fit A) and the quark magnetic moments taken as independent parameters (fit B).

	Exp. $[e\hbar/2M_pc]$	Quark Model	A	B
μ_p	$2.79284739(6)$	$\frac{4\,\mu_u - \mu_d}{3}$	2.79	2.79
μ_n	$-1.9130428(5)$	$\frac{4\,\mu_d - \mu_u}{3}$	-1.86	-1.91
μ_Λ	$-0.613(4)$	μ_s	-0.93	-0.61
μ_{Σ^+}	$2.46(1)$	$\frac{4\,\mu_u - \mu_s}{3}$	2.79	2.67
$\lvert\mu_{\Sigma^0\Lambda}\rvert$	$1.61(8)$	$\frac{\lvert\mu_u - \mu_d\rvert}{\sqrt{3}}$	1.61	1.63
μ_{Σ^-}	$-1.16(2)$	$\frac{4\,\mu_d - \mu_s}{3}$	-0.93	-1.09
μ_{Ξ^0}	$-1.25(1)$	$\frac{4\,\mu_s - \mu_u}{3}$	-1.86	-1.44
μ_{Ξ^-}	$-0.651(3)$	$\frac{4\,\mu_s - \mu_d}{3}$	-0.93	-0.49

Axial current matrix elements

Another fundamental property of the nucleon is its axial vector coupling constant, g_A, which determines the matrix element for neutron beta decay (as discussed in Section 3.3). In quark models g_A is expressed in terms of the axial current contributions from each constituent quark. Non-relativistically, the space components of the quark axial current, $A_a^\mu = \overline{\psi}\gamma^\mu\gamma_5\frac{\lambda_a}{2}\psi$, reduce to the following spin-flavor operator:

$$\vec{A}_a \;=\; \sum_{q,q'\ldots} g_{A,q}\,\langle q'\nu'|\,\vec{\sigma}\,\frac{\lambda_a}{2}|q\nu\rangle\; b^\dagger_{q'\nu',i}\, b_{q\nu,i}, \tag{8.20}$$

where for bare quarks $g_{A,q} = 1$. For constituent quarks with their non-trivial substructure the axial constant could well deviate from one but this implies a certain arbitrariness and lack of predictivity which we prefer not to introduce, at least not in this section.

It is now a straightforward exercise to evaluate the beta decay matrix element

$$\langle p \uparrow |\vec{A}_+|n \uparrow\rangle \;=\; g_A\langle\uparrow|\,\vec{\sigma}\,|\uparrow\rangle \tag{8.21}$$

with $\vec{A}_+ = \vec{A}_1 + i\vec{A}_2$, using the explicit neutron and proton wave functions from Section 8.2.1. The result is:

$$g_A = \frac{5}{3}, \tag{8.22}$$

to be compared with the empirical $g_A = 1.26$. Later we shall see that relativistic effects from the lower components of the quark Dirac wave functions tend to lower the non-relativistic quark model result significantly.

8.2.3 Potential Model

Assume that the motion of constituent quarks inside the baryon is described by a non-relativistic Hamiltonian

$$H = \sum_{i=1}^{3} \left(m_i + \frac{\vec{p}_i^{\,2}}{2m_i} \right) + \frac{1}{2} \sum_{i,j} V(\vec{r}_{ij}), \tag{8.23}$$

with a potential V depending on the distances, $\vec{r}_{ij} = \vec{r}_i - \vec{r}_j$, between the quarks. Here, in particular, we consider a harmonic oscillator potential

$$V(\vec{r}_{ij}) = \frac{K}{2} r_{ij}^2. \tag{8.24}$$

For nucleons and their non-strange excited states one can start with equal quark masses $m = m_u = m_d$. It is convenient to introduce a center-of-mass coordinate vector $\vec{R}$ and internal coordinates $\vec{\lambda}$ and $\vec{\rho}$ by

$$\begin{aligned} \vec{R} &= \frac{1}{3} \left(\vec{r}_1 + \vec{r}_2 + \vec{r}_3 \right), \\ \vec{\lambda} &= \frac{1}{\sqrt{2}} \left(\vec{r}_1 - \vec{r}_2 \right), \\ \vec{\rho} &= \frac{1}{\sqrt{6}} \left(\vec{r}_1 + \vec{r}_2 - 2\,\vec{r}_3 \right). \end{aligned} \tag{8.25}$$

Writing $H = 3m + H_0$ and removing the center-of-mass coordinates we are left with the intrinsic Hamiltonian

$$H_0^{\text{int}} = \frac{\vec{p}_\rho^{\,2}}{2m} + \frac{3K}{2} \vec{\rho}^{\,2} + \frac{\vec{p}_\lambda^{\,2}}{2m} + \frac{3K}{2} \vec{\lambda}^2 \tag{8.26}$$

of two independent oscillators, each with frequency

$$\omega_0 = \sqrt{\frac{3K}{m}}. \tag{8.27}$$

The zero-order baryon spectrum is then

$$E_N = E_0 + N\omega_0, \tag{8.28}$$

where E_0 is the ground state energy and $N = N_\rho + N_\lambda$ with the numbers $N_{\rho,\lambda} = 0, 1, 2, \ldots$ of λ and ρ oscillator quanta. The total angular momentum $\vec{J} = \vec{L} + \vec{S}$ is the sum of the total quark spin $\vec{S} = \sum \vec{s}_i$ and the orbital angular momentum $\vec{L} = \vec{l}_\rho + \vec{l}_\lambda$ associated with the coordinates $\vec{\rho}$ and $\vec{\lambda}$. The parity of a given state is $P = (-1)^{l_\rho + l_\lambda}$.

The oscillator frequency, ω_0, can be estimated from the observed masses of the lowest negative parity (p-shell) N^* resonances as compared to the average mass of the nucleon and $\Delta(1232)$. The p-shell has several N^* states with spins $\frac{1}{2}^-$, $\frac{3}{2}^-$ and $\frac{5}{2}^-$ as shown in Fig. 8.1. These resonances have their average mass around $\overline{M}^* \simeq 1.6$ GeV. The average N and Δ

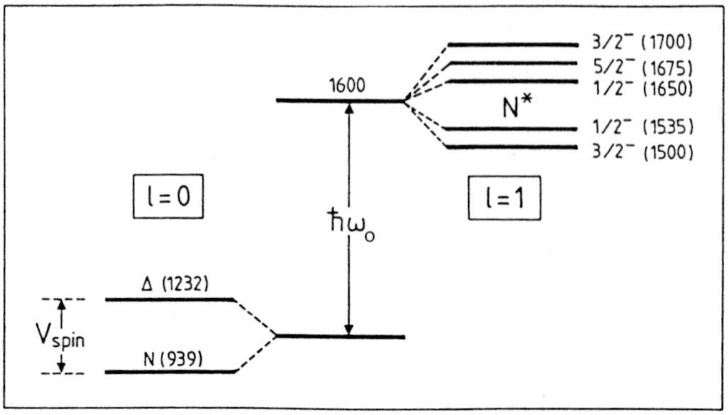

Figure 8.1: Spectrum of low-lying nucleon and N^* resonance states. The s- and p-states are separated by the oscillator frequency ω_0. The hyperfine ("$V_{\rm spin}$") splitting between N and Δ is also indicated.

mass is $\overline{M} = \frac{1}{2}(M_N + M_\Delta) \simeq 1.1$ GeV, so that the characteristic oscillator energy should be roughly

$$\omega_0 = \overline{M}^* - \overline{M} \simeq 0.5 \text{ GeV}. \tag{8.29}$$

The mean square radius, calculated using the intrinsic ground state wave function:

$$\Psi_0 = \left(\frac{m\omega_0}{\pi}\right)^{3/2} \exp\left[-\frac{m\omega_0}{2}\left(\vec{\rho}^{\,2} + \vec{\lambda}^2\right)\right] \tag{8.30}$$

becomes

$$\langle r^2 \rangle = (m\omega_0)^{-1}. \tag{8.31}$$

With $m \simeq 0.34$ GeV the resulting $\sqrt{\langle r^2 \rangle} \simeq 0.5$ fm is too small when compared with the measured proton charge radius – i.e., 0.86 fm, as given in Section 2.2.1. However, this should not be so surprising, given the extreme simplicity of the model.

In practice the harmonic oscillator quark model fails in direct confrontation with details of the observed baryon spectrum, but it provides a useful calculational basis for further refinements to which we shall now turn.

8.2.4 Baryon Spectroscopy

The quark shell model with $N_f = 3$ predicts a systematic spectral pattern for baryons with states clustering in $SU(6) \times O(3)$ supermultiplets. This structure follows from the group products when combining three quarks, each represented by a fundamental triplet of $SU(3)_{\rm flavor}$ and a doublet of $SU(2)_{\rm spin}$, with the $O(3)$ group describing the orbital motion. The classification of the supermultiplet bands makes use of the oscillator quantum numbers $N = 0, 1, 2, \ldots$ introduced in the previous section. The ground state ($N = 0$) multiplet arises as the $[56]$

representation of $SU(6)$ which contains the spin $\frac{1}{2}$ octet and the spin $\frac{3}{2}$ decuplet as shown previously. The "$1\hbar\omega_0$" excitations ($N = 1$) involve orbital angular momentum $L = 1$ and form a $[70]$ multiplet with negative parity which includes octets and decuplets with total spins $\frac{1}{2}$, $\frac{3}{2}$ and $\frac{5}{2}$.

The book-keeping of states is conveniently done using the notation (r, L_N^P) where the first entry refers to the $SU(6)$ representation $[r]$, the index N labels the oscillator quanta, P the parity, and L is the orbital angular momentum. The systematics of the resulting low-lying baryon spectrum is shown in Fig. 8.2. In a pure harmonic oscillator shell model the multiplets within the $N = 0$ and $N = 1$ bands, as well as the supermultiplets in the $N = 2$ band, would all be degenerate. Deviations of the inter-quark potential from a harmonic oscillator, together with spin dependent interactions, remove these degeneracies and cause the splitting of supermultiplets in the $N = 2$ bands as well as the hyperfine splitting of the $N = 0, 1$ bands into multiplets with different total angular momentum J, as illustrated schematically in Fig. 8.2.

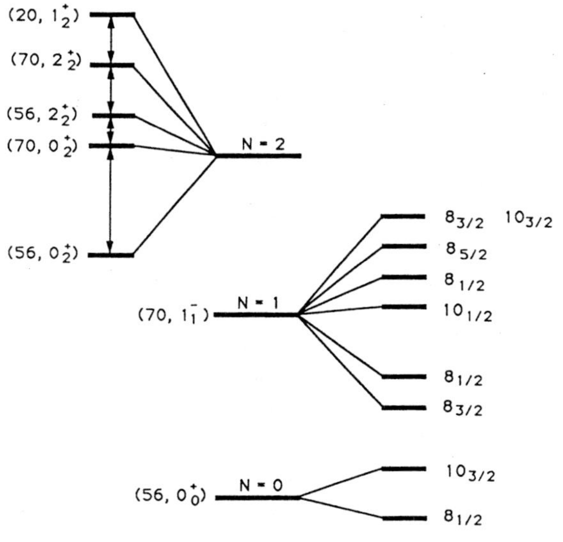

Figure 8.2: Schematic picture of the low-lying baryon spectrum. The labels (r, L_N^P) to the left indicate the $SU(6) \times O(3)$ supermultiplets. The hyperfine splittings of the $N = 0$ and $N = 1$ bands are denoted by the $SU(3)$ flavor multiplets and the total spin J attached to each multiplet.

8.2.5 The Model of deRujula, Georgi and Glashow

The interest in the non-relativistic quark model was revived in the mid seventies when deRujula, Georgi and Glashow [RGG 75] introduced aspects of QCD into the description of constituent quark dynamics. The basic assumptions of the RGG model are the following:

(i) the starting point is the non-relativistic quark model with $SU(6)$ spin-flavor symmetry;

(ii) $SU(3)$ flavor breaking occurs only via the quark masses, as in QCD;

(iii) there exists a long-range confining force which is independent of flavor and spin;

(iv) short-range spin and flavor dependent interactions arise solely from one-gluon exchange between quarks.

The model does not make attempts to justify the concept of constituent quarks from QCD, and the assumption (iv), of keeping only one-gluon exchange as a residual interaction between quarks, is certainly an idealization. In essence, the hypothesis (i) of non-relativistic quark dynamics permits one to reduce one-gluon exchange to the Breit interaction familiar from the electrodynamics of two slowly moving spin $\frac{1}{2}$ particles. In QCD the corresponding interaction involves the color coupling $(\alpha_s/4)\vec{\lambda}_1 \cdot \vec{\lambda}_2$ between two quarks, with $SU(3)_{\text{color}}$ generators $\vec{\lambda}/2$. Each pair of quarks within a color singlet baryon must be in a $[\bar{3}]$ (antitriplet) representation for which $\langle \vec{\lambda}_1 \cdot \vec{\lambda}_2 \rangle = -\frac{8}{3}$. With inclusion of the resulting color coupling factor $-\frac{2}{3}\alpha_s$ which replaces the fine structure constant in the electromagnetic case, the gluon exchange Breit interaction between two quarks in a baryon has the following form when expanded to order m^{-2} in the quark masses:

$$
\begin{aligned}
V_{ij} = & -\frac{2\,\alpha_s}{3r} + \frac{\alpha_s}{3m_i m_j}\left[\frac{\vec{p}_i \cdot \vec{p}_j}{r} + \frac{(\vec{r}\cdot\vec{p}_i)(\vec{r}\cdot\vec{p}_j)}{r^3}\right] + \frac{\pi\alpha_s}{3}\delta^3(\vec{r})\left(\frac{1}{m_i^2} + \frac{1}{m_j^2}\right) \\
& + \frac{2\,\alpha_s}{3m_i m_j}\left[\frac{8\,\pi}{3}\,\delta^3(\vec{r})\,\vec{s}_i\cdot\vec{s}_j + \frac{3(\vec{s}_i\cdot\hat{r})(\vec{s}_j\cdot\hat{r}) - \vec{s}_i\cdot\vec{s}_j}{r^3}\right] \\
& + \frac{\alpha_s}{3r^3}\left[\frac{\vec{r}\times\vec{p}_i\cdot\vec{s}_i}{m_i^2} - \frac{\vec{r}\times\vec{p}_j\cdot\vec{s}_j}{m_j^2} + \frac{2}{m_i m_j}\,(\vec{r}\times\vec{p}_i\cdot\vec{s}_j - \vec{r}\times\vec{p}_j\cdot\vec{s}_i)\right],
\end{aligned}
\tag{8.32}
$$

with $\vec{r} = \vec{r}_i - \vec{r}_j$ and $\hat{r} = \vec{r}/r$. Apart from the Coulomb, Darwin and contact terms, this interaction has spin-spin, tensor and spin-orbit contributions, listed in this order.

The Hamiltonian of the three-quark system is then written as follows:

$$
H = U_{\text{conf}} + \sum_{i=1}^{3}\left(m_i + \frac{\vec{p}_i^{\,2}}{2m_i}\right) + \sum_{i<j} V_{ij}, \tag{8.33}
$$

where U_{conf} is a confining potential. This Hamiltonian is conveniently rewritten as

$$
H = H^{(0)} + \Delta V, \tag{8.34}
$$

where we have introduced the $SU(3)$ symmetric Hamiltonian

$$
H^{(0)} = U_{\text{conf}} + 3m + \sum_{i=1}^{3}\frac{\vec{p}_i^{\,2}}{2m}, \tag{8.35}
$$

with the effective non-strange quark mass $m = m_u$. The eigenstates of $H^{(0)}$ are the usual $SU(6)$ multiplets. The ground state baryon octet and decuplet wave functions, which satisfy

$$
H^{(0)}|\Psi_B\rangle = M_0|\Psi_B\rangle, \tag{8.36}
$$

are degenerate at that level. The degeneracy is removed by ΔV, which includes $SU(3)$ breaking by the strange quark mass and spin dependent interactions. For quarks in their lowest s-wave orbitals the tensor and spin-orbit interactions do not contribute. To first order in ΔV, the hyperfine splitting between the spin-$\frac{3}{2}$ decuplet and the spin-$\frac{1}{2}$ octet is then determined by the spin-spin part of the one-gluon exchange interaction alone. For $B = N, \Delta$ we have the mass formula:

$$M_B = M_0 + \frac{16\,\pi}{9m^2}\alpha_s \langle \Psi_B | \sum_{i<j} \delta^3(\vec{r}_{ij})\, \vec{s}_i \cdot \vec{s}_j | \Psi_B \rangle. \tag{8.37}$$

We note that for baryons with total spin, $\vec{S} = \vec{s}_1 + \vec{s}_2 + \vec{s}_3$, the spin-spin interaction operator reduces to

$$\sum_{i<j} \vec{s}_i \cdot \vec{s}_j = \frac{\vec{S}^2}{2} - \frac{9}{8}, \tag{8.38}$$

with eigenvalues $\pm\frac{3}{4}$ for $S = \frac{3}{2}$ and $\frac{1}{2}$, respectively. Using the harmonic oscillator wave function (8.30) with the usual frequency parameter $\omega_0 (\simeq 0.5\,\mathrm{GeV})$, the mass splitting between the Δ and the nucleon becomes:

$$M_\Delta - M_N = \frac{4}{3}\alpha_s\,\omega_0 \left(\frac{\omega_0}{2\,\pi m}\right)^{1/2}. \tag{8.39}$$

The hyperfine splitting induced by one-gluon-exchange evidently goes in the right direction, but the observed $\Delta - N$ mass difference, of about 0.3 GeV, requires a strong coupling constant $\alpha_s \simeq 0.9$ (using a constituent quark mass $m \simeq 0.34$ GeV). With such a large value of α_s, one suspects that the original assumption about *perturbative* one-gluon-exchange being the *only* source of the hyperfine splitting, should be taken with caution.

8.2.6 The Isgur-Karl Model

Extensive calculations of the baryon spectra have been performed by Isgur, Karl and collaborators [IK 78, IK 79] starting from a harmonic oscillator Hamiltonian combined with an anharmonic perturbation, U, and a hyperfine interaction V^{hyp}:

$$H = \sum_{i=1}^{3}\left(m_i + \frac{\vec{p}_i^{\,2}}{2m_i}\right) + \frac{K}{2}\sum_{i<j}\vec{r}_{ij}^{\,2} + \sum_{i<j}\left(U(\vec{r}_{ij}) + V_{ij}^{\mathrm{hyp}}\right). \tag{8.40}$$

The clue to the successful phenomenology of the IK model is the omission of the spin-orbit forces, retaining only the spin-spin and tensor terms from the Breit interaction:

$$V_{ij}^{\mathrm{hyp}} = \frac{2\,\alpha_s}{3m_i m_j}\left[\frac{8\,\pi}{3}\,\delta^3(\vec{r})\,\vec{s}_i \cdot \vec{s}_j + \frac{3(\vec{s}_i \cdot \hat{r})(\vec{s}_j \cdot \hat{r}) - \vec{s}_i \cdot \vec{s}_j}{r^3}\right]. \tag{8.41}$$

The Fermi contact term in the spin-spin interaction operates only in quark pairs (ij) with zero orbital angular momentum, whereas the tensor term acts only when the orbital angular momentum of the pair is non-zero. The anharmonic perturbation, U, is introduced primarily to break the degeneracy of the five $SU(6)$ multiplets in the $N = 2$ band as shown in Fig. 8.2. The theoretical justification for dropping the spin-orbit force is not obvious. However, its omission definitely improves the quality of the fit to the baryon resonance spectra.

The IK model adequately describes the systematics of the baryon spectrum, with few notable exceptions. We summarize here the results for the non-strange baryon sector which includes the nucleon and the sequence of its excited states. Their empirical masses and widths are deduced mainly from the analysis of pion-nucleon scattering and photoproduction data. The $N = 1$ negative parity states are displayed in Fig. 8.3, and the results for the $N = 2$ positive parity resonances are shown in Fig. 8.4. The IK model predicts more states than those actually observed. However, the model also permits calculations of the couplings of these resonances to the πN channel, which is the one most readily accessible to experiment. Indeed, it turns out that most of the unobserved states are weakly coupled to the πN channel.

The IK analysis uses the following input:

(i) The zeroth order mass of the $N = 1$ multiplet $(70, 1_1^-)$ is chosen to be 1610 MeV, corresponding to a harmonic oscillator spacing

$$\omega_0 \simeq 0.52 \text{ GeV}. \tag{8.42}$$

(ii) The non-strange constituent quark mass is

$$m \simeq 0.4 \text{ GeV}, \tag{8.43}$$

somewhat larger than the value which reproduces the nucleon magnetic moments. The ratio of non-strange and strange quark masses is $m/m_s \simeq 0.6$.

(iii) The strength of the hyperfine interaction is essentially determined by the $N - \Delta$ mass splitting of about 300 MeV. This requires α_s to be somewhat larger than 0.9 (see also Eq. (8.39)).

(iv) The anharmonic perturbation, U, acts only in the $N = 2$ band. Its main role is to remove the degeneracy of this band and to reduce the spacing between $N = 1$ and $N = 2$ zero order masses to a value below 0.5 GeV.

An interesting consequence of the hyperfine interaction is to mix the $N = 2$ configurations into the $N = 0$ ground state wave function. Explicit calculations suggest the following composition of the physical nucleon wave function:

$$\begin{aligned} |N\rangle \simeq\ & 0.90\,|N(56, 0_0^+)\rangle - 0.34\,|N(56, 0_2^+)\rangle \\ & - 0.27\,|N(70, 0_2^+)\rangle - 0.06\,|N(70, 2_2^+)\rangle . \end{aligned} \tag{8.44}$$

This means that the nucleon in this picture has about $80\ \%$ probability to be found in the $N = 0$ ground state configuration, while it exists for about $20\ \%$ of the time in one of its "$2\hbar\omega_0$" excitations. The primary component of the $N = 2$ admixture is the Roper resonance $N(1440)$, the $J = \frac{1}{2}$ "breathing mode" of the nucleon.

A related phenomenon is the possible admixture of d-states into the nucleon and $\Delta(1232)$ wave functions by residual tensor forces between quarks. Such admixtures should have observable consequences in the electromagnetic $N \rightarrow \Delta$ transition. While this process is dominated by a strong magnetic dipole $(M1)$ amplitude, it can also proceed as an electric quadrupole $(E2)$ transition. Simple versions of quark models, without configuration mixing, predict that the $E2$ amplitude should vanish since the quarks are in s-states. This changes in the presence of non-zero d-wave components. As we saw in Section 2.3.4, measurements of the $E2$ amplitude in the $N \rightarrow \Delta$ transition have shown that this matrix element is indeed non-vanishing, but small. This result effectively rules out the possibility of a large d-wave admixture in the nucleon wave function.

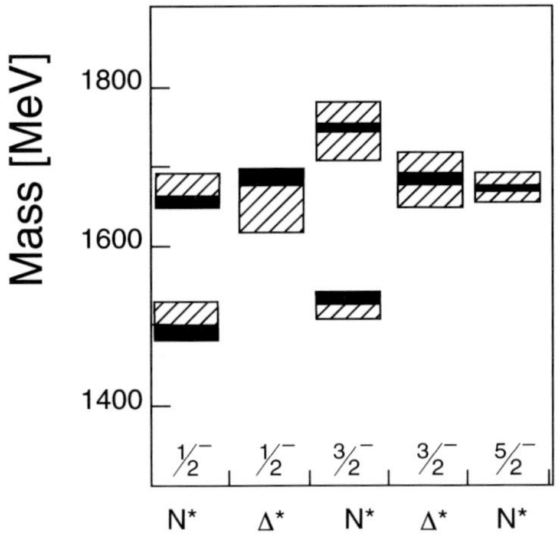

Figure 8.3: Spectrum of "$1\hbar\omega_0$" negative parity nucleon and delta resonances. The shaded regions indicate the empirical positions and widths [PDG 98/00]. The solid bars are the predictions of the Isgur-Karl model [IK 78, IK 79].

Whereas the phenomenology of the IK model indicates that spin-orbit forces between constituent quarks are unimportant in the baryon mass spectrum, the QCD Breit interaction certainly yields such forces. This raises a conceptual problem about the detailed properties of residual interactions beyond one-gluon-exchange, which has been intensely discussed in the literature [He 83, Gr 76]. The issues are relativistic corrections, $\vec{L} \cdot \vec{S}$ forces from a Lorentz scalar confining potential via Thomas precession, genuine three-body interactions between quarks, and the like. We do not enter into those discussions here, instead referring the reader to the more specialized literature [Is 00, GR 96]. The role of Goldstone boson exchange, which is of considerable interest in its own right, will be deferred until we discuss chiral quark models in Sections 8.5 and 8.6.

Basic questions about the deeper foundations of the non-relativistic quark models remain unanswered. For example, we note the kinematical inconsistency that the average momentum of a constituent quark in a nucleon is comparable to its mass. To a certain extent this has been addressed by relativistic extensions of the Isgur-Karl model by Godfey and Isgur [GI 85] and others [CR 00]. Nevertheless, in spite of their conceptual problems, the non-relativistic

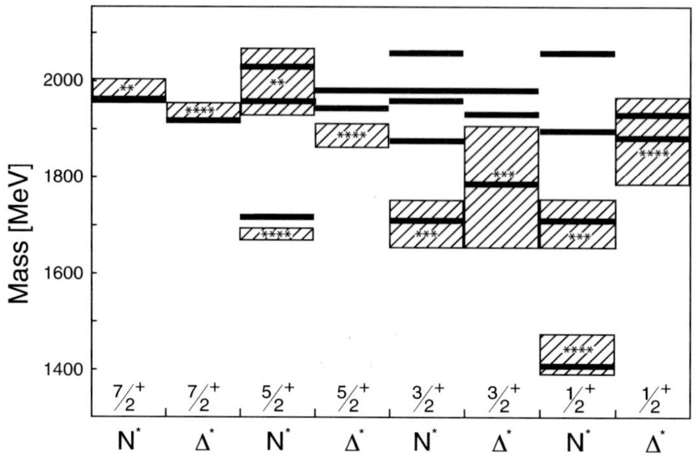

Figure 8.4: Spectrum of "$2\hbar\omega_0$" positive parity nucleon and delta resonances. The shaded regions indicate the empirical positions and widths [PDG 98/00]. The solid bars are the predictions of the Isgur-Karl model [IK 78, IK 79].

models have a major virtue, namely their phenomenological simplicity. From this perspective, the Isgur-Karl model presents a prototype Hamiltonian which successfully combines the group theoretical framework of $SU(6) \times O(3)$ symmetry with minimal dynamical input and gives a surprisingly good description of a large amount of baryon data.

8.2.7 Transition Amplitudes

The quark model provides useful guidance for evaluating electromagnetic transition amplitudes connecting the nucleon and its excited states. In Section 2.3.1 we have seen that such $\gamma N \leftrightarrow N^*$ transitions are conveniently described using helicity amplitudes, $A_{\frac{1}{2}}$ and $A_{\frac{3}{2}}$. As explained in that earlier section, these amplitudes represent helicity non-flip and flip processes induced by a circularly polarized real or virtual photon.

In the quark model the basic hypothesis is that the photon is emitted or absorbed by a single constituent quark in the baryon. A calculation of $\gamma N \leftrightarrow N^*$ amplitudes then involves a sum over the three constituent quarks in the initial and final baryon states. For a real photon the electromagnetic interaction Hamiltonian, $\mathcal{H}_{\text{int}} = \sum_{i=1}^{3} \vec{J}_i \cdot \vec{\mathcal{A}}(\vec{r}_i)$, of the constituent quarks reduces to:

$$H_{\text{int}} = \sum_{i=1}^{3} \frac{e_i}{2m_i} \left[(\vec{p}_i + \vec{p}_i\,') \cdot \vec{\mathcal{A}}(\vec{r}_i) - i\vec{\sigma}_i \cdot \vec{q} \times \vec{\mathcal{A}}(\vec{r}_i) \right], \tag{8.45}$$

where $\vec{p}_i$ and $\vec{p}_i\,'$ are the quark momenta in the initial and final state, e_i is the quark charge and m_i its mass. The photon has momentum $\vec{q}$ and energy ω. Its electromagnetic vector potential is

$$\vec{\mathcal{A}}(\vec{r}) = \sqrt{\frac{2\pi}{\omega}}\, \vec{\varepsilon} \left[a_q^\dagger\, e^{i\vec{q}\cdot\vec{r}} + a_q\, e^{-i\vec{q}\cdot\vec{r}} \right]. \tag{8.46}$$

The polarization vector is chosen, without loss of generality, to be that of a right-handed photon (helicity $+1$) moving in the z-direction:

$$\vec{\varepsilon} = -\frac{1}{\sqrt{2}}\,(1, i, 0). \tag{8.47}$$

For photoabsorption the interaction Hamiltonian then reduces to

$$H_{\text{int}} = \sqrt{\frac{\pi}{\omega}} \sum_{i=1}^{3} \frac{e_i}{2m_i} \left[q\sigma_+(i) - 2p_+(i) \right] e^{-iqz_i}, \tag{8.48}$$

with $q = |\vec{q}\,|$, $\sigma_+ = \sigma_x + i\sigma_y$ and $p_+ = p_x + ip_y$. The relevant helicity amplitudes for $N \rightarrow N^*$ transitions induced by the absorption of a real photon are

$$A_{\frac{1}{2}} = \langle N^*_{j,+\frac{1}{2}} |\, H_{\text{int}}\, | N_{\frac{1}{2},-\frac{1}{2}} \rangle, \tag{8.49}$$

$$A_{\frac{3}{2}} = \langle N^*_{j,+\frac{3}{2}} |\, H_{\text{int}}\, | N_{\frac{1}{2},+\frac{1}{2}} \rangle. \tag{8.50}$$

The cross section for photo-excitation of an N^* resonance is

$$\sigma(\gamma p \rightarrow N^*) = \frac{2M_p}{\Gamma_R M_R} \left[|A_{\frac{1}{2}}|^2 + |A_{\frac{3}{2}}|^2 \right], \tag{8.51}$$

where Γ_R is the resonance width and M_R its mass. The radiative decay width of a resonance with total angular momentum j is given by

$$\Gamma(N^* \rightarrow N\gamma) = \frac{\omega^2}{\pi} \frac{2M_N}{(2j+1)M_R} \left[|A_{\frac{1}{2}}|^2 + |A_{\frac{3}{2}}|^2 \right]. \tag{8.52}$$

The photo- and electroproduction cross sections in the resonance region (see section 2.3.1) display three pronounced peaks. They belong to the $\Delta_{33}(1232)$ (isospin $I = \frac{3}{2}$, spin and parity $j^P = \frac{3}{2}^+$), the $N^*_{13}(1520)$ ($I = \frac{1}{2}$, $j^P = \frac{3}{2}^-$) and the $N^*_{15}(1680)$ ($I = \frac{1}{2}$, $j^P = \frac{5}{2}^+$) resonances. Less prominently seen are the $N^*_{11}(1440)$ ($I = \frac{1}{2}$, $j^P = \frac{1}{2}^+$) and the $N^*_{11}(1535)$ ($I = \frac{1}{2}$, $j^P = \frac{1}{2}^-$). In the quark model the simplest case is that of the Δ resonance, which is reached by a pure spin-flip, magnetic dipole transition. The excitation of the higher N^* resonances involves transitions induced by both the spin-flip (σ_+) and orbital (p_+) transition operators at the quark level.

Detailed applications of the non-relativistic quark model to the calculation of such transition amplitudes are given in [Cl 79]. Here we briefly summarize and discuss some typical results of such calculations. In order to be specific, we start with the amplitudes $A^{p,n}_{1/2}$ for the proton and neutron and consider the excitation of a resonance with spin $j = \frac{1}{2}$ or $\frac{3}{2}$ and orbital angular momentum $L = 1$. The resulting expressions for the proton and neutron amplitudes are:

$$A^p_{\frac{1}{2}} = \sqrt{\frac{2\,\pi}{\omega}} \mu_N \left[C^j_{11}\mathcal{R}_{11} - q\, C^j_{10}\mathcal{R}_{10} \right], \tag{8.53}$$

$$A^n_{\frac{1}{2}} = \sqrt{\frac{2\,\pi}{\omega}} \mu_N \left[-C^j_{11}\mathcal{R}_{11} + \frac{1}{3} q\, C^j_{10}\mathcal{R}_{10} \right], \tag{8.54}$$

with the Clebsch-Gordon coefficients $C^j_{1\,\lambda} = (j\,\frac{1}{2}|1\,\lambda\,\frac{1}{2} - \frac{1}{2})$. The amplitudes are given in units of the nuclear magneton, $\mu_N = \frac{e}{2M}$. The quantities $\mathcal{R}_{L,m_L}$ refer to the orbital transition form factors which involve Fourier integrals over products of the quark model wave functions in the initial and final states. For a spin-flip transition the orbital angular momentum projection is not altered and we have $m_L = 0$. An orbital transition via the operator p_+ leads to $m_L = 1$ in the final state. In the harmonic oscillator model with quark mass m and oscillator spacing ω_0 one finds, explicitly:

$$\mathcal{R}_{10}(q) = \frac{iq}{\sqrt{3m\omega_0}} \exp\left(-\frac{q^2}{6m\omega_0}\right), \tag{8.55}$$

$$\mathcal{R}_{11}(q) = i\sqrt{\frac{2m\omega_0}{3}} \exp\left(-\frac{q^2}{6m\omega_0}\right). \tag{8.56}$$

For the example of $N^*_{13}(1520)$ resonance excitation, the previous amplitudes are then given as:

$$A^p_{\frac{1}{2}} = \frac{2i}{3}\sqrt{\frac{\pi m\omega_0}{\omega}}\mu_N \left(1 - \frac{q^2}{m\omega_0}\right) \exp\left(-\frac{q^2}{6m\omega_0}\right), \tag{8.57}$$

$$A^n_{\frac{1}{2}} = -\frac{2i}{3}\sqrt{\frac{\pi m\omega_0}{\omega}}\mu_N \left(1 - \frac{1}{3}\frac{q^2}{m\omega_0}\right) \exp\left(-\frac{q^2}{6m\omega_0}\right). \tag{8.58}$$

These amplitudes hold for real photons at resonance, with $\omega = q$ matching the difference between the nucleon and resonance masses in the static limit. For virtual photons with $q > \omega$, as provided by inelastic electron scattering, one should be aware of the rapid dependence of these amplitudes on the transferred photon momentum. Characteristic sign changes are expected to occur when q^2 equals some multiple of $m\omega_0$, the inverse mean square oscillator radius.

In order that the amplitude $A_{\frac{3}{2}}$ be non-zero the resonance which is excited must have $j = \frac{3}{2}$, at least (since one requires $m_j = \frac{3}{2}$ in the excited state). For the N^*_{13} this cannot be done by quark spin-flip. Only the orbital p_+ term contributes to $A_{\frac{3}{2}}$ in this case. One finds:

$$A^p_{\frac{3}{2}} = \sqrt{\frac{2\,\pi}{\omega}}\mu_0\mathcal{R}_{11} = -A^n_{\frac{3}{2}}. \tag{8.59}$$

An interesting quantity to investigate is the asymmetry

$$\mathcal{A} = \frac{|A_{\frac{1}{2}}|^2 - |A_{\frac{3}{2}}|^2}{|A_{\frac{1}{2}}|^2 + |A_{\frac{3}{2}}|^2} \tag{8.60}$$

at resonance. It shows the rapid change in helicity structure when going from photo- to electro-production of a given nucleon resonance. Let us discuss again the example of the $N^*_{13}(1520)$ and notice first that the characteristic q^2 dependence of $A^p_{\frac{1}{2}}$ translates into the ratio

$$\frac{A^p_{\frac{1}{2}}}{A^p_{\frac{3}{2}}} = \frac{1}{\sqrt{3}}\left(1 - \frac{q^2}{m\omega_0}\right). \tag{8.61}$$

The helicity asymmetry $\mathcal{A}^p$ as a function of $Q^2 = q^2 - \omega_R^2$, taken at the excitation energy ω_R of the N^*_{13}, is shown in Fig. 8.5. The qualitative pattern seen there, with its typical sign change at low Q^2 reflecting the size of the transition form factor, is actually observed experimentally, as has been discussed in Section 2.3.4.

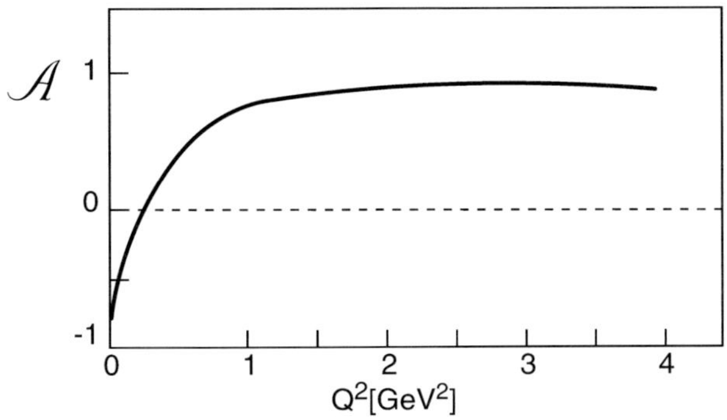

Figure 8.5: The helicity asymmetry, $\mathcal{A} = (|A_{\frac{1}{2}}|^2 - |A_{\frac{3}{2}}|^2)/(|A_{\frac{1}{2}}|^2 + |A_{\frac{3}{2}}|^2)$, for $\gamma^* p \to N^*_{13}(1520)$ as a function of squared photon four-momentum, Q^2, calculated in the simple harmonic oscillator quark model. For comparison with data see Fig. 2.19 of Section 2.3.1.

For the $N^*_{15}(1680)$ resonance similar features occur, except that now the zero of the asymmetry is shifted to slightly higher Q^2. The reason is that this resonance requires $L = 2$, and the transition formfactor which enters $A^p_{1/2}$ now has its zero at $q^2 = 2m\omega_0$, reflecting the fact that this is a "$2\hbar\omega_0$" excitation in the oscillator quark model.

8.3 Soliton Models

While the long-range potentials introduced in the non-relativistic quark models certainly have the effect of confining the quarks, they do not provide any physical picture of the mechanism for confinement within QCD. Moreover, since QCD is a quantum field theory one would actually expect that the phenomenon might be a many-body effect, more analogous to a phase transition between the interior and exterior of the hadron. Reasoning such as this led T. D. Lee [Lee 81] to suggest that quark confinement should be understandable, at least phenomenologically, through a quasi-classical microscopic theory.

In order to understand how a quasi-classical theory might lead to a rather natural description of confinement, we suppose that the QCD vacuum can be treated as a color dia-electric medium (with dielectric constant $\kappa < 1$). Any color charge, ϵ, inserted into such a medium will produce a hole (vacuum) with $\kappa = 1$ (as illustrated in Fig. 8.6). The repulsion between ϵ and the surface charges means that work has to be done to shrink the hole. The continuity of the displacement, $\vec{D}$:

$$\vec{D}_{\text{in}}(= \vec{E}_{\text{in}}) = \vec{D}_{\text{out}}(= \kappa \vec{E}_{\text{out}})\Big|_{r=R}, \qquad (8.62)$$

means that the color electric field is given by

$$\begin{aligned}\vec{E}_{\text{in}} &= \frac{\epsilon}{r^2}\hat{r} \qquad\qquad r \le R, \\ \vec{E}_{\text{out}} &= \frac{\epsilon}{\kappa r^2}\hat{r} \qquad\qquad r \ge R. \end{aligned} \tag{8.63}$$

Thus we find that the energy of the charge in the cavity (relative to the energy of the same charge in free space) has the form

$$\sim \epsilon^2 \frac{\kappa^{-1} - 1}{R} \longrightarrow \infty \quad \text{as} \quad \kappa \to 0, \tag{8.64}$$

so that a perfect dia-electric is *confining*.

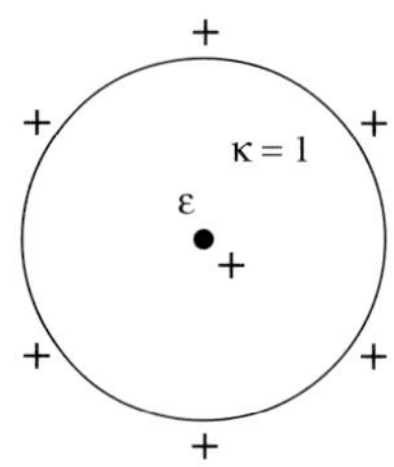

Figure 8.6: If the non-perturbative vacuum is a dia-electric (i.e., $\kappa \ll 1$), it is anti-shielding. As a result a cavity of perturbative vacuum ($\kappa = 1$) will form around any color charge, ϵ, placed in it.

Now we suppose that the dia-electric vacuum is the *lowest* energy state, so that it costs energy to make this hole with $\kappa = 1$. In general the energy required to make such a hole is given by

$$\mathcal{U}_{\text{hole}} = BV + CS, \tag{8.65}$$

where B and C are constants, V is the volume of the hole and S is the surface area. Therefore we find the equilibrium radius, R_{equil}, when

$$\frac{d}{dR}\left(\mathcal{U}_{\text{hole}} - \mathcal{U}_{\text{electric}}\right)\bigg|_{R=R_{\text{equil}}} = 0. \tag{8.66}$$

Clearly this equilibrium radius will tend to ∞ when $\kappa \to 0$, so the only solutions with finite mass are those of the form illustrated in Fig. 8.7, where the total charge inside the cavity is zero.

Figure 8.7 illustrates the close analogy between the superconducting state, which excludes magnetic fields (the Meissner effect), and the QCD vacuum which excludes the color electric field, leading to confinement of color charges. Recent work based on an Abelian projection of lattice QCD tends to support this picture of the QCD vacuum [Kom 00, BWS 91], which appears to contain a condensate of color magnetic monopoles. This is a very active area of research (see, for example, Ref. [ER 00]), which would take us too far from the phenomenological emphasis of this chapter. We turn now to the quark models which have been built upon the idea of a color dia-electric vacuum.

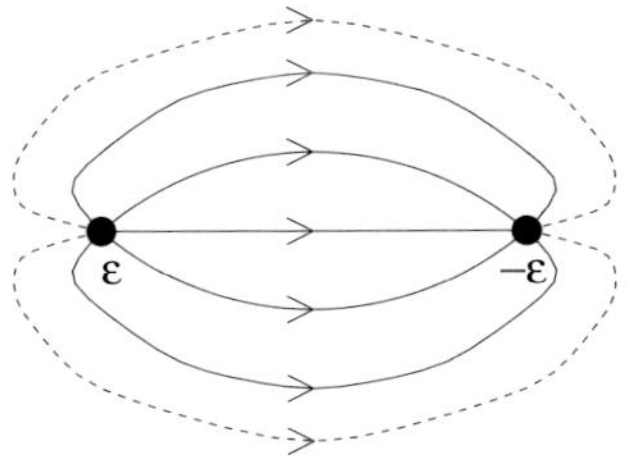

Figure 8.7: Schematic illustration of a color dipole in a dia-electric medium. Here the dashed lines can be visualized as the self-consistent "bag surface"

8.3.1 The Friedberg-Lee Soliton Model

The Friedberg-Lee soliton model [FL 77] in its simplest form is implemented through the Lagrangian density:

$$\begin{aligned}\mathcal{L}_{\mathrm{FL}} &= \bar{\psi}\,(i\gamma_\mu D^\mu - m - f\sigma)\,\psi - \frac{1}{4}\,\kappa\,G_a^{\mu\nu}\,G^a_{\mu\nu} \\ &+ \frac{1}{2}\,(\partial_\mu\sigma)^2 - \mathcal{U}(\sigma),\end{aligned} \tag{8.67}$$

where σ is a new scalar field, and the term involving κ is a dielectric function that we treat perturbatively for colorless states. The classical field, σ, experiences a potential $\mathcal{U}(\sigma)$, the behavior of which is illustrated in Fig. 8.8.

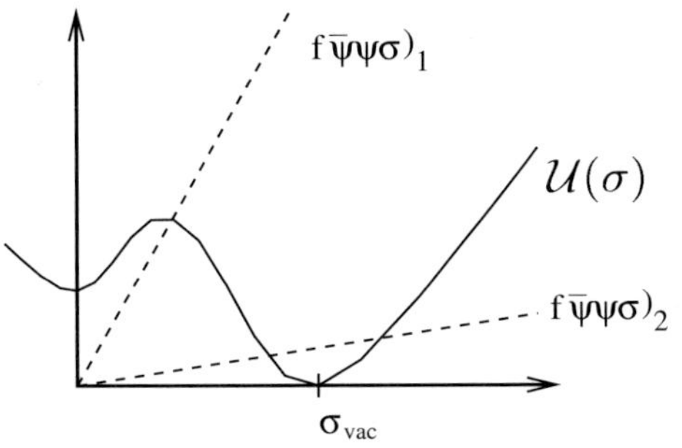

Figure 8.8: Illustration of the scalar field and quark-scalar-field contributions to the total energy for two choices of the valence quark scalar density.

How such a model leads to a localized, self-consistent soliton[1] is illustrated in Fig. 8.8. When $\psi = 0$, which corresponds to the non-perturbative vacuum with no valence quarks, the lowest energy state has $\sigma = \sigma_{\mathrm{vac}}$. If, on the other hand, we consider the case where the quark density is high, $f\bar{\psi}\psi)_1$, we find that σ becomes small. This means that when the scalar, valence quark density is high the perturbative vacuum is restored. Finally, in the case

[1]Note that the term "soliton" used here refers to a so-called "non-topological soliton", for which there is no topological index or winding number.

that the quark field is small, $f\bar{\psi}\psi)_2$, we see that $\sigma \rightarrow \sigma_{\rm vac}$ again – corresponding to the non-perturbative vacuum.

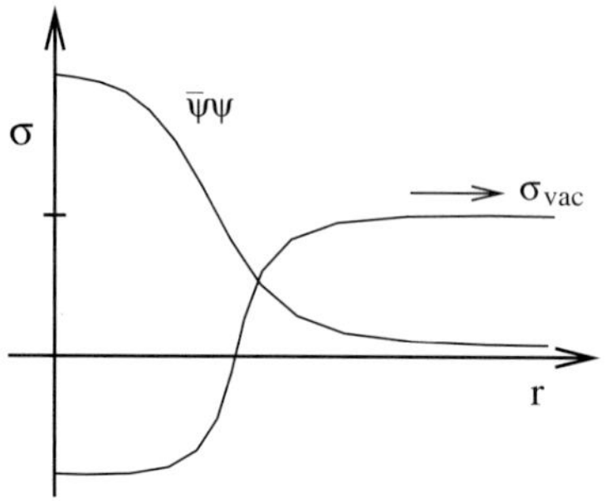

Figure 8.9: Formation of a non-topological soliton as a self-consistent combination of localized valence quark density and a "hole" in the non-perturbative vacuum field configuration.

In summary, within the Friedberg-Lee soliton model:

- the quarks dig a self-consistent "hole" in the scalar field;
- asymptotically quarks have an effective mass, $m_{\rm eff} = m + f\sigma_{\rm vac}$, where m is the perturbative QCD mass;
- quarks are "confined" in the sense that if, for example, $\kappa \sim (1 - \sigma/\sigma_{\rm vac})^n$ then $\kappa \rightarrow 0$ outside the soliton. Hence the total color electric energy is infinite, unless the system is in a colorless state. This is shown by using the same argument as for a charge in a dia-electric cavity.

8.3.2 The Color Dielectric Model

A slightly more sophisticated non-topological soliton model, which was motivated by a block-spin treatment of lattice QCD, is the color dielectric model [NP 82, Pir 87, Pir 92]. This model is most simply expressed through the Lagrangian density:

$$\mathcal{L} = i\bar{\psi}\gamma_\mu\partial^\mu\psi - \frac{m}{\chi}\,\bar{\psi}\psi - U(\chi) + \frac{1}{2}\,\sigma_{\rm V}{}^2\,(\partial_\mu\chi)^2 - \frac{\kappa(\chi)}{4}\,G^a_{\mu\nu}\,G^{\mu\nu}_a\,, \tag{8.68}$$

where $\sigma_{\rm V}\chi$ is a confining scalar field which is meant to mimic the non-perturbative effect of the gluons, the constant, $\sigma_{\rm V}$, sets the scale and χ is a dimensionless scalar field from which the soliton is self-consistently constructed. Note that when $\chi \rightarrow 1$ (inside the soliton) the second term goes to the usual mass term, $m\bar{\psi}\psi$, while as $\chi \rightarrow 0$ the effective mass of the quark goes to infinity. This certainly amounts to quark confinement as

$$\chi \rightarrow 0 \quad \Longrightarrow \quad \frac{m}{\chi} \rightarrow \infty. \tag{8.69}$$

In the simplest case, the potential term, $U(\chi)$, may be taken to be quadratic in χ:

$$U(\chi) \approx \frac{1}{2}\,m^2_{\rm GB}\sigma_{\rm V}{}^2\chi^2, \tag{8.70}$$

where the parameter m_{GB} should be of the order of the mass of a glueball, the characteristic gluonic energy scale.

There have been extensive numerical studies of this model [DWT 87, BCB 87] (and other variations of the non-topological soliton idea [BB 86, Bir 90]), which because of its non-linear structure can produce an interesting variety of solutions. In particular, for some values of the parameters the solutions look very much like a constituent quark model, with χ small and slowly varying in the interior of the hadron. For other parameters the lowest energy solution has two phases with χ near unity in the interior but dropping rapidly to zero outside. In this case the quarks are confined but behave almost like free particles inside the hadron. Since solutions of this kind exhibit behavior like asymptotic freedom inside the hadron and confinement at large distances they are attractive candidates for modelling hadron structure. As the results of these studies are very similar to those of the MIT bag model, which has the advantage that it can be solved analytically, rather than purely numerically, we turn immediately to the MIT bag.

8.4 Bag Models

Despite extensive searches in the late 60's (continuing to the present day), no free quarks were found. It was therefore clear, even before QCD, that one needed some form of confinement mechanism. The most obvious was to make the quarks extremely massive, but have them bound in an extremely deep potential so that the observed hadronic states could still have reasonable masses. A simple model of this kind was constructed by Bogoliubov [Bog 67]. He set the quark masses to m, with $m \to \infty$, but placed them in a spherical volume of radius R, within which they felt an attractive scalar field of strength m. This extremely simple model led to an accurate prediction for the ratio of the mass of the Roper resonance to that of the nucleon, to a simple explanation of the deviation of g_A from the naive $SU(6)$ result of $\frac{5}{3}$ and many other interesting results.

However, the real renaissance of interest in such models came only after it was realized that QCD is asymptotically free, at short distances, as well as (probably) confining at large distances. The MIT bag model [Ch 74, DeG 75] was constructed in such a way as to build these properties into a simple phenomenological model. Space was divided into two regions, the interior of the bag in which the quarks had very small (current) masses and felt only weak forces and the exterior in which the quarks were not allowed to propagate and which had a different (lower) vacuum energy. It was soon realized that as far as the motion of the quarks was concerned, at least for a static, spherical cavity, the wave functions in the MIT bag were identical to those in the model of Bogoliubov in the limit where the parameter $m \to \infty$. Bearing this in mind, we begin our discussion of bag models with a review of the model of Bogoliubov which is mathematically very simple.

8.4.1 The Model of Bogoliubov

Consider the Dirac equation for a particle of mass m moving in a spherical cavity of radius R within which it feels a constant scalar potential $-V_s$:

$$[\vec{\alpha} \cdot \vec{p} + \beta(m - V_s)]\, \psi = E\psi. \tag{8.71}$$

The operators representing the total angular momentum, $\vec{j} = \vec{l} + \vec{\sigma}/2$, and also $K = \beta(\vec{\sigma}\cdot\vec{l}+1)$ commute with this Hamiltonian and may therefore be used to classify the eigenstates of Eq. (8.71). Taking the eigenvalues of $\vec{j}^2$, j_z and K to be $j(j+1)$, μ and $-\kappa$, respectively, one can show that κ is equal to $\pm(j+\frac{1}{2})$ and it is therefore sufficient to label ψ by κ and μ, ψ^{μ}_{κ}. The 4×4 matrix operator K obviously has the 2×2 matrices $\pm(\vec{l}\cdot\vec{\sigma}+1) \equiv \pm k$ down the diagonal. Taking the eigenfunctions of k and j_z to be χ^{μ}_{κ} we may therefore write:

$$\psi^{\mu}_{\kappa} = \begin{pmatrix} g(r)\chi^{\mu}_{\kappa} \\ i\,f(r)\chi^{\mu}_{-\kappa} \end{pmatrix}. \tag{8.72}$$

Using the operator identity

$$\vec{\nabla} = \hat{r}\frac{\partial}{\partial r} - i\frac{\hat{r}}{r}\times\vec{l}, \tag{8.73}$$

we find

$$\vec{\alpha}\cdot\vec{p} = -i\vec{\alpha}\cdot\hat{r}\frac{\partial}{\partial r} + \frac{i}{r}\vec{\alpha}\cdot\hat{r}(\beta K - 1). \tag{8.74}$$

It is then simple to write down the two coupled differential equations for $g(r)$ and $f(r)$:

$$\begin{aligned} (E+V_s-m)g &= -\left(\frac{df}{dr}+\frac{f}{r}\right)+\kappa\frac{f}{r}, \\ (E-V_s+m)f &= \left(\frac{dg}{dr}+\frac{g}{r}\right)+\kappa\frac{g}{r}. \end{aligned} \tag{8.75}$$

If we now take $V_s = m$ for $r < R$ and $V_s = 0$ for $r > R$ it is easy to solve these coupled equations. For $\kappa = -1$ (in which case the upper component is in s-wave) we find:

$$g(r) = \begin{cases} A\dfrac{\sin Er}{r} & r < R, \\ A\dfrac{\sin Er}{r}e^{-\sqrt{m^2-E^2}(r-R)} & r > R, \end{cases} \tag{8.76}$$

where the eigenenergy is obtained by matching the first derivatives at $r = R$.

The particular case of interest here is the limit where $m \to \infty$, so that the quarks behave as though they were free (with zero effective mass) inside R, but have infinite mass outside – and are therefore confined. In that limit, matching the derivative inside and outside gives the condition

$$j_0(ER) = j_1(ER), \tag{8.77}$$

where j_0 and j_1 are the spherical Bessel functions

$$j_0(x) = \frac{\sin x}{x}; \qquad j_1(x) = \frac{\sin x}{x^2} - \frac{\cos x}{x}. \tag{8.78}$$

Equation (8.77) has the solutions $ER = 2.04, 5.40, \ldots$, so that if we label the eigenenergies by the principal quantum number, n, and κ we have $E_{n,\kappa} = \omega_{n,\kappa}/R$ with, for example, $\omega_{n,-1} = 2.04, 5.40, \ldots$. Using the property

$$\chi^{\mu}_{-\kappa} = -\vec{\sigma}\cdot\hat{r}\chi^{\mu}_{\kappa}, \tag{8.79}$$

we can now write down the wave functions for the confined quark in this simple problem

$$\psi^{\mu}_{n,-1} = \frac{N_{n,-1}}{\sqrt{4\pi}} \begin{bmatrix} j_0(\omega_{n,-1}\frac{r}{R}) \\ i\vec{\sigma}\cdot\hat{r}j_1(\omega_{n,-1}\frac{r}{R}) \end{bmatrix} \chi^{\mu}_{-1}, \tag{8.80}$$

for $r < R$. (It is, of course, zero for $r > R$.) In the present case, the upper component is in an s-state and the spin-angle function, χ^{μ}_{-1}, is just a Pauli spinor. The normalization constant is given by:

$$N_{n,-1} = \frac{\omega^3_{n,-1}}{2R^3(\omega_{n,-1} - 1)\sin^2(\omega_{n,-1})}. \tag{8.81}$$

The first excited state

The immediate success of this simple model, which was observed by Bogoliubov, is that the first excited state of the nucleon (with positive parity) has two quarks in the 1s-state and one in the 2s-state. The ratio of the mass of this excited state to that of the nucleon is therefore $(5.40 + 2 \times 2.04)/(3 \times 2.04) = 1.55$, which is in remarkable agreement with the ratio of the mass of the Roper resonance to that of the nucleon, namely 1.56. Of course, this close agreement cannot now be regarded as anything more than a happy coincidence as there are many subtleties ignored in this extremely simple model. Nevertheless, it is an indication that the model is not grossly incorrect and certainly served to encourage early work.

The axial coupling constant

Within the non-relativistic $SU(6)$ quark model we have already seen that the axial coupling constant of the nucleon is predicted to be $\frac{5}{3}$, in disagreement with the experimental value of 1.26. In a relativistic treatment the axial current involves $\bar{\psi}\gamma^{\mu}\gamma_5\frac{\lambda_a}{2}\psi$ and for the case of massless, confined quarks one can easily see that, in the Bogoliubov model, the non-relativistic result is just multiplied by a factor

$$\frac{\int_0^R drr^2[j_0^2 - \frac{1}{3}j_1^2]}{\int_0^R drr^2[j_0^2 + j_1^2]} = 1 - \frac{1}{3}\left(\frac{2\omega_{1,-1} - 3}{\omega_{1,-1} - 1}\right) = 0.65. \tag{8.82}$$

This reduces g_A from $\frac{5}{3}$ to 1.09 – somewhat below the experimental value but much closer to it.

8.4.2 The MIT Bag Model

Although, from the modern point of view, the model of Bogoliubov was quite attractive, building in asymptotic freedom and confinement in a simple way, it was not very useful for spectroscopy. In particular, the radius of the spherical cavity was put in by hand, rather than being determined dynamically. Indeed, minimizing the energy of the hadron would lead us to favour $R \to \infty$. In addition, one would prefer to formulate a model of hadron structure covariantly, even if, in practice, one needed to make approximations to find a solution.

The MIT bag model satisfies all of these criteria, while retaining an elegant simplicity. In its most general form, for massless quarks, the model involves a three-dimensional volume V – the "bag" – with surface S. If $\theta_{\rm V}$ is one inside the volume and zero outside the Lagrangian density for the MIT bag is

$$\mathcal{L} = \left(i\bar{\psi}\gamma^\mu\partial_\mu\psi - B\right)\theta_{\rm V} - \frac{1}{2}\,\bar{\psi}\psi\delta_{\rm S}, \tag{8.83}$$

where $\delta_{\rm S}$ is a δ function at the bag surface. The additional constant, B, is supposed to be the extra energy per unit volume required to create a region of space where the vacuum is perturbative – as opposed to the region outside the bag, where there are no free quarks and the vacuum is non-perturbative.

Requiring that the action associated with $\mathcal{L}$ be stationary with respect to variations in the field and the bag surface, S, leads to three equations. The first is the free Dirac equation for a massless quark inside the bag

$$i\gamma^\mu\partial_\mu\psi = 0. \tag{8.84}$$

The other two equations are boundary conditions for ψ on S, a linear boundary condition (l.b.c.):

$$i\gamma^\mu\eta_\mu\psi(x) = \psi(x), \qquad x \in {\rm S}, \tag{8.85}$$

and a non-linear boundary condition (n.l.b.c.):

$$B = -\frac{1}{2}\,\eta_\mu\partial^\mu\left(\bar{\psi}\psi(x)\right), \qquad x \in {\rm S}, \tag{8.86}$$

where η^μ is a normal to the surface (with $\eta^2 = -1$).

The l.b.c. can be used to show that the component of the quark current, $j^\mu = \bar{\psi}\gamma^\mu\psi$, normal to the bag surface is zero:

$$\eta_\mu\bar{\psi}\gamma^\mu\psi(x) = 0, \qquad x \in {\rm S}. \tag{8.87}$$

The proof is easy. Taking the Hermitian conjugate of Eq. (8.85) and multiplying to the right by γ^0 we find

$$\bar{\psi} = -i\bar{\psi}\gamma\cdot\eta. \tag{8.88}$$

Hence, we may write

$$\begin{aligned} i\eta_\mu j^\mu &= \left(\bar{\psi}i\gamma\cdot\eta\right)\psi = -\bar{\psi}\psi \\ &= \bar{\psi}\left(i\gamma\cdot\eta\psi\right) = +\bar{\psi}\psi \\ &= 0. \end{aligned} \tag{8.89}$$

As there is no component of the current normal to the surface the quarks are confined to the interior of the bag. In the case of a static, spherical bag, $\eta^\mu = (0,\hat{r})$, and substituting

Eq. (8.80), which for arbitrary $\omega_{n,-1}$ is the solution of the free Dirac equation (in s-wave), into Eq. (8.85), the l.b.c., we find (at $r = R$):

$$j_0(\omega_{n,-1}) = j_1(\omega_{n,-1}). \tag{8.90}$$

This is exactly the same eigenvalue condition as we found in the model of Bogoliubov.

On the other hand, the n.l.b.c., Eq. (8.86), is new. It involves a relationship between the normal derivative of $\bar{\psi}\psi$ at the surface of the bag and B. This has a simple interpretation as a stability condition because, if we consider the energy-momentum tensor for the model

$$T^{\mu\nu} = \left(\frac{i}{2}\bar{\psi}(\overrightarrow{\partial}_\mu - \overleftarrow{\partial}_\mu)\psi + Bg^{\mu\nu}\right)\theta_V, \tag{8.91}$$

we find, using the relation

$$\delta_S = -\eta \cdot \partial\theta_V, \tag{8.92}$$

that the n.l.b.c. ensures that $\partial_\mu T^{\mu\nu} = 0$ on the surface (it is guaranteed by the Dirac equation elsewhere). Thus B is essential to ensure energy momentum conservation. From the energy momentum tensor we identify $-\frac{1}{2}\eta \cdot \partial(\bar{\psi}\psi)$ as the normal component of the pressure exerted by the gas of free Dirac particles inside the bag. The n.l.b.c. ensures that this is exactly balanced at the surface by the difference in the energy density inside and outside the bag – and hence that the system is stable.

Using the energy momentum tensor we can construct the energy of the bag as

$$P^0 = \int d^3x\, T^{00}. \tag{8.93}$$

Specializing again to the simplest case of a static spherical bag of radius R, with three quarks in the 1s-state, this yields

$$P^0 = 3\,\frac{\omega_{1-1}}{R} + \frac{4\,\pi}{3}\,BR^3, \tag{8.94}$$

and the n.l.b.c. requires

$$\frac{\partial P^0}{\partial R} = 0. \tag{8.95}$$

Thus the internal energy of the quarks determines the radius of the cavity if B is taken to be a universal constant – independent of the hadron being considered. One can now discuss excited states with confidence.

Static spherical cavity

The MIT bag model, as formulated above, is a very complicated problem in quantum field theory. It has only been solved in one space and one time dimension [Jaf 81]. For practical purposes it has mainly been used in a much simpler form where the cavity is taken to be a static, spherical cavity of radius R, fixed by satisfying the n.l.b.c. on average over the surface.

(One can only satisfy the n.l.b.c. exactly over a spherical surface for states with quantum numbers $s_{\frac{1}{2}}$ and $p_{\frac{1}{2}}$.) As we have seen this is the model of Bogoliubov, but with the radius determined dynamically.

The MIT bag model is constructed to model QCD and one must therefore also include the glue required to ensure local gauge invariance. Of course, the bag itself is supposed to be generated by the non-perturbative interactions between gluons so there is some danger of double counting. For this reason, while the gluons are included in the Lagrangian density describing the bag model, they are typically treated only in low order perturbation theory. (An exception to the rule is the treatment of glueball or hybrid states, but those are not of direct concern to us in the context of the structure of the nucleon.) Taking the volume to be V and the surface S in this case, the Lagrangian density is

$$L = \left(i\bar{\psi}\gamma^\mu D_\mu \psi - \frac{1}{4} G^a_{\mu\nu} G^{\mu\nu}_a - B \right) \theta_V - \frac{1}{2} \bar{\psi}\psi\delta_S. \tag{8.96}$$

In Eq. (8.96) $D^\mu = \partial^\mu - igA^\mu$ (with $A^\mu = \frac{\lambda_a}{2} A^\mu_a$), the usual covariant derivative in QCD.

If we choose η^μ to be the outward normal for a static spherical bag

$$\eta^\mu = (0, \hat{r}), \tag{8.97}$$

minimizing the action associated with Eq. (8.96) leads to the quark and gluon field equations:

$$\left[i\gamma^\mu \partial_\mu + g\gamma^\mu A_\mu \right] \psi = 0, \qquad r \leq R \tag{8.98}$$

$$\partial_\nu G^{\mu\nu}_a = g \left[\bar{\psi}\gamma^\mu \frac{\lambda_a}{2} \psi + f_{abc} G^{\mu\nu}_b A_{c\nu} \right], \qquad r \leq R, \tag{8.99}$$

linear boundary conditions (l.b.c.) for the quarks and gluons:

$$-i\vec{\gamma} \cdot \hat{r}\psi = \psi, \qquad r = R, \tag{8.100}$$

$$\hat{r} \cdot \vec{E}_a = 0, \qquad \hat{r} \times \vec{B}_a = 0, \qquad r = R, \tag{8.101}$$

and finally a generalization of the n.l:b.c. to include the pressure of the confined gluons:

$$B = -\frac{1}{2} \frac{\partial}{\partial r} (\bar{\psi}\psi) - \frac{1}{4} G^a_{\mu\nu} G^{\mu\nu}_a, \qquad r = R. \tag{8.102}$$

If we treat the effect of the gluons in low order perturbation theory, for the reasons explained earlier, Eqs. (8.98) and (8.100) are just the free Dirac equation and the linear boundary condition for which we obtained the s-wave solution above. The solutions with arbitrary angular momentum are (recall $\kappa = \pm(j + \frac{1}{2})$)

$$\psi^\mu_{n,\kappa} = N_{n,\kappa} \begin{pmatrix} j_{-\kappa-1}(\omega_{n,\kappa} \frac{r}{R}) \\ i\vec{\sigma} \cdot \hat{r}\, j_{-\kappa}(\omega_{n,\kappa} \frac{r}{R}) \end{pmatrix} \chi^\mu_\kappa, \qquad \text{for} \quad \kappa < 0 \tag{8.103}$$

and

$$\psi^\mu_{n,\kappa} = N_{n,\kappa} \begin{pmatrix} i\vec{\sigma} \cdot \hat{r}\, j_\kappa(\omega_{n,\kappa} \frac{r}{R}) \\ j_{\kappa-1}(\omega_{n,\kappa} \frac{r}{R}) \end{pmatrix} \chi^\mu_\kappa, \qquad \text{for} \quad \kappa > 0. \tag{8.104}$$

In general χ^{μ}_{κ} is given by:

$$\chi^{\mu}_{\kappa} = \sum_{m,\mu} C^{m\nu\mu}_{l\frac{1}{2}j} Y^{m}_{l} \chi^{\nu}_{\frac{1}{2}}. \tag{8.105}$$

As we have already seen the eigenvalues, $\omega_{n,\kappa}$, are determined by the l.b.c., Eq. (8.100), which implies

$$j_{-\kappa-1}(\omega_{n,\kappa}) = j_{-\kappa}(\omega_{n,\kappa}); \qquad j_{\kappa}(\omega_{n,\kappa}) = -j_{\kappa-1}(\omega_{n,\kappa}). \tag{8.106}$$

(Of course, the first and second cases in Eq. (8.106) correspond to κ less than or greater than zero, respectively.) For example, for $ns_{1/2}$ one has $\omega_{n,-1} = 2.043,\ 5.396,\ 8.578,\ldots$ and for $np_{\frac{1}{2}}$, $\omega_{n,+1} = 3.812,\ 7.002,\ldots$, etc.

A final feature of the bag which needs some mention is that in practical applications it is common to introduce an additional correction to the bag energy of the form $-Z_0/R$. This was originally introduced by DeGrand et al.[DeG 75] as an approximation to the Casimir energy associated with gluon and $q\bar{q}$ fluctuations in an empty bag. In practice there does seem to be a need for such a term, with $Z_0 \sim 1.8$. It has the effect of raising the value of B needed to fit the nucleon mass and hence making the nucleon bag radius somewhat smaller. However, its origin in the Casimir effect is rather unclear. We simply note that the center of mass correction, which will be discussed in more detail below, also tends to lower the total energy of the hadronic states in the bag by an amount proportional to $1/R$. From the practical point of view, the term $-Z_0/R$ may be viewed as including both the center of mass and Casimir corrections in a phenomenological way.

Spectroscopy

The key additions to the model required in order that it can be applied to the lowest mass baryons and vector mesons are: (a) a non-zero mass for the strange quark, and (b) the spin-dependent energy shift arising from perturbative gluon exchange. Introducing a quark mass term into Eqs. (8.96) and (8.98) adds no significant complications to what we have already discussed. For example, for an s-state ($\kappa = -1$) we find

$$\psi_{n,-1} = N_{n,-1} \begin{pmatrix} \alpha_{+}\, j_0(\frac{xr}{R}) \\ \alpha_{-}\, i\vec{\sigma}\cdot\hat{r}\, j_1(\frac{xr}{R}) \end{pmatrix} \chi^{\mu}_{-1}, \tag{8.107}$$

where the quark energy is related to the spatial frequency, x, by

$$E = \frac{\Omega}{R}; \qquad \Omega = \sqrt{x^2 + (mR)^2}, \tag{8.108}$$

and x is determined by the l.b.c.

$$\tan x = \frac{x}{1 - mR - \Omega}. \tag{8.109}$$

The coefficients $\alpha_{\pm}$ are $\sqrt{(E \pm m)/m}$ and the normalization constant is

$$N^2 = \frac{\Omega(\Omega - mR)}{R^3 j_0^2(x)\left[2\,\Omega(\Omega - 1) + mR\right]}. \tag{8.110}$$

By introducing $m_s \neq m_{u,d}$ one can fit the $SU(3)$ breaking in the hadron mass spectrum, just as in the non-relativistic quark model.

The treatment of gluon exchange is somewhat more idiosyncratic, as the total Coulomb energy is supposed to be zero because the physical hadrons are color singlets. Thus the only term retained in second order perturbation theory is the spin-dependent one gluon exchange between different quarks

$$\Delta E_g^M = -\frac{\alpha_s}{2} \sum_{i<j} \int d^3x\, \vec{B}_i^a \cdot \vec{B}_j^a. \tag{8.111}$$

Here $\vec{B}_i^a$ is the solution of the classical field equations for the gluon field strength inside the cavity, including the boundary conditions, generated by the color current of quark i. (Note that each $\vec{B}_i^a$ separately satisfies the l.b.c. $\hat{r} \times \vec{B}_i^a = 0$.) Equation (8.111) leads to

$$\Delta E_g^M = -3\frac{\alpha_s}{R} \sum_{i<j} (\vec{\sigma}_i \lambda_i^a) \cdot (\vec{\sigma}_j \lambda_j^a) M(m_i, m_j, R), \tag{8.112}$$

where M can be obtained in closed form. Clearly Eq. (8.112) has exactly the same spin-flavor structure as the hyperfine interaction in the non-relativistic quark model and since α_s is treated as an adjustable parameter one can fit the $N - \Delta$ mass difference and the $\Sigma - \Lambda$ splitting – as was discussed in Section 8.2.6.

In summary, the energy of a bag, including all the effects we have discussed, is

$$E(R) = \sum_i \frac{\Omega_i}{R} + \frac{4\pi}{3} R^3 B + \Delta E_g^M - \frac{Z}{R} \tag{8.113}$$

and the radius is determined by the n.l.b.c., $\partial E(R)/\partial R = 0$. By adjusting B, α_s, Z and m_s one can obtain a very good fit to the masses of the ground state baryon octet and decuplet as well as the vector mesons. B is typically (146 MeV)$^{\frac{1}{4}}$, or roughly 58 MeV/fm^3, and the nucleon bag radius is about 1 fm. (The other parameters are $m_s = 279$ MeV and $Z_0 \sim 1.84$, while $\alpha_s \sim 2.2$ is even larger than in the non-relativistic quark model.)

One major difference between the non-relativistic oscillator model and the bag is that in the latter the spin and spatial wave functions are inextricably mixed. For the [56] representation of $SU(6)$, which contains the nucleon spin-$\frac{1}{2}$ octet and the delta spin-$\frac{3}{2}$ decuplet, this is not a complication as the wave function is a totally anti-symmetric product of flavor, color and spin wave functions:

$$|\Psi\rangle \sim SU(3)_{\text{flavor}} \times SU(3)_{\text{color}} \times SU(2)_{\text{“spin”}}. \tag{8.114}$$

Here $SU(2)_{\text{“spin”}}$ is the full Dirac wave function for the $1s_{1/2}$ state, but in any case the spatial wave function is totally symmetric. However, for excited states, for example, where one quark is in a $p_{\frac{1}{2}}$ state, the situation becomes much more complicated.

Whereas one can remove the center of mass motion exactly in the oscillator model, leaving two internal, spatial degrees of freedom, this is not possible in the bag. Equation (8.114) corresponds to three, independent spatial degrees of freedom in the fixed cavity. This is physically incorrect and a great deal of effort has gone into attempts to remove spurious center of mass motion leading to unphysical excitations. For example, whereas it is trivial to see that the [56] 1^- is purely a spurious excitation of the center of mass in the non-relativistic harmonic oscillator model, this is much less obvious in the bag model. In particular, one could imagine a genuine excitation in which the quarks oscillate with respect to the non-perturbative glue making up the bag. In order to address this one must go beyond the static bag approximation and allow at least small surface excitations of the bag. Finally, we note that, because the spin and orbital angular momentum are linked through the Dirac wave function, the natural coupling scheme for the bag is j-j coupling. On the other hand, the success of the non-relativistic harmonic oscillator model strongly suggests that L-S coupling is the better scheme within which to tackle the spectrum of excited states.

In summary, the discussion of excited states within the bag model is technically complicated because of the spurious center of mass motion and the fact that j-j coupling is most natural. The overall description of the baryon spectrum is certainly less impressive in the bag model than in the non-relativistic harmonic oscillator model, but it is not clear to what extent this is a consequence of the technical difficulties, rather than a failing of the model. Certainly the relativistic quark wave functions in the bag model have led to dramatic improvements in the predictions for some photoproduction amplitudes (e.g., the [70] 1^-) in comparison with the non-relativistic models [He 83]. The new feature, which could be extremely important, is the possibility of genuinely new states, not present in the oscillator model, in which the non-perturbative glue and the quarks move relative to each other or in which the glue itself is excited. As an important example of the latter phenomenon, it has been suggested that the Roper resonance may involve a breathing mode (0^+) excitation of the bag itself [Gui 85]. These are all important topics for future research.

Charge and current densities

The bag model Lagrangian density is clearly invariant under the phase transformations

$$\begin{aligned} \psi &\rightarrow \psi + i\varepsilon\psi, \\ \bar{\psi} &\rightarrow \bar{\psi} - i\varepsilon\bar{\psi}. \end{aligned} \tag{8.115}$$

and hence the vector current, $j^\mu = \bar{\psi}\gamma^\mu\psi\theta_{\rm V}$ is conserved. Noting that the upper and lower components of the quark wave functions must be equal at the bag boundary and that the charge density is

$$\begin{aligned} j^0 &\equiv \rho = \bar{\psi}\gamma^0\psi \\ &\propto \left[j_0{}^2\left(\frac{\omega r}{R}\right) + j_1{}^2\left(\frac{\omega r}{R}\right) \right], \end{aligned} \tag{8.116}$$

we can easily understand the forms shown in Fig. 8.10.

The calculation of the magnetic moment is a little more interesting because massless quarks have no Dirac moment. This is completely changed by the confinement as we now illustrate. Suppose that the bag is immersed in a constant magnetic field $\vec{B}$, with corresponding vector potential $\vec{A} = \frac{1}{2}\vec{B} \times \vec{r}$. Then the confined quark will have a magnetic moment given by

$$\vec{\mu} \cdot \vec{B} = \int d^3r\, \vec{j} \cdot \vec{A},$$

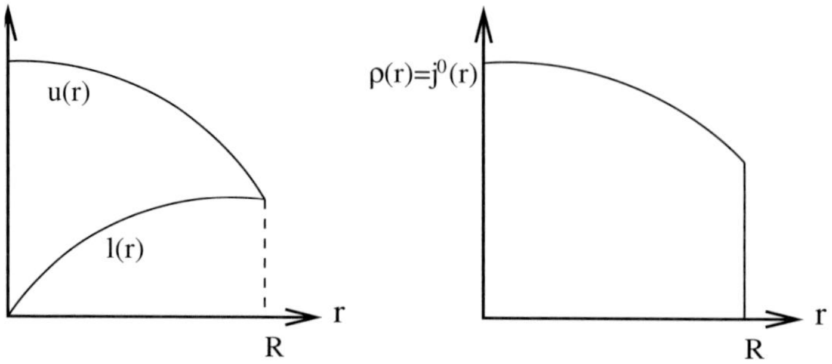

Figure 8.10: Illustration of the bag boundary condition that the upper and lower components of the Dirac wavefunction (u(r) and l(r), respectively) should be equal at $r = R$, together with the corresponding charge distribution.

where the integral is taken over the bag volume. Substituting for $\vec{j}$ the spatial piece of the conserved vector current this gives

$$\begin{aligned}\vec{\mu} \cdot \vec{B} &= \frac{e}{2} \int d^3r \, (\psi^\dagger \vec{\alpha} \psi) \cdot \left(\vec{B} \times \vec{r}\right) \\ &= \frac{e}{2} \int d^3r \, \vec{r} \times (\psi^\dagger \vec{\alpha} \psi) \cdot \vec{B},\end{aligned} \tag{8.118}$$

and hence

$$\begin{aligned}\vec{\mu} &= \frac{e}{2} \int d^3r \, \vec{r} \times \psi^\dagger \vec{\alpha} \psi \\ &= \mu_{\rm conf} \vec{\sigma},\end{aligned} \tag{8.119}$$

where

$$\mu_{\rm conf} = \frac{eR}{2\Omega} \cdot \left\{ \frac{4\Omega - 3}{6(\Omega - 1)} \right\}. \tag{8.120}$$

Thus the energy of the confined quark, Ω/R, acts something like a constituent quark mass. We note that the famous relationship $\frac{\mu_p}{\mu_n} = -\frac{3}{2}$ is preserved here too.

8.5 Chiral Quark Models

Whereas the vector current is quite satisfactory in the bag model, as we have just seen, the axial current is a different matter. Assuming flavor $SU(2)$, with the quark field operator, ψ, a two component vector for up and down quarks, we consider the chiral transformation familiar from Section 6.1.1, under which the QCD Lagrangian is invariant:

$$\begin{aligned}\psi &\rightarrow \psi - i \frac{\boldsymbol{\tau} \cdot \boldsymbol{\varepsilon}}{2} \gamma_5 \psi, \\ \bar{\psi} &\rightarrow \bar{\psi} - i \bar{\psi} \frac{\boldsymbol{\tau} \cdot \boldsymbol{\varepsilon}}{2} \gamma_5.\end{aligned} \tag{8.121}$$

For the bag, the surface term, $-\frac{1}{2}\bar{\psi}\psi\delta_S$, which is essential to confine the quarks, breaks the invariance under the chiral transformation, Eq. (8.121). Instead of being conserved, the axial current

$$\boldsymbol{A}^\mu = \bar{\psi}\gamma^\mu\gamma_5\frac{\boldsymbol{\tau}}{2}\psi\theta_V, \tag{8.122}$$

satisfies

$$\partial_\mu \boldsymbol{A}^\mu = -i\bar{\psi}\gamma_5\frac{\boldsymbol{\tau}}{2}\psi\delta_S. \tag{8.123}$$

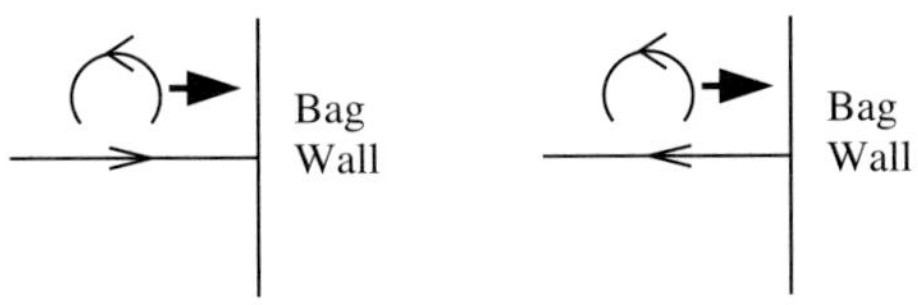

Figure 8.11: Violation of chiral symmetry at the bag surface.

The physical origin of this symmetry breaking, which was first clearly explained in the context of chiral bag models by Brown and Rho [BR 79], is illustrated in Fig. 8.11. Because the confinement is provided by a Lorentz scalar interaction, when a quark is reflected from the bag surface its spin is not flipped and therefore its helicity *does* change. This violates chiral symmetry.

Of course, as we discussed in detail in Chapters 6 and 7, even within QCD itself the naive expectations based on chiral symmetry quickly run into trouble. The hadronic spectrum shows no degeneracy between positive and negative parity states and therefore the exact chiral symmetry of massless QCD must be dynamically broken leading to a vacuum containing massless Goldstone bosons – the pions in two-flavor QCD. This realization provides a natural solution to the problem which we have just encountered and a large number of phenomenological models have been constructed on the basis of this solution. The idea is to add terms involving a pion field to the basic bag model Lagrangian density. This field should be coupled to the confined quarks in such a way that under a chiral transformation of *both* the quark and pion fields the new Lagrangian density is invariant and therefore yields a conserved axial current.

There are many choices possible for the Lagrangian density of such a chiral quark model. The following has been the basis of an enormous amount of work, beginning in the late 70's [Th 80]:

$$\mathcal{L}_{\text{CBM}} = \left[i\bar{\psi}\gamma_\mu\partial^\mu\psi - B\right]\theta_V - \frac{1}{2}\bar{\psi}e^{i\boldsymbol{\tau}\cdot\boldsymbol{\phi}\gamma_5/f}\psi\delta_S + \frac{1}{2}\left(D_\mu\boldsymbol{\phi}\right)^2. \tag{8.124}$$

Here the covariant derivative on the pion field is:

$$D_\mu\boldsymbol{\phi} = (\partial_\mu\phi)\,\hat{\boldsymbol{\phi}} + f\sin\left(\frac{\phi}{f}\right)\partial_\mu\hat{\boldsymbol{\phi}}, \tag{8.125}$$

$$= \partial_\mu\boldsymbol{\phi} + \mathcal{O}(\phi^3), \tag{8.126}$$

where $\hat{\phi}$ is a unit vector in isospin space pointing in the direction of $\vec{\phi}$. By construction the Lagrangian density is invariant under the chiral transformations:

$$\begin{aligned}
\psi &\rightarrow \psi - i\frac{\boldsymbol{\tau}\cdot\boldsymbol{\varepsilon}}{2}\gamma_5\psi,\\
\bar{\psi} &\rightarrow \bar{\psi} - i\bar{\psi}\frac{\boldsymbol{\tau}\cdot\boldsymbol{\varepsilon}}{2}\gamma_5,\\
\boldsymbol{\phi} &\rightarrow \boldsymbol{\phi} + \boldsymbol{\varepsilon} f + f\left(\boldsymbol{\varepsilon}\times\hat{\phi}\right)\times\hat{\phi}\left[1 - \frac{\phi}{f}\cot\left(\frac{\phi}{f}\right)\right].
\end{aligned} \tag{8.127}$$

As a consequence there is a conserved axial current $\boldsymbol{A}^{\mu}$, which has the form

$$\begin{aligned}
\boldsymbol{A}^{\mu} &= \bar{\psi}\gamma^{\mu}\gamma_5\frac{\boldsymbol{\tau}}{2}\psi\theta_V + \left[f\hat{\phi}(\partial^{\mu}\phi) + \frac{f^2}{2}\partial^{\mu}\hat{\phi}\sin\left(\frac{2\,\phi}{f}\right)\right]\\
&= \boldsymbol{A}^{\mu}_{\text{quark}} + f\partial^{\mu}\boldsymbol{\phi} + \mathcal{O}(\phi^3).
\end{aligned} \tag{8.128}$$

In examining models of this kind we need to bear in mind a number of points:

- These models are effective theories, *not* QCD. They are only designed to give us insight into some aspects of hadron structure and spectroscopy.
- A prime example of the effective nature of such a model is the fact that the pion appears as an elementary field. The idea is that the Goldstone boson is a highly collective state and for many purposes it should act like an independent degree of freedom. That is, we should be able to neglect its internal structure, at least in the long wavelength limit. Of course, this will break down whenever we consider processes that involve very high energy or momentum transfer.
- Attempts to understand bag formation are usually based on the instability of the perturbative vacuum, in which chiral symmetry is realized in the Wigner mode, with respect to the Goldstone phase. This suggests naively that one should exclude the pions from the interior of the bag. On the other hand, recent lattice simulations of hadron structure suggest a much more complicated structure with a snapshot of a nucleon revealing something far from a static spherical region of Wigner phase. Indeed the quarks are linked by flux tubes with a radius of typically 0.5 fm [SBS 99]. On average, for the ground state of the system, the region which the valence quarks occupy must be spherical. However, there is no reason that the pions, being created in some dynamical process involving flux tube breaking, should be excluded from this spherical region.

Rather than attempting an exhaustive summary of all the chiral quark models we shall concentrate on one example, the cloudy bag model. This has as its basic hypothesis the idea that the pion cloud required by chiral symmetry should only perturb the MIT bag in a minor way. Such an approach has the advantage that it allows a very transparent physical picture of the observed hadrons. Moreover, the approach has proven quite successful phenomenologically.

8.5.1 The Cloudy Bag Model

If we suppose that the basic valence structure of the nucleon is reasonably well described by the MIT bag and that the pion field required by chiral symmetry only modifies it perturbatively, it is natural to expand the general chiral quark model, given in Eq. (8.124), in powers of the pion field. Keeping terms up to $\mathcal{O}(\phi^2)$ only this yields [Th 80, TTM 81, Th 84]:

$$\begin{aligned}\mathcal{L}_{\text{CBM}} &\cong \left(i\bar{\psi}\gamma_\mu\partial^\mu\psi - m\bar{\psi}\psi - B\right)\theta_{\text{V}} - \frac{1}{2}\bar{\psi}\psi\delta_{\text{S}} - \\ &\quad \frac{i}{2f}\bar{\psi}\gamma_5\frac{\boldsymbol{\tau}}{2}\psi\cdot\boldsymbol{\phi}\delta_{\text{S}} + \frac{1}{2}\left(\partial_\mu\boldsymbol{\phi}\right)^2 - \frac{1}{2}m_\pi^2\boldsymbol{\phi}^2, \end{aligned} \tag{8.129}$$

where the additional terms in m (the current quark mass) and m_π (the pion mass) break the exact chiral symmetry, leading to PCAC.

We first note from Eq. (8.129) that chiral symmetry requires (at least) linear coupling of pions to the bag through the $\boldsymbol{\tau}\cdot\boldsymbol{\phi}$ term. That is, the CBM Lagrangian reduces to the MIT bag plus a free pion field with interactions between the two dictated by chiral symmetry. In this case the full Lagrangian density may be written as

$$\mathcal{L}_{\text{CBM}} \cong \mathcal{L}_{\text{MIT Bag}} + \mathcal{L}_{\text{Free}-\pi} + \mathcal{L}_{\text{int}}, \tag{8.130}$$

which implies that the Hamiltonian must have the form

$$H = H_{\text{MIT}} + H_\pi + H_{\text{int}} \tag{8.131}$$

$$\quad = \sum_B \varepsilon_B B^\dagger B + \sum_k \omega_k a_k^\dagger a_k + \sum_{B,\tilde{B},k}\left[\tilde{B}^\dagger B a_k v_k^{\tilde{B}B} + \text{h.c.}\right], \tag{8.132}$$

where $|B\rangle = B^\dagger\,|0\rangle$ is a bag model state (analogous to the non-relativistic, quark model state given in Eq. (8.4)) and $a_k^\dagger$ creates a pion with momentum $\vec{k}$ and isospin i ($k = \vec{k}, i$). The matrix elements for the pion couplings to the composite baryon states are:

$$v_{\vec{k},i}^{\tilde{B}B} = \frac{i}{2f}\frac{1}{\sqrt{2\omega_k}}\int d^3r\, e^{-i\vec{k}\cdot\vec{r}}\delta(r-R)\langle\tilde{B}|\bar{\psi}(\vec{r})\gamma_5\tau_i\psi(\vec{r})|B\rangle. \tag{8.133}$$

We stress that the pion-baryon couplings involve form factors which arise naturally as a consequence of the internal structure of the baryons and which can be calculated using the underlying quark model – be it a simple $(1\text{s})^3$ bag configuration or something more sophisticated.

Taking as the simplest example the states $B = |N\rangle$ and $\tilde{B} = |N'\rangle$, we can establish the vertex function for the absorption of a pion of momentum, $\vec{k}$, and isospin, i, by a nucleon:

$$v_{\vec{k},i}^{N'N} = -\left(2\,\omega_k\right)^{-1/2}\frac{i}{2f}\frac{\Omega}{(\Omega-1)}\frac{j_1(kR)}{kR} \times {}_{SI}\left\langle N'\left|\sum_{q=1}^{3}\tau_{qi}\vec{\sigma}_q\cdot\vec{k}\right|N\right\rangle, \tag{8.134}$$

where the matrix element on the right hand side of Eq. (8.134) involves only the spin-isospin (SI) parts of the wavefunction. Using the fact that the axial charge of the nucleon in the bag model is $g_{\text{A}}^{\text{Bag}} = \left(\frac{\Omega}{\Omega-1}\right)\frac{5}{9}$, we see that

$$v_{\vec{k},i}^{N'N} = -\frac{i}{(2\,\omega_k)^{1/2}}\left(\frac{g_{\text{A}}^{\text{Bag}}}{2f}\right)u(k)\,\langle N'|\,\tau_i\vec{\sigma}\cdot\vec{k}\,|N\rangle\,, \tag{8.135}$$

where

$$u(k) \equiv \frac{3j_1(kR)}{kR} \quad \left(\stackrel{k\to 0}{\longrightarrow} 1\right). \tag{8.136}$$

Comparing this with the usual πNN coupling, we find

$$\frac{g_{\pi NN}}{2M_N} = \frac{g_\mathrm{A}}{2f}, \tag{8.137}$$

and hence

$$f g_{\pi NN} = g_\mathrm{A} M_N, \tag{8.138}$$

which is the Goldberger-Treiman relation.

In a similar way one can calculate the pion couplings between any bag model states – e.g., $\Delta N\pi, \Delta\Delta\pi, \Sigma\Lambda\pi$ etc. To the extent that the Δ has the same quark wavefunction as the nucleon the form factors for the former two vertices will have the same shape as the $N \to N\pi$ vertex, Eq. (8.136), and the coupling constants will satisfy the usual $SU(6)$ relations. By extending the symmetry to $SU(3)_L \times SU(3)_R$ one can also calculate both the coupling constants and form factors for kaon emission and absorption [VTJ 85].

The physical nucleon

Perhaps the most important consequence of the cloudy bag model is the physical insight it gives us concerning the structure of the physical hadrons. We take the nucleon as the example of direct interest, but bear in mind that similar considerations apply to all hadrons.

The form of the Hamiltonian given in Eq. (8.132) already makes it clear that the structure of a physical hadron will consist of a valence quark core "dressed" by a cloud of pions. Indeed, the physical nucleon, $|\tilde{N}\rangle$, is an eigenstate of the CBM Hamiltonian with eigenvalue M_N:

$$H|\tilde{N}\rangle = M_\mathrm{N}|\tilde{N}\rangle. \tag{8.139}$$

This should be compared with the eigenstate of the MIT bag piece of the Hamiltonian:

$$\begin{aligned} H_\mathrm{MIT}\,|N\rangle &= M_\mathrm{N}^{(0)}\,|N\rangle\,, \\ H_\mathrm{MIT}\,|\Delta\rangle &= M_\Delta^{(0)}\,|\Delta\rangle\,, \text{ etc.,} \end{aligned} \tag{8.140}$$

where the one gluon exchange mass shift is included in the $M^{(0)}$ terms. In first order perturbation theory:

$$\begin{aligned} |\tilde{N}\rangle &= \sqrt{Z}\left\{|N\rangle + \sum_{B,k} \frac{\langle B\pi_k|\,H_\mathrm{int}\,|N\rangle}{M_\mathrm{N} - w_k - M_B}\,|B\pi_k\rangle\right\} \\ &\cong \sqrt{Z}\left\{|N\rangle + \sum_k c_1(k)\frac{|N\pi_k\rangle}{-\omega_k} + \sum_k c_2(k)\frac{|\Delta\pi_k\rangle}{M_\mathrm{N} - M_\Delta - \omega_k}\right\}. \end{aligned} \tag{8.141}$$

A crucial feature of the model is that this perturbative expansion can be proven to be extremely good as long as the bag radius, R, is large (≥ 0.7 fm) – i.e., the form-factor is soft [DTA 81]. As for the non-relativistic oscillator model, with its restriction to a model space (e.g., $2\hbar\omega$) it is also necessary to restrict the model space here to the low-lying bag states (e.g. $N, \Delta, N^*, \ldots$). (If one were to include states that require too much excitation energy, or for which the typical momentum transfer is too high, the treatment of the pion as elementary would break down.) Thus for many purposes it is possible to think of the physical proton as consisting, some of the time (probability Z) of three valence quarks and some of the time as a pion plus a nucleon core (probability $P_{N\pi}$) and some of the time a pion and three valence quarks with the quantum numbers of a Δ (i.e., a "bare" Δ) – with probability $P_{\Delta\pi}$. There is considerable phenomenological evidence pointing to values for these probabilities of order $P_{N\pi} \sim 20\,\%$ and $P_{\Delta\pi} \sim 10\,\%$ [ST 97, Th 84].

The charge distribution of the neutron

If we ignore configuration mixing in the bag, so that the neutron consists of one up and two down quarks in the 1s-state, the "bare" neutron has no charge distribution. On the other hand, once we take account of the chiral structure of the nucleon, its wave function has the form:

$$|\tilde{n}\rangle = \sqrt{Z}\left\{|n\rangle + \sum_k c_1(k)\left[\left(\frac{2}{3}\right)^{1/2}|p\pi_k^-\rangle + \left(\frac{1}{3}\right)^{1/2}|n\pi_k^0\rangle\right] + \cdots\right\}. \tag{8.142}$$

Since $|n\rangle$ contributes nothing to the charge distribution and nor does $|n\pi^0\rangle$, the *leading order contribution* to the neutron charge distribution comes from the term $|p\pi_k^-\rangle$. (Note that the coefficients $\frac{2}{3}$ and $\frac{1}{3}$ in Eq. (8.142) are just the Clebsch-Gordon coefficients for isospin.)

It is easy to see that, as the bag has a sharp surface, this simple model will produce a zero in the neutron charge density at the bag radius [TMT 82]. Detailed calculations of the neutron charge distribution are complicated by the need to project a state of good total momentum and to remove spurious center of mass motion (which cannot be treated exactly, as it was in the oscillator model). Nevertheless, after all the necessary corrections are made this feature persists to quite good accuracy. In Fig. 8.12, we see the results of a calculation [Lu 98], which illustrate the importance of accurate measurements of the neutron charge form factor.

8.5.2 Corrections to Hadronic Masses

To lowest, non-trivial order (second order in H_{int}) we find that the shift in the mass of the nucleon is given by the two processes shown in Fig. 8.13, that is:

$$\begin{aligned} M_N = M_N^{(0)} &- \frac{3}{16\,\pi^2 f_\pi^2}\, g_A^2 \int_0^\infty dk\, \frac{k^4\, u_{NN}^2(k)}{w^2(k)} \\ &- \frac{3}{16\,\pi^2 f_\pi^2}\,\frac{32}{25}\, g_A^2 \int_0^\infty dk\, \frac{k^4\, u_{N\Delta}^2(k)}{w(k)\,(\Delta M + w(k))}, \end{aligned} \tag{8.143}$$

where $\Delta M = M_\Delta^{(0)} - M_N$ and we have used the $SU(6)$ relation to express the $N \to \Delta\pi$ coupling in terms of g_A. For simplicity we have also used the static approximation for the

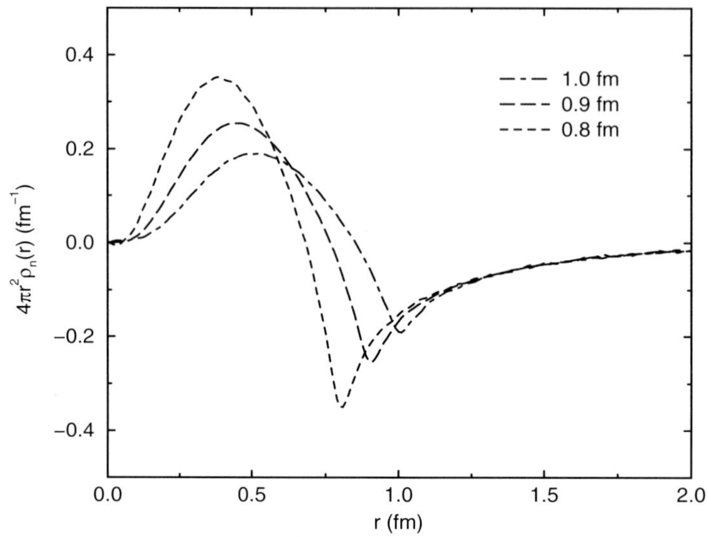

Figure 8.12: Neutron charge density calculated in the CBM for several choices of bag radius. Note that the peak in the negative charge density always occurs very near the chosen value of R.

propagator of the heavy baryon. This nevertheless preserves the correct chiral behavior as $m_\pi \to 0$ – for example, the leading non-analytic behavior in the quark mass ($m_\pi \sim \sqrt{m_q}$), arising from Fig. 8.13(a), is:

$$\delta M_N^{LNA} = -\frac{3}{32\pi f_\pi^2}\, g_A^2\, m_\pi^3, \tag{8.144}$$

as it should be. Since both of the processes shown in Fig. 8.13 are attractive they can give a sizeable correction to the nucleon mass, varying from about -150 MeV (for $R \sim 1.0$ fm or a dipole form factor of order 700 MeV) to -900 MeV (for $R \sim 0.6$ fm or a dipole of mass around 1.2 GeV).

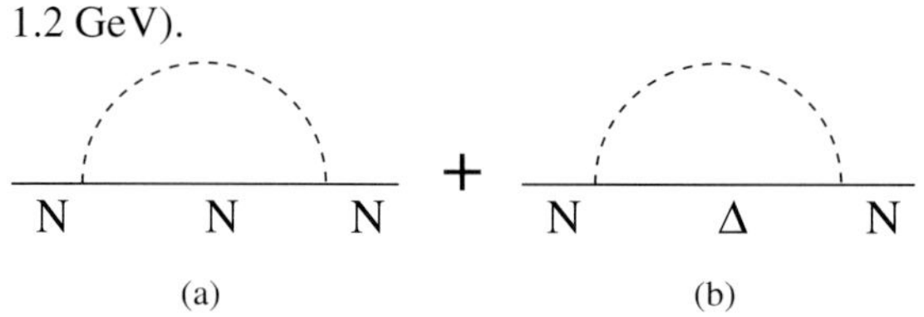

Figure 8.13: Pion self-energy corrections for the nucleon.

We consider next the corresponding self-energy corrections for the Δ, shown in Fig. 8.14, that is:

$$\begin{aligned} M_\Delta = M_\Delta^{(0)} &+ \frac{3}{16\,\pi^2 f_\pi^2}\,\frac{8}{25}\, g_A^2 \int_0^\infty dk\, \frac{k^4\, u_{N\Delta}^2(k)}{w(k)\,(\Delta M - w(k))} \\ &- \frac{3}{16\,\pi^2 f_\pi^2}\, g_A^2 \int_0^\infty dk\, \frac{k^4\, u_{\Delta\Delta}^2(k)}{w^2(k)}. \end{aligned} \tag{8.145}$$

Clearly the contributions from Fig. 8.13(a) and Fig. 8.14(b) (the second and third terms on the right of Eqs. (8.143) and (8.145), respectively) are equal in magnitude (and have the same leading non-analytic chiral behavior, Eq. (8.144)), while that from Fig. 8.14(a) is small. The latter is small not only because of the small coupling constant but also because the integral corresponds to an open channel (the decay $\Delta \rightarrow N\pi$) and the real part is therefore a principal value integral. This varies between -250 and -100 MeV for R between 0.6 and 1.0 fm.

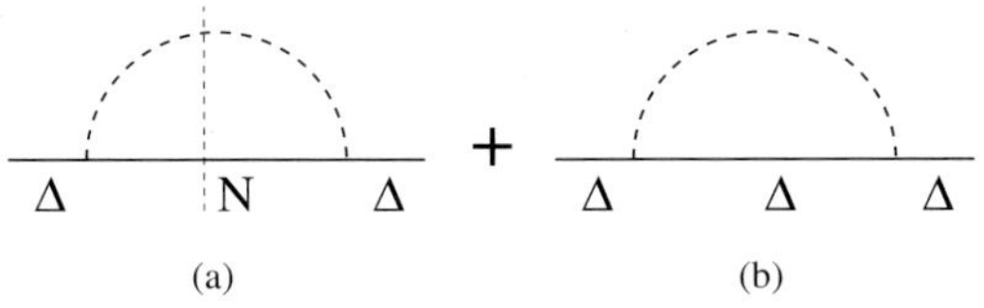

Figure 8.14: Pion self-energy corrections for the Δ. The short-dashed line in part (a) indicates that this term will have an imaginary part.

Clearly the difference between the pionic self-energy corrections to the N and Δ masses contributes significantly to the observed mass splitting [TTM 81, Myh 81, GR 96]. One finds a mass splitting of between 300 and 30 MeV for R between 0.6 and 1.0 fm. The price of getting the full $\Delta - N$ mass splitting from pions is a very large absolute shift in their masses – of order 1 GeV [TK 00]. It seems most likely that the observed mass splitting is the result of a linear combination of the pionic self-energy and one gluon exchange. Any pionic contribution has the desirable effect of reducing the size of the strong coupling constant needed for gluon exchange, hence ameliorating the spin-orbit problem discussed in Section 8.2.6.

Before completing this discussion we need to make one more point which is far more general than the Δ itself. Resonances like the Δ are unstable against decay to other hadrons and thus we must face two issues.

- Is there an underlying (stable) "bare" quark state upon which the observed resonance is based? This is quite fundamental to any spectroscopic studies and there are quite compelling suggestions in the case of the Roper resonance[Kre 00] and the $\Lambda(1405)$ [VJT 85, KSW 95] that the resonances observed are predominantly the result of channel coupling effects, with no corresponding "bare" quark state near those masses.

- In the case where there is an underlying "bare" quark state involved, there can be quite large mass shifts corresponding to the effects of channel coupling. Thus one is faced with a serious problem in identifying the mass that one needs to fit in a detailed spectroscopic study. It may be quite misleading to simply use the mass of the observed resonance as the mass to be fit in a shell-model type of study. In general one should really treat each resonance as a coupled channel problem, within a consistent chiral quark model framework.

8.5.3 Appearances can be deceptive

Within the framework that we used to introduce the pion field in the CBM it may have (erroneously) appeared to be a way to deal with a problem specific to the MIT bag – namely the violation of chiral symmetry at the bag surface. Since the evidence from heavy quark systems is that the hyperfine structure is consistent with a confining interaction that is a Lorentz scalar, there is clearly nothing special about the bag; any acceptable mechanism for confinement would violate chiral symmetry and would require a similar resolution. Another feature which is a little unsatisfactory is the fact that surface γ_5 coupling implies that to recover some low energy theorems, one needs to sum over a large number of strongly coupled negative energy states. This in turn can hide the fundamental simplicity of the structure of the nucleon with its three valence quarks.

One can resolve the technical problem just described and at the same time remove any specific connection to the bag surface by making a unitary transformation to a new, dressed quark field, Q [Th 81]:

$$Q = U(\phi)\psi; \qquad U(\phi) = e^{i\boldsymbol{\tau}\cdot\boldsymbol{\phi}\gamma_5/2f}. \tag{8.146}$$

Then using the identity

$$U\partial_\mu U^\dagger = \int_0^1 d\lambda U^\lambda \partial_\mu \left(\ln U^\dagger\right) (U^\dagger)^\lambda, \tag{8.147}$$

we find that the CBM Lagrangian density becomes

$$\begin{aligned}\mathcal{L}'_{\text{CBM}}(x) = \left[i\bar{Q}\not{D}Q - m\bar{Q}Q - B\right]\theta_{\text{V}} - \frac{1}{2}\bar{Q}Q\delta_S + \frac{1}{2}\left(D_\mu\boldsymbol{\phi}\right)^2 \\ + \frac{1}{2f}\bar{Q}\gamma^\mu\gamma_5\boldsymbol{\tau}Q\cdot\left(D_\mu\boldsymbol{\phi}\right)\theta_{\text{V}}.\end{aligned} \tag{8.148}$$

Clearly the troubling pseudoscalar coupling of the pions to the quarks at the bag surface has been replaced in Eq. (8.148) by pseudovector coupling throughout the bag volume. This is exactly as one expects from current algebra and chiral perturbation theory. On the other hand, it is reassuring that the coupling constant (for massless quarks) is again given by the Goldberger-Treiman relation – i.e., it is essentially $\frac{g_A}{2f}$ for the nucleon – because the pion couples to the longitudinal component of the axial current. Indeed, it is relatively simple to show by integration by parts that, again in the chiral limit, the momentum dependence of the $NN\pi$ form factor is identical to that obtained in the previous version of the model (c.f. Eq. (8.136)). Thus the results described earlier, such as the contribution to the $\Delta - N$ mass splitting, remain unchanged.

Next we turn to the covariant derivative on the Q-field in Eq. (8.148)

$$D_\mu Q = \partial_\mu Q - \frac{i}{2}\left[\cos\left(\frac{\phi}{f}\right) - 1\right]\boldsymbol{\tau}\cdot\left(\hat{\boldsymbol{\phi}}\times\partial_\mu\hat{\boldsymbol{\phi}}\right)Q. \tag{8.149}$$

This term contains the low energy theorems in a very elegant and compact form. For example, let us consider low energy pion scattering in s-wave. Then to $\mathcal{O}(\phi^2)$ the covariant derivative in Eq. (8.149) leads to the following interaction:

$$\mathcal{L}_{\text{int}} = -\frac{1}{2f^2}\left[\bar{Q}\gamma^0\frac{\boldsymbol{\tau}}{2}Q\right]\cdot\left(\boldsymbol{\phi}\times\partial_0\boldsymbol{\phi}\right)\theta_{\text{V}}. \tag{8.150}$$

We easily recognize that the term in square brackets is just the isospin density for the target bag, while the term in round brackets is the usual pion isospin operator. For pions of zero momentum the latter term is independent of x and hence we obtain the Hamiltonian density for low energy pion scattering

$$\mathcal{H}_{\mathrm{WT}} = \frac{\boldsymbol{T}_{\mathrm{target}} \cdot \boldsymbol{t}_{\pi}}{2f^2}, \tag{8.151}$$

where $\boldsymbol{T}_{\mathrm{target}}$ is the isospin of the target bag. Since the vector current is conserved in the model this generalizes easily to the total isospin of the target when the bag is "dressed" by its pion field. Equation (8.151) is the famous Weinberg-Tomozawa relation for low energy pion scattering established on the basis of current algebra – see Section 7.3.2.

We note that Eq. (8.148) is readily generalized to $SU(3)_L \times SU(3)_R$. The octet of low lying Goldstone bosons are then emitted and absorbed through pseudovector coupling throughout the volume of the corresponding bag. In addition, the interaction in Eq. (8.150) is generalized to include processes such as $\pi\Sigma \to \pi\Sigma$ and $\pi\Sigma \to \bar{K}N$, which are so important to understanding the nature of the $\Lambda(1405)$. (For example, from Eq. (8.151) we easily see that the low energy $\pi\Sigma$ interaction is extremely attractive in the channel with total isospin zero.)

Relation to other chiral quark models

More important than the specific results which we have presented is the fact that this new (unitarily equivalent) representation presents what looks like a completely new physical picture. In particular, it is now clear that the bag itself plays no special role. One could in fact replace it with any other quark model, such as the constituent quark model discussed earlier in this chapter. The only difference is that one would then identify the quark mass in Eq. (8.148) as a constituent quark mass. Apart from that the model has all the features of models like the chiral quark model of Georgi and Manohar [MG 84], with pseudovector coupling throughout the body of the hadron. A complete treatment of hadron structure would then include **both** configuration mixing, as described in Section 8.2.6, **and** the pion cloud required by chiral symmetry.

In this context we should also mention the spectroscopic studies of Glozman and Riska [GR 96], which have played a major role in re-emphasizing the importance of Goldstone boson exchange, versus one-gluon-exchange, in reproducing the fine structure of the hadron spectrum – c.f. our earlier discussion of the $\Delta - N$ mass splitting, which they applied to all major shells in the oscillator quark model. In connection with studies of this kind we finish with two cautions. First to ensure the correct chiral behavior of the pionic corrections, as given for example in Eq. (8.144), one must project the Hamiltonian involving quark and meson fields onto states with allowed baryon quantum numbers [TK 00, Th 84]. It is very difficult to ensure the correct chiral behavior by working only with quarks. Secondly, as we explained earlier, one cannot avoid the necessity of dealing with channel coupling effects and cannot simply compare the quark model results directly with the observed spectrum. While in some cases it may be a reasonable approximation (such as the Δ) in others it can be quite misleading.

8.6 Chiral Solitons

The previous discussions have emphasized the important role of spontaneous chiral symmetry breaking and pions as Goldstone bosons in the context of nucleon structure. In this section we briefly describe the conceptually interesting idea that the nucleon may be viewed as a chiral soliton, a non-linear dynamical system in which quarks are bound and localized by their interaction with chiral fields.

Quarks bound by a scalar field have already been encountered in the form of the Friedberg-Lee soliton model (see section 8.3.1). The fact that QCD has a spontaneously broken chiral symmetry implies that scalar and pion fields should appear in a combination such that the resulting effective Lagrangian, apart from small quark mass terms, is symmetric under chiral transformations. If one chooses to introduce these fields in a minimal way, as for example in the linear sigma model, with no other scalar fields present, one can find chiral solitons for certain choices of parameters.

The concept of the chiral soliton is based on the observation that there exists an energetically favoured configuration of chiral fields which binds valence quarks in a well localized object, a soliton, of baryon number $B = 1$. As one approaches the soliton from outside and proceeds to its center, the field reveals a characteristic twist, or winding number, which distinctly separates the chiral field configuration of a baryon ("winding number one") from purely mesonic field configurations ("winding number zero"). Thus the winding number acquires an interesting topological meaning of its own. One can even go one step further and eliminate the quarks altogether. The (topological) soliton which emerges in this way is called a *Skyrmion*, to be identified with a baryon by virtue of its winding number which takes over the role of the baryon number.

While the mathematics surrounding these developments has its own appeal, we make no attempt to cover Skyrmions and topological concepts here since this has already been done extensively in a previous textbook [Bha 88]. The interested reader is referred to the original literature [Wit 83, Adk 84, Me 86]. The conceptual framework and many applications can also be found in [Ri 97] and several reviews [ZB 86, Schw 89, Hol 93, ARW 96].

8.6.1 Quarks Bound by Chiral Fields

The prototype of a model which describes the interaction of Dirac particles with a chiral field is the linear sigma model. In our following brief exposition, we will use this model in its $SU(2)_L \times SU(2)_R$ version and identify the Fermion field ψ with the doublet of massless u- and d-quarks. The fermionic part of the sigma model (see Eq. (7.19))is

$$\mathcal{L}_F = \bar{\psi}\Big[i\gamma_\mu\partial^\mu - g(\sigma - i\gamma_5\, \boldsymbol{\tau}\cdot\boldsymbol{\phi})\Big]\psi, \tag{8.152}$$

with the scalar σ and the pseudoscalar Goldstone boson $\boldsymbol{\phi}$ treated as elementary fields. A closely related structure is realized in the Nambu and Jona-Lasinio (NJL) model (see section 6.1.5), where the scalar and pseudoscalar fields are quark-antiquark composites, with the replacements $g\sigma = -(G/2)\bar{\psi}\psi$ and $g\boldsymbol{\phi} = (G/2)\bar{\psi}i\gamma_5\boldsymbol{\tau}\psi$ in Eq. (8.152).

For convenience we introduce the scalar and pseudoscalar fields as $S(x) = g\sigma(x)$ and $P_a(x) = -g\phi_a(x)$ (with isospin index $a = 1, 2, 3$) and write the interaction part of Eq. (8.152) as

$$\mathcal{L}_{\text{int}} = -\bar{\psi}[S + i\gamma_5\tau^a P_a]\psi. \tag{8.153}$$

A time-independent field, which turns out to be particularly suitable for building solitons, is the so-called *hedgehog* configuration [CT 75], which has the following special shape:

$$\begin{aligned} S(\vec{r}) + i\gamma_5\tau^a P_a(\vec{r}) &= M(r)\,\exp\big[i\gamma_5\tau^a\hat{r}_a\theta(r)\big] \\ &= M(r)\,\big[\cos\theta(r) + i\gamma_5\tau^a\hat{r}_a\,\sin\theta(r)\big]. \end{aligned} \tag{8.154}$$

It combines the components of the radial unit vector, $\hat{r} = \vec{r}/r$, in coordinate space with the three Pauli matrices, τ^a, in isospin space in a peculiar way, such that Eq. (8.154) is invariant under a simultaneous rotation in *both* spaces: it represents a mapping of $SU(2)$ onto R^3.

The hedgehog field is special because it turns out that it makes the actions of the linear sigma model, the NJL model and the Skyrme model stationary. It is subject to the following boundary conditions for the $chiral\ angle$, $\theta(r)$:

$$\theta(r \to \infty) = 0, \qquad \theta(r \to 0) = n\pi, \tag{8.155}$$

where the integer n is called the *winding number* of the chiral field. At $r \to \infty$ the field is entirely scalar, and the effective (or constituent) quark mass $M(r)$ approaches a constant, M. As one moves to smaller r, the chiral angle $\theta(r)$ increases and the pseudoscalar field gets involved. The square of the chiral field,

$$S^2(r) + P_a^2(r) = M^2(r), \tag{8.156}$$

defines a circle of radius $M(r)$, the $chiral\ circle$.

The Dirac Hamiltonian for quarks moving in the chiral hedgehog field is

$$H = -i\vec{\alpha}\cdot\vec{\nabla} + \beta M(r)\,[\cos\theta(r) + i\gamma_5\tau^a\hat{r}_a\,\sin\theta(r)], \tag{8.157}$$

and their energy spectrum is determined by solving the equation

$$H|\lambda\rangle \;=\; E_\lambda|\lambda\rangle. \tag{8.158}$$

The eigenstates $|\lambda\rangle$ have quantum numbers reflecting the symmetries of H. With rotations in isospin- and r-space being tied together, the grand spin $\boldsymbol{K} = \boldsymbol{t} + \boldsymbol{j}$ commutes with H, where $\boldsymbol{t} = \boldsymbol{\tau}/2$ is the isospin and $\boldsymbol{j} = \boldsymbol{l} + \boldsymbol{s}$ is the total angular momentum of the quark.

A schematic example of the spectrum of quark orbits in such a chiral field is shown in Fig. 8.15. In this example the function $M(r)$ is held constant at an input mass parameter M so that changes of the scalar and pseudoscalar fields are restricted to the chiral circle $S^2+P_a^2 = M^2$ with constant radius. The chiral angle profile with winding number $n = 1$ is chosen as

$$\theta(r) \;=\; \pi\,\exp\left(-\frac{r}{R}\right). \tag{8.159}$$

The quark energy eigenvalues are given as the ratio E_λ/M in Fig. 8.15. The chiral field is controlled by a single dimensionless parameter, $X = MR$. As X grows, so does the effect of the chiral field. An orbit with grand spin $K = 0$ and positive parity (corresponding to a quark

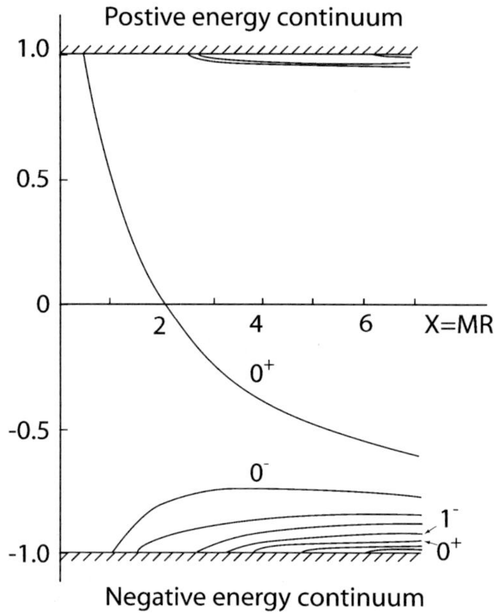

Figure 8.15: The energy spectrum of the Dirac Hamiltonian (8.157), for the chiral field profile (8.159), plotted as a function of the dimensionless parameter $X = MR$. The energy is given as the ratio E/M. Adapted from [Ri 97].

state with $s = t = \frac{1}{2}$ and $l = 0$) dives into the gap between the positive and negative energy continuum and becomes a bound state. Other states emerge from the Dirac sea. According to the Pauli principle, the 0^+ orbit can be occupied maximally by N_c quarks, where N_c is the number of colours. Then the soliton is a color singlet.

Together with the filled Dirac sea, the soliton formed out of the occupied 0^+ level and the chiral field has baryon number $B = 1$. Its baryon density can be computed using the quark wave functions, $\varphi_\lambda(\vec{r}) = \langle \vec{r}\,|\lambda\rangle$, as

$$\rho_B(\vec{r}) = \sum_{\lambda(\text{occ})} |\varphi_\lambda(\vec{r})|^2 - \sum_{\text{vacuum}} |\varphi_\lambda^{(0)}(\vec{r})|^2, \tag{8.160}$$

where the first sum is over all occupied orbits (the 0^+ valence orbit and the negative energy Dirac sea), and the second (vacuum) sum extends over the unperturbed Dirac sea with $\theta = 0$ (i.e. with $S = M$ and $P_a = 0$). The total baryon number thus undergoes a fractioning into contributions from the valence orbital and from the Dirac sea.

8.6.2 Nucleons as Chiral Solitons

So far the chiral soliton just described is a baryon but not a nucleon. The specific hedgehog form of the chiral field is not invariant under *separate* rotations in the coordinate and isospin spaces. The soliton has a definite grand spin $\boldsymbol{K} = \boldsymbol{J} + \boldsymbol{I}$ but it does not have definite angular momentum quantum numbers (J, J_z) *and* isospin quantum numbers (I, I_3) as it should in order to be a *hadronic* eigenstate. In fact the lowest grand spin 0^+ eigenstate occupied by N_c valence quarks is a linear combination of baryon states with $J = I = \frac{1}{2}, \frac{3}{2}, \ldots$ etc. While these states would indeed all be degenerate in the limit $N_c \to \infty$, this is not so in the real world with $N_c = 3$.

For $N_c = 3$ the hedgehog soliton includes both the nucleon with $J = I = \frac{1}{2}$ and the Δ isobar with $J = I = \frac{3}{2}$. To remove their degeneracy, techniques familiar from the quantization of collective motion in nuclear physics are applied to the soliton. Suppose we let the soliton rotate slowly in coordinate space so that its internal structure is not changed. The energy required is then just the rotational kinetic energy, $J(J+1)/2\Theta$, where Θ is the moment of inertia. The low lying excited states form a rotational band. The mass difference between the $\Delta(1232)$ and the $N(939)$, to this order, is simply given by $M_\Delta - M_N = 3/2\Theta$. Since the moment of inertia is proportional to N_c, just like the baryon mass itself, the $N - \Delta$ mass difference is an effect of order $1/N_c$.

Further effects must be taken into account when relaxing the assumption of a rigid rotor. The rotation produces a Coriolis force which tends to align the quark spins with the rotation axis and deform the soliton. In addition, however, the rotating pion field also contributes to the angular momentum so that only a fraction of the nucleon spin ends up being carried by the quarks, their precise portion depending on the size of the soliton [Bl 88].

The actual calculation of nucleon properties requires a rather complicated technical machinery which we shall not pursue, instead referring to [Ri 97, Kim 96] for details. In short, the procedure is as follows. One starts from the hedgehog soliton with constant mass parameter M. The form of the chiral profile $\theta(r)$, with appropriate boundary conditions, is computed self-consistently by variational methods, demanding that the soliton energy be minimized. The quantization of the system involves quark loops which need to be regularized. The cutoff is determined by the input mass M, once the pion decay constant f_π is fixed at its empirical value. Small current quark masses are also introduced by fitting the pion mass m_π. Nucleon properties are then evaluated for different values of the constituent quark mass M by exposing the (quantized) soliton to electro-weak perturbing operators.

Table 8.2: Nucleon properties calculated in a chiral soliton model for different values of the constituent quark mass M – see [Kim 96].

M [MeV]	370	420	450	Exp.
$\langle r^2\rangle_p$ [fm^2]	0.85	0.70	0.66	0.74
$\langle r^2\rangle_n$ [fm^2]	-0.22	-0.18	-0.18	-0.11
μ_p $[\frac{e\hbar}{2M_pc}]$	2.12	2.03	1.86	2.79
μ_n $[\frac{e\hbar}{2M_pc}]$	-1.44	-1.41	-1.29	1.46
$M_\Delta - M_N$ [MeV]	213	280	314	293
g_A	1.26	1.21	1.13	1.26

Typical results are summarized in Table 8.2. While orders of magnitude for radii, magnetic and axial moments come out reasonably, it is also evident that such models have their limitations, despite the formidable technical effort that they require. One important restriction is that the pion and scalar fields are permitted to vary only along the chiral circle, keeping the constituent quark mass essentially constant. This is a serious constraint. It enforces a role on the scalar field quite different from that in color dielectric models where bag formation is accompanied by the reduction of the quark mass inside the baryon. Clearly, the very nature of the scalar field and its relationship to confinement and the gluon sector of QCD is the crucial issue.

9 Final Remarks

As the lowest mass excitation of the QCD vacuum with one unit of baryon number, the nucleon is of fundamental importance to our understanding of the strong interaction. It exhibits all of the phenomena associated with QCD, notably confinement, spontaneous chiral symmetry breaking and asymptotic freedom. In addition to extensive numerical simulations based on lattice QCD, there is an impressive array of models which have proven helpful in developing our understanding. Furthermore, there are certain limits in which one can obtain model independent constraints. At very low momentum transfer these come from chiral symmetry, while at very high momentum transfer we can rigorously analyze data in terms of perturbative QCD. Complementary to these theoretical considerations is a wealth of experimental information obtained with electroweak probes. One of our prime purposes in preparing this book has been to collect a representative sample of this vast amount of information in a digestible form.

Based on the experimental evidence and the various attempts to interpret it theoretically, we have an understanding of some of the key features of the structure of the nucleon. From its excitation spectrum it is clear that valence quarks play a crucial role in its structure. The local gauge invariance of QCD demands that these valence quarks are connected by gauge strings, but from deep inelastic data we see that these strings do not carry a large fraction of the energy of the system, at least when viewed with low resolution – above, say, 0.1 fm. Studies of dynamical chiral symmetry breaking have given some guidance on the origin of the mass of the constituent quarks. They suggest that the quark "mass" is a function of momentum, which takes values of order 300-400 MeV at momenta characteristic of a quark in a nucleon. At high momentum this function rapidly decreases to a few MeV, for the u and d quarks. The same process yields the long range pion cloud of the nucleon about which we have so much phenomenological information. Finally, this object sits, and the pseudo-Goldstone pions move, in a complex soup of quark and gluon condensates known as the QCD vacuum.

Yet for all the beauty and consistency of this picture there remain more issues to resolve than we have so far understood. We do not know to what extent the interior of the nucleon is swept clean of the condensates characteristic of the QCD vacuum. This is directly related to how chiral symmetry is realized at non-zero baryon density and, in turn, impacts on our understanding of the relevance of the constituent quark picture given above. There are tantalizing hints of the role of topological structures such as instantons, but much remains to be learnt. We still have little solid information on the role of the strange quark and still less for heavier quarks. There is little evidence concerning the distribution of the total angular momentum of the nucleon between its constituent quarks and gluons and in turn their spin and orbital angular momentum, and so on.

For those who are fascinated by the strong interaction and, in particular, the structure of the nucleon there can be no more exciting period than that currently facing us. On the theoretical side, developments in lattice QCD, including:

- clever improvements in gluon and quark actions which permit larger lattice spacing;
- tremendous increases in computing speed;
- improved techniques for chiral extrapolation;

promise direct information on a variety of observables, including masses and electroweak form factors, of an accuracy that was hitherto unthinkable. One is beginning to imagine models that bear a direct, quantitative connection to QCD and respect fundamental requirements such as Lorentz covariance.

In addition, we have at our disposal experimental facilities which will provide a vast improvement in our knowledge of nucleon observables. Baryon and meson spectroscopy is undergoing a revolution with high intensity, high duty factor machines equipped with multi-particle detectors. Along with this spectroscopic information will come precise data on a variety of transition form factors to well defined final hadronic states. New probes such as virtual Compton scattering will provide complementary information on the amplitudes for exciting the nucleon. Precise measurements of parity violation, together with new, high intensity neutrino beams, will reveal more about the role of strangeness in nucleon structure.

Complementing the exclusive measurements just described, we also look forward to important new insights from deep inelastic probes. Traditional inclusive scattering experiments at higher precision and in new kinematic regimes will provide guidance on the spin and flavor structure of the nucleon, including the role of heavy quarks – for example, whether or not the nucleon has any intrinsic charm. As well, a variety of completely new experiments, such as deeply virtual Compton scattering, will yield qualitatively new information.

While we have constrained ourselves to the properties of the free nucleon, one must also acknowledge with anticipation, the promise of information on the temperature and density dependence of hadron properties from a variety of experiments, especially from relativistic heavy ion collisions.

The understanding of the strong interaction is one of the most challenging problems in subatomic physics and every piece of information we can gather, from whatever source, must become part of the rich tapestry that we eventually complete.

A Notations and Conventions

A.1 Isospin

The proton and neutron form an isospin $SU(2)$ doublet

$$|\text{proton}\rangle = \begin{pmatrix} 1 \\ 0 \end{pmatrix}, \quad |\text{neutron}\rangle = \begin{pmatrix} 0 \\ 1 \end{pmatrix} \tag{A.1}$$

We work with standard isospin matrices

$$\vec{\tau} = (\tau_1, \tau_2, \tau_3), \quad \tau_\pm = \frac{1}{2}(\tau_1 \pm i\tau_2), \tag{A.2}$$

$$\tau_1 = \begin{pmatrix} 0 & 1 \\ 1 & 0 \end{pmatrix}, \quad \tau_2 = \begin{pmatrix} 0 & -i \\ i & 0 \end{pmatrix}, \quad \tau_3 = \begin{pmatrix} 1 & 0 \\ 0 & -1 \end{pmatrix},$$

so that

$$\tau_3\,|p\rangle = |\,p\rangle, \quad \tau_3\,|n\rangle = -|\,n\rangle, \quad \tau_-\,|p\rangle = |\,n\rangle, \quad \tau_+\,|n\rangle = |\,p\rangle. \tag{A.3}$$

The electric charge of the nucleon is

$$Q = \frac{1}{2}(B + \tau_3) \tag{A.4}$$

with baryon number $B = 1$.

Analogous relations hold for the isospin doublet of u- and d-quarks, with the proton replaced by the u-quark, the neutron by the d-quark and baryon number $B = 1/3$.

A.2 Gell-Mann matrices

We use the standard Gell-Mann forms of the $SU(3)$ matrices representing flavor or color degrees of freedom:

$$\lambda_1 = \begin{pmatrix} 0 & 1 & 0 \\ 1 & 0 & 0 \\ 0 & 0 & 0 \end{pmatrix}, \quad \lambda_2 = \begin{pmatrix} 0 & -i & 0 \\ i & 0 & 0 \\ 0 & 0 & 0 \end{pmatrix}, \quad \lambda_3 = \begin{pmatrix} 1 & 0 & 0 \\ 0 & -1 & 0 \\ 0 & 0 & 0 \end{pmatrix},$$

$$\lambda_4 = \begin{pmatrix} 0 & 0 & 1 \\ 0 & 0 & 0 \\ 1 & 0 & 0 \end{pmatrix}, \quad \lambda_5 = \begin{pmatrix} 0 & 0 & -i \\ 0 & 0 & 0 \\ i & 0 & 0 \end{pmatrix}, \quad \lambda_6 = \begin{pmatrix} 0 & 0 & 0 \\ 0 & 0 & 1 \\ 0 & 1 & 0 \end{pmatrix},$$

$$\lambda_7 = \begin{pmatrix} 0 & 0 & 0 \\ 0 & 0 & -i \\ 0 & i & 0 \end{pmatrix}, \qquad \lambda_8 = \frac{1}{\sqrt{3}} \begin{pmatrix} 1 & 0 & 0 \\ 0 & 1 & 0 \\ 0 & 0 & -2 \end{pmatrix}.$$

They satisfy the commutation relations $[\lambda_a, \lambda_b] = 2\,i f_{abc}\,\lambda_c$ with the (totally antisymmetric) $SU(3)$ structure constants f_{abc}. The non-zero elements (up to permutations) are:

$$f_{123} = 2f_{147} = 2f_{246} = 2f_{257} = 2f_{345} = -2f_{156} = -2f_{367} = \frac{2}{\sqrt{3}} f_{458} = \frac{2}{\sqrt{3}} f_{678} = 1$$

A.3 Dirac matrices

We use the standard representation of Dirac matrices:

$$\gamma^\mu = (\gamma^0, \vec{\gamma}), \quad \gamma^0 = \gamma_0 = \begin{pmatrix} \mathbf{1} & 0 \\ 0 & -\mathbf{1} \end{pmatrix}, \quad \vec{\gamma} = \begin{pmatrix} 0 & \vec{\sigma} \\ -\vec{\sigma} & 0 \end{pmatrix} \tag{A.5}$$

with the 2×2 unit matrix $\mathbf{1}$ and the Pauli spin matrices

$$\vec{\sigma} = (\sigma_x, \sigma_y, \sigma_z),$$

$$\sigma_x = \begin{pmatrix} 0 & 1 \\ 1 & 0 \end{pmatrix}, \quad \sigma_y = \begin{pmatrix} 0 & -i \\ i & 0 \end{pmatrix}, \quad \sigma_z = \begin{pmatrix} 1 & 0 \\ 0 & -1 \end{pmatrix}.$$

Further important combinations of γ matrices are

$$\gamma^5 = i\gamma^0\gamma^1\gamma^2\gamma^3 = \gamma_5 = \begin{pmatrix} 0 & \mathbf{1} \\ \mathbf{1} & 0 \end{pmatrix}, \tag{A.6}$$

and the commutator

$$\sigma^{\mu\nu} = \frac{i}{2}[\gamma^\mu, \gamma^\nu] \equiv \frac{i}{2}(\gamma^\mu\gamma^\nu - \gamma^\nu\gamma^\mu). \tag{A.7}$$

The Dirac matrices satisfy $\{\gamma^\mu, \gamma^\nu\} = 2g^{\mu\nu}$ with the metric tensor

$$g^{\mu\nu} = \begin{pmatrix} 1 & 0 & 0 & 0 \\ 0 & -1 & 0 & 0 \\ 0 & 0 & -1 & 0 \\ 0 & 0 & 0 & -1 \end{pmatrix}.$$

A.4 Dirac fields and spinors

Nucleons

The state vector of a free nucleon is

$$\big|N_s(P)\rangle = b^\dagger(P, s)\big|\,0\rangle, \tag{A.8}$$

where the normalization of the vacuum state is $\langle 0|0\rangle = 1$; the nucleon four-momentum is $P^\mu = (E, \vec{P})$, with $E = \sqrt{\vec{P}^{\,2} + M^2}$, and $s = \pm\frac{1}{2}$ is the spin projection. We use the normalization

$$\langle N_{s'}(P')|N_s(P)\rangle = 2E(2\pi)^3\delta^3(\vec{P}\,' - \vec{P}\,)\delta_{s's}. \tag{A.9}$$

The nucleon field satisfies the free Dirac equation

$$(i\gamma_\mu\partial^\mu - M)\Psi(x) = 0 \tag{A.10}$$

The solution for $E > 0$ is

$$\Psi^{(+)}(x) = \sum_{s=\pm 1/2}\int \frac{d^3P}{2E(2\pi)^3}\, b(P,s)u_s(P)e^{-iP\cdot x}, \tag{A.11}$$

where the Dirac spinor, $u(P)$, satisfies

$$(\gamma_\mu P^\mu - M)u_s(P) = 0. \tag{A.12}$$

Its explicit form is

$$u_s(P) = \sqrt{E+M}\begin{pmatrix} \chi_s \\ \frac{\vec{\sigma}\cdot\vec{P}}{E+M}\chi_s \end{pmatrix}, \tag{A.13}$$

with two-component Pauli spinors

$$\chi_{s=+1/2} = \begin{pmatrix} 1 \\ 0 \end{pmatrix}, \quad \chi_{s=-1/2} = \begin{pmatrix} 0 \\ 1 \end{pmatrix}, \tag{A.14}$$

and the normalization

$$u^\dagger_{s'}(P)u_s(P) = 2E\delta_{s's}, \quad \bar{u}_{s'}(P)u_s(P) = u^\dagger_{s'}(P)\gamma_0 u_s(P) = 2M\delta_{s's}. \tag{A.15}$$

Leptons and quarks

The following relations are written for leptons, but identical relations hold for the quarks. State vectors:

$$\big|l_s(p)\rangle = b^\dagger(p,s)\big|\,0\rangle, \tag{A.16}$$

with four-momentum $p^\mu = (\epsilon, \vec{p})$ and $\epsilon = \sqrt{\vec{p}^{\,2} + m^2}$, and spin projection $s = \pm\frac{1}{2}$. Normalization:

$$\langle l_{s'}(p')|l_s(p)\rangle = 2\,\epsilon\,(2\pi)^3\delta^3(\vec{p}\,' - \vec{p})\,\delta_{s's}. \tag{A.17}$$

The lepton field satisfies the free Dirac equation

$$(i\gamma_\mu\partial^\mu - m)\psi(x) = 0, \tag{A.18}$$

for which the solution is

$$\psi(x) = \sum_{s=\pm 1/2} \int \frac{d^3p}{2\,\epsilon(2\,\pi)^3} \left[b(p,s)u_s(p)e^{-ip\cdot x} + d^\dagger(p,s)v_s(p)e^{ip\cdot x} \right]. \tag{A.19}$$

The explicit form of the Dirac spinor is

$$u_s(p) = \sqrt{\epsilon + m} \begin{pmatrix} \chi_s \\ \frac{\vec{\sigma}\cdot\vec{p}}{\epsilon+m}\chi_s \end{pmatrix}, \quad v_s(p) = \sqrt{\epsilon + m} \begin{pmatrix} \frac{\vec{\sigma}\cdot\vec{p}}{\epsilon+m}\chi_s \\ \chi_s \end{pmatrix}, \tag{A.20}$$

with the normalization

$$u_{s'}^\dagger(p)u_s(p) = v_{s'}^\dagger(p)v_s(p) = 2\,\epsilon\,\delta_{s's}. \tag{A.21}$$

B Cross Sections for Lepton- and Photon-Nucleon Scattering

Here we summarize the main steps in the derivation of basic cross sections for the scattering of leptons and photons from nucleons.

B.1 Electron and Muon Scattering

The process under consideration is

$$l(p) + N(P) \to l(p') + X(P'), \tag{B.1}$$

in which a charged lepton l (electron e or muon μ) with four-momentum $p^\mu = (\epsilon, \vec{p})$ scatters from a nucleon N with four-momentum $P^\mu = (E, \vec{P})$ and $E = \sqrt{M^2 + \vec{P}^2}$. The scattered lepton has outgoing four-momentum $p'^\mu = (\epsilon', \vec{p'})$. The hadronic final state X is left with a four-momentum $P'^\mu = (E', \vec{P'})$. Elastic scattering corresponds to $X \equiv N$.

Kinematic variables

energy transfer:

$$\nu = \epsilon - \epsilon' = E' - E \tag{B.2}$$

momentum transfer:

$$\vec{q} = \vec{p} - \vec{p'} = \vec{P'} - \vec{P} \tag{B.3}$$

squared four-momentum transfer:

$$q^2 = \nu^2 - \vec{q}^{\,2} \equiv -Q^2 \equiv t \tag{B.4}$$

squared total center-of-mass energy:

$$s = (p + P)^2 \equiv W^2 \tag{B.5}$$

squared invariant mass of the residual hadronic system:

$$s^* = M_X^2 = (P + q)^2 = M^2 + 2P \cdot q - Q^2 \tag{B.6}$$

Bjorken scaling variable:

$$x = \frac{Q^2}{2P \cdot q} \tag{B.7}$$

fraction of lepton energy loss in lab. frame:

$$y = \frac{q \cdot P}{p \cdot P} \tag{B.8}$$

Lab frame relations

Consider the laboratory frame in which $P^\mu = (M, 0)$. Let ϵ, ϵ' be the lepton energies, ν the energy transfer and θ the lepton scattering angle in that frame. We work with relativistic leptons and neglect their mass in comparison with their energy ($\epsilon \simeq |\vec{p}|$). We have

$$Q^2 \equiv -q^2 \simeq 4\,\epsilon\epsilon' \sin^2\frac{\theta}{2}, \tag{B.9}$$

$$s \equiv W^2 \simeq 2M\epsilon + M^2, \tag{B.10}$$

$$s^* = M^2 + 2M\nu - Q^2, \tag{B.11}$$

$$x = \frac{Q^2}{2M\nu}, \tag{B.12}$$

$$y = \frac{\nu}{\epsilon}. \tag{B.13}$$

For elastic scattering, $x = 1$ and $\nu = \frac{Q^2}{2M}$.

T-Matrix

The T-matrix for the scattering process is related to the S-matrix by

$$S_{fi} = \langle f|i\rangle + (2\,\pi)^4 i\delta^4(p + P - p' - P')T_{fi}. \tag{B.14}$$

In the one-photon exchange approximation, which is excellent throughout the kinematic region of interest, this T-matrix

$$T_{fi} = \frac{e^2}{q^2}\langle l(p')|j^\mu(0)|l(p)\rangle\langle X(P')|J_\mu(0)|N(P)\rangle, \tag{B.15}$$

where $j^\mu(x)$ and $J_\mu(x)$ are the lepton and quark electromagnetic currents as given by the Standard Model (see Chapter 1). The lepton current matrix elements are

$$\langle l(p')|j^\mu(0)|l(p)\rangle = \bar{u}(p')\gamma^\mu u(p) \tag{B.16}$$

in terms of the electron or muon Dirac spinors $u(p)$. Since the high-energy leptons behave as almost massless particles, it is convenient to choose a normalization for their spinors such that

$$\sum_{s=\pm\frac{1}{2}} u_s(p)\bar{u}_s(p) = \gamma \cdot p \tag{B.17}$$

as detailed in Appendix A.4.

Scattering of unpolarized leptons from a point-like Dirac particle

Here we sketch the derivation of the cross section for the scattering of unpolarized high energy electrons or muons on a pointlike Dirac particle of mass M. In this case only elastic scattering is possible. The relevant current matrix elements of the nucleon are

$$\langle N(P')|J_\mu(0)|N(P)\rangle = \bar{u}(P')\gamma_\mu u(P), \tag{B.18}$$

and we adopt the normalization of spinors as in Appendix A.4, with

$$\sum_{s=\pm\frac{1}{2}} u_s(P)\bar{u}_s(P) = \gamma \cdot P + M. \tag{B.19}$$

The differential cross section for this process is

$$d\sigma = \frac{\overline{|T_{fi}|^2}}{\mathcal{J}} \frac{d^3p'}{2\,\epsilon'(2\,\pi)^3} \frac{d^3P'}{2E'(2\,\pi)^3}\,(2\,\pi)^4\delta^4(p - p' + P - P'). \tag{B.20}$$

The calculation is carried out in the lab frame where the flux factor $\mathcal{J} = 4P \cdot p$ reduces to $4M\epsilon$. The factors $(2\,\epsilon')^{-1}$ and $(2E')^{-1}$ are related to the normalizations of the lepton and nucleon wave functions in the final state.

For unpolarized scattering the spin average over the squared T-matrix gives

$$\overline{|T_{fi}|^2} = \frac{e^4}{q^4}\frac{1}{4}\sum_{\text{spins}} |\bar{u}(p')\gamma_\mu u(p)\;\bar{u}(P')\gamma^\mu u(P)|^2 = \frac{e^4}{q^4}\,L_{\mu\nu}N^{\mu\nu}, \tag{B.21}$$

with the (spin-averaged) lepton and point nucleon tensors

$$L_{\mu\nu} = \frac{1}{2}tr[\gamma \cdot p'\gamma_\mu\gamma \cdot p\,\gamma_\nu] = 2[p_\mu p'_\nu + p'_\mu p_\nu - g_{\mu\nu}p \cdot p'], \tag{B.22}$$

and

$$\begin{aligned} N_{\mu\nu} &= \frac{1}{2}tr\,[(\gamma \cdot P' + M)\gamma_\mu(\gamma \cdot P + M)\gamma_\nu] \\ &= 2[P_\mu P'_\nu + P'_\mu P_\nu - g_{\mu\nu}(P \cdot P' - M^2)], \end{aligned} \tag{B.23}$$

where we have neglected the lepton mass. Using $P' = P + p - p'$ one finds, in the lab frame:

$$\begin{aligned} L_{\mu\nu}N^{\mu\nu} &= 8\left[2P \cdot p'\;P \cdot p + p \cdot p'(P \cdot q - M^2)\right] \\ &= 16M^2\epsilon\epsilon'\left(\cos^2\frac{\theta}{2} - \frac{q^2}{2M^2}\sin^2\frac{\theta}{2}\right). \end{aligned} \tag{B.24}$$

Next we integrate over d^3P' and insert $d^3p' = p'^2dp'd\Omega \simeq \epsilon'^2d\epsilon'd\Omega$. The differential cross section can be written in the form

$$\frac{d^2\sigma}{d\Omega\,d\epsilon'} = \frac{\alpha^2}{4\,\epsilon^2\sin^4\frac{\theta}{2}}\left(\cos^2\frac{\theta}{2} - \frac{q^2}{2M^2}\sin^2\frac{\theta}{2}\right)\delta(\nu + \frac{q^2}{2M}), \tag{B.25}$$

with $\alpha = e^2/4\pi$. Here we have used the fact that in this elastic scattering process, energy and momentum conservation require $\nu = \epsilon - \epsilon' = E' - E = Q^2/2M$.

Finally, the integration over ϵ', with

$$\epsilon' = \frac{\epsilon}{1 + \frac{2\epsilon}{M}\sin^2\frac{\theta}{2}}, \tag{B.26}$$

gives

$$\frac{d\sigma}{d\Omega} = \left(\frac{d\sigma}{d\Omega}\right)_{\text{Mott}} \left[1 - \frac{q^2}{2M^2}\tan^2\frac{\theta}{2}\right], \tag{B.27}$$

where the Mott cross section, with inclusion of the recoil factor ϵ'/ϵ, is:

$$\left(\frac{d\sigma}{d\Omega}\right)_{\text{Mott}} = \frac{\alpha^2\,\cos^2\frac{\theta}{2}}{4\,\epsilon^2\sin^4\frac{\theta}{2}\left(1 + \frac{2\epsilon}{M}\sin^2\frac{\theta}{2}\right)}. \tag{B.28}$$

(We note that the traditional form of the Mott cross section is commonly written in the limit $M \to \infty$ or $\frac{\epsilon'}{\epsilon} \to 1$, omitting recoil.)

Rosenbluth cross section

The nucleon matrix element of the electromagnetic current with inclusion of the Dirac form factor $F_1(Q^2)$ and the Pauli form factor $F_2(Q^2)$ is

$$\langle N(P')|J^\mu(0)|N(P)\rangle = \bar{u}(P')\left[\gamma^\mu F_1(Q^2) + i\sigma^{\mu\nu}\frac{q_\nu}{2M}F_2(Q^2)\right]u(P), \tag{B.29}$$

with $q_\nu = (P' - P)_\nu$, $Q^2 = -q^2$. Alternatively, one uses the Sachs form factors

$$G_E = F_1 + \frac{q^2}{4M^2}F_2, \quad G_M = F_1 + F_2. \tag{B.30}$$

The nucleon tensor $N^{\mu\nu}$ of Eq. (B.23) is now replaced by

$$N_{\mu\nu} = \frac{1}{2}tr\left[(\gamma\cdot P' + M)\Gamma_\mu(\gamma\cdot P + M)\Gamma_\nu\right], \tag{B.31}$$

with

$$\Gamma_\mu = G_M\gamma_\mu - F_2\frac{(P + P')_\mu}{2M}, \tag{B.32}$$

where the Gordon identity $i\bar{u}(P')\sigma_{\mu\nu}u(P)q^\nu = 2M\bar{u}(P')\gamma_\mu u(P) - (P + P')_\mu$ has been used.

The product of lepton and nucleon tensors becomes

$$L_{\mu\nu}N^{\mu\nu} = 16M^2\epsilon\epsilon'\left[\left(F_1^2 - \frac{q^2}{4M^2}F_2^2\right)\cos^2\frac{\theta}{2} - \frac{q^2}{2M^2}G_M^2\sin^2\frac{\theta}{2}\right]. \tag{B.33}$$

The resulting elastic differential cross section which replaces Eq. (B.27) is

$$\frac{d\sigma}{d\Omega} = \left(\frac{d\sigma}{d\Omega}\right)_{\text{Mott}}\left[\frac{G_E^2 - \frac{q^2}{4M^2}G_M^2}{1 - \frac{q^2}{4M^2}} - \frac{q^2}{2M^2}G_M^2\tan^2\frac{\theta}{2}\right]. \tag{B.34}$$

This is the Rosenbluth formula [Ros 50].

Inclusive inelastic scattering

Here we summarize the general differential cross section for the inelastic scattering of charged leptons on nucleons without detection of the hadronic final state, but including the possibility that both leptons and nucleons in the initial state are polarized.

The lepton tensor that replaces the spin-averaged one of Eq. (B.22) is now:

$$\begin{aligned} L_{\mu\nu} &= \sum_{s'} \bar{u}_s(p)\gamma_\mu u_{s'}(p')\bar{u}_{s'}(p')\gamma_\nu u_s(p) \\ &= 2\left[p_\mu p'_\nu + p'_\mu p_\nu - g_{\mu\nu} p\cdot p' + i\epsilon_{\mu\nu\alpha\beta}s^\alpha q^\beta\right], \end{aligned} \tag{B.35}$$

where $s_\mu = \bar{u}\gamma_\mu\gamma_5 u$ is the lepton spin vector and we have neglected again the small lepton mass. Our convention for the Levi-Civita tensor is $\epsilon_{0123} = +1$. The antisymmetric part of this tensor vanishes when averaging over the initial spin.

The inelastic scattering process involves all possible transitions of the nucleon from its ground state, $|N(P)\rangle$, to any excited state, $|X(P')\rangle$. The term $\int d^3P'(\pi/E')\delta^4(p-p'+P-P')N_{\mu\nu}$, from Eqs. (B.23) and (B.20), is now to be replaced by the hadronic current tensor

$$\begin{aligned} W_{\mu\nu}(P,q) = \frac{1}{2M}\sum_X &\langle N_s(P)|J_\mu(0)|X(P')\rangle\langle X(P')|J_\nu(0)|N_s(P)\rangle \\ &\cdot(2\pi)^3\delta^4(q+P-P'), \end{aligned} \tag{B.36}$$

The initial nucleon spin state is denoted by $s = \pm\frac{1}{2}$. The summation over unobserved final states X is understood to include the integration $\int d^3P'/2E'(2\pi)^3$ over their total momenta. The lab frame differential cross section is now

$$\frac{d^2\sigma}{d\Omega d\epsilon'} = \frac{\alpha^2}{q^4}\frac{\epsilon'}{\epsilon}L_{\mu\nu}W^{\mu\nu}. \tag{B.37}$$

The tensor $W_{\mu\nu}$ can be presented in a more compact form using the completeness of the states $|X\rangle$:

$$W_{\mu\nu} = \frac{1}{4\pi M}\int d^4z\, e^{iq\cdot z}\langle N_s(P)|J_\mu(z)J_\nu(0)|N_s(P)\rangle. \tag{B.38}$$

It is sometimes convenient to rewrite this as a current commutator,

$$W_{\mu\nu} = \frac{1}{4\pi M}\int d^4z\, e^{iq\cdot z}\langle N_s(P)|[J_\mu(z),J_\nu(0)]|N_s(P)\rangle, \tag{B.39}$$

where the extra term, $J_\nu(0)J_\mu(z)$, gives a vanishing matrix element since it produces a delta function $\delta^4(q-P+P')$ which cannot be satisfied because of energy conservation as the nucleon is the ground state baryon (there is no intermediate state with $E' = M - \nu \le M$).

One can further split $W_{\mu\nu}$ into symmetric and antisymmetric parts:

$$W_{\mu\nu} = W^S_{\mu\nu} + W^A_{\mu\nu}. \tag{B.40}$$

Lorentz and gauge invariance together with parity conservation of the electromagnetic interaction imply the most general forms of these terms:

$$\begin{aligned} W^S_{\mu\nu} = & \; W_1(\nu, Q^2)\left(\frac{q_\mu q_\nu}{q^2} - g_{\mu\nu}\right) \\ & + \frac{W_2(\nu, Q^2)}{M^2}\left(P_\mu - \frac{P\cdot q}{q^2}q_\mu\right)\left(P_\nu - \frac{P\cdot q}{q^2}q_\nu\right), \end{aligned} \tag{B.41}$$

and

$$W^A_{\mu\nu} = i\epsilon_{\mu\nu\alpha\beta}q^\alpha\left[G_1(\nu, Q^2)S^\beta + \frac{G_2(\nu, Q^2)}{M^2}\left(S^\beta P\cdot q - P^\beta S\cdot q\right)\right]. \tag{B.42}$$

This defines four response functions $W_{1,2}(\nu, Q^2)$ and $G_{1,2}(\nu, Q^2)$. The first two are observed in the scattering of unpolarized particles, while the latter two require scattering of polarized leptons on polarized nucleons for their determination.

In Eq. (B.42) we have introduced the nucleon spin vector

$$S^\mu = \bar{u}(P)\gamma^\mu\gamma_5 u(P)/2M. \tag{B.43}$$

For a nucleon at rest, $S^\mu = (0, \chi^\dagger_S\vec{\sigma}\chi_S)$.

Structure functions

In the description of deep-inelastic scattering processes the response functions $W_{1,2}(\nu, Q^2)$ and $G_{1,2}(\nu, Q^2)$ are often replaced by the dimensionless structure functions $F_{1,2}(x, Q^2)$ and $g_{1,2}(x, Q^2)$, expressed in terms of the Bjorken variable x together with Q^2:

$$F_1(x, Q^2) = MW_1(\nu, Q^2), \tag{B.44}$$

$$F_2(x, Q^2) = \nu W_2(\nu, Q^2), \tag{B.45}$$

and

$$g_1(x, Q^2) = M\nu G_1(\nu, Q^2), \tag{B.46}$$

$$g_2(x, Q^2) = \nu^2 G_2(\nu, Q^2). \tag{B.47}$$

Unpolarized differential cross section

Averaging over the spins of the incident lepton and target nucleon, one obtains the unpolarized differential cross section in the laboratory frame, to be compared with Eq. (B.25):

$$\frac{d^2\sigma}{d\Omega\, d\epsilon'} = \frac{\alpha^2}{4\,\epsilon^2\sin^4\frac{\theta}{2}}\left(W_2(\nu, Q^2)\cos^2\frac{\theta}{2} + 2W_1(\nu, Q^2)\sin^2\frac{\theta}{2}\right). \tag{B.48}$$

The following invariant cross sections are also frequently used:

$$\frac{d^2\sigma}{dx\, dy} = x(s - M^2)\frac{d^2\sigma}{dx\, dQ^2} = \frac{2\,\pi M\nu}{\epsilon'}\frac{d^2\sigma}{d\Omega\, d\epsilon'}. \tag{B.49}$$

For example,

$$\frac{d^2\sigma}{dx\, dQ^2} = \frac{4\,\pi\alpha^2}{Q^4}\left[\left(\frac{1-y}{x} - \frac{M^2 y}{s - M^2}\right)F_2(x, Q^2) + y^2 F_1(x, Q^2)\right]. \tag{B.50}$$

Elastic scattering

In the elastic scattering limit the response functions reduce to combinations of squared electromagnetic form factors, as follows:

$$W_1(\nu, Q^2) = \frac{Q^2}{4M^2} G_M^2(Q^2)\, \delta\left(\nu - \frac{Q^2}{2M}\right), \tag{B.51}$$

$$W_2(\nu, Q^2) = \frac{G_E^2(Q^2) + \frac{Q^2}{4M^2} G_M^2(Q^2)}{1 + \frac{Q^2}{4M^2}}\, \delta\left(\nu - \frac{Q^2}{2M}\right). \tag{B.52}$$

Insertion of these expressions in Eq. (B.48) leads back to the Rosenbluth formula (B.34).

Virtual photon cross sections

The differential cross section (B.48) can be expressed in terms of the cross sections for absorption of transverse and longitudinal virtual photons on a nucleon. We denote these cross sections by σ_T and σ_L, respectively. The lepton scattering cross section in the lab frame is expressed as

$$\frac{d^2\sigma}{d\Omega\, d\epsilon'} = \Gamma \left[\sigma_T(\nu, Q^2) + \varepsilon \sigma_L(\nu, Q^2)\right], \tag{B.53}$$

where

$$\varepsilon = \left[1 + 2\left(1 + \frac{\nu^2}{Q^2}\right) \tan^2 \frac{\theta}{2}\right]^{-1} \tag{B.54}$$

is the ratio of longitudinal to transverse polarization of the virtual photon and

$$\Gamma = \frac{\alpha}{2\,\pi^2 Q^2} \cdot \frac{\epsilon'}{\epsilon} \cdot \frac{\nu^*}{1 - \varepsilon}, \tag{B.55}$$

with

$$\nu^* = \frac{s^* - M^2}{2M} = \nu - \frac{Q^2}{2M} = \nu(1 - x) \tag{B.56}$$

a kinematic factor associated with our choice for the flux of virtual photons. Here s^* is the squared center-of-mass energy of the excited hadronic system. For real photons, i.e. in the limit $Q^2 \to 0$, one has $\nu^* = \nu$, $\sigma_L \to 0$ and $\sigma_T \to \sigma_\gamma(\nu)$, the total photoabsorption cross section.

The following relations hold between the virtual photon cross sections and the response functions:

$$W_1(\nu, Q^2) = \frac{\nu^*}{4\,\pi^2\alpha} \sigma_T(\nu, Q^2), \tag{B.57}$$

$$W_2(\nu, Q^2) = \frac{\nu^*}{4\,\pi^2\alpha} \frac{Q^2}{Q^2 + \nu^2} \left[\sigma_T(\nu, Q^2) + \sigma_L(\nu, Q^2)\right]. \tag{B.58}$$

The ratio of longitudinal to transverse cross sections is:

$$R \equiv \frac{\sigma_L}{\sigma_T} = \frac{W_2}{W_1}\left(1 + \frac{\nu^2}{Q^2}\right) - 1. \tag{B.59}$$

Polarized inclusive scattering

Consider a measurement with both the lepton beam and the nucleon target polarized. Let $\sigma^{\rightrightarrows}$ be the cross section for lepton and target spins parallel, $\sigma^{\rightleftarrows}$ the corresponding one for antiparallel spins. In terms of the spin dependent response functions $G_{1,2}$, the differential cross sections in the laboratory frame are:

$$\frac{d^2\sigma^{\rightrightarrows}}{d\Omega d\epsilon'} = \frac{d^2\sigma}{d\Omega d\epsilon'} - \frac{2\,\alpha^2}{Q^2}\frac{\epsilon'}{\epsilon}\left[(\epsilon+\epsilon'\cos\theta)G_1(\nu,Q^2) - \frac{Q^2}{M}G_2(\nu,Q^2)\right],$$
$$\frac{d^2\sigma^{\rightleftarrows}}{d\Omega d\epsilon'} = \frac{d^2\sigma}{d\Omega d\epsilon'} + \frac{2\,\alpha^2}{Q^2}\frac{\epsilon'}{\epsilon}\left[(\epsilon+\epsilon'\cos\theta)G_1(\nu,Q^2) - \frac{Q^2}{M}G_2(\nu,Q^2)\right], \tag{B.60}$$

where

$$\frac{d^2\sigma}{d\Omega d\epsilon'} = \frac{1}{2}\left(\frac{d^2\sigma^{\rightrightarrows}}{d\Omega d\epsilon'} + \frac{d^2\sigma^{\rightleftarrows}}{d\Omega d\epsilon'}\right) \tag{B.61}$$

is the unpolarized cross section of Eq. (B.48). The lepton asymmetry is

$$A = \frac{d\sigma^{\rightleftarrows} - d\sigma^{\rightrightarrows}}{d\sigma^{\rightleftarrows} + d\sigma^{\rightrightarrows}}. \tag{B.62}$$

The cross section difference expressed in terms of the dimensionless spin structure functions $g_{1,2}$ is

$$\frac{d^2\sigma^{\rightleftarrows}}{d\Omega d\epsilon'} - \frac{d^2\sigma^{\rightrightarrows}}{d\Omega d\epsilon'} = \frac{4\,\alpha^2}{Q^2}\frac{\epsilon'}{M\nu}\left[\left(1+\frac{\epsilon'}{\epsilon}\cos\theta\right)\,g_1(x,Q^2) - \frac{Q^2}{\epsilon\nu}\,g_2(x,Q^2)\right]. \tag{B.63}$$

It is often convenient to work with the virtual photon cross sections defined in Eq. (B.87). Introducing $\sigma_{\frac{1}{2}}$ and $\sigma_{\frac{3}{2}}$ for the cross sections with photon and nucleon helicities antiparallel and parallel, respectively, the photon asymmetry is

$$A_1 = \frac{\sigma_{\frac{1}{2}} - \sigma_{\frac{3}{2}}}{2\sigma_T}, \tag{B.64}$$

where

$$\sigma_T = \frac{1}{2}\left[\sigma_{\frac{1}{2}} + \sigma_{\frac{3}{2}}\right] \tag{B.65}$$

is the transverse cross section of Eq. (B.57). The relation to the spin structure functions is:

$$\sigma_{\frac{1}{2}} - \sigma_{\frac{3}{2}} = \frac{8\,\pi^2\alpha}{M\nu^*}\left[g_1(x,Q^2) - \frac{Q^2}{\nu^2}\,g_2(x,Q^2)\right], \tag{B.66}$$

with $\nu^* = \nu - \frac{Q^2}{2M} = \nu(1-x)$.

B.2 Neutrino Scattering

The processes under consideration are now

$$\nu_\mu(p) + N(P) \to \mu^-(p') + X(P'),$$
$$\bar{\nu}_\mu(p) + N(P) \to \mu^+(p') + X(P'),$$

in which a neutrino or antineutrino with four-momentum $p^\mu = (\epsilon, \vec{p})$ scatters from a nucleon N with four-momentum $P^\mu = (E, \vec{P})$ and $E = \sqrt{M^2 + \vec{P}^2}$. The outgoing muon has four-momentum $p'^\mu = (\epsilon', \vec{p'})$. The hadronic final state X is left with a four-momentum $P'^\mu = (E', \vec{P'})$.

All steps carry through as before in electron or muon scattering, apart from the following modifications:

a) The factor α/Q^2 is replaced by $\dfrac{G_F\, M_W^2}{2\,\pi\sqrt{2}(Q^2 + M_W^2)}$;

b) The response functions $W_{1,2}$ are replaced by $W_{1,2}^{\nu N}$ or $W_{1,2}^{\bar{\nu} N}$;

c) A third response function W_3 enters due to the parity violating weak interaction.

Differential cross sections

The results for the differential cross sections are:

$$\frac{d^2\sigma(\nu N)}{d\Omega\, d\epsilon'} = \frac{G_F^2\, M_W^4\, \epsilon'^2}{2\,\pi^2(Q^2 + M_W^2)^2} \tag{B.67}$$
$$\left(2W_1^{\nu N}(\nu, Q^2)\sin^2\frac{\theta}{2} + W_2^{\nu N}(\nu, Q^2)\cos^2\frac{\theta}{2} + W_3^{\nu N}(\nu, Q^2)\frac{\epsilon + \epsilon'}{M}\sin^2\frac{\theta}{2}\right),$$

$$\frac{d^2\sigma(\bar{\nu} N)}{d\Omega\, d\epsilon'} = \frac{G_F^2\, M_W^4\, \epsilon'^2}{2\,\pi^2(Q^2 + M_W^2)^2} \tag{B.68}$$
$$\left(2W_1^{\bar{\nu} N}(\nu, Q^2)\sin^2\frac{\theta}{2} + W_2^{\bar{\nu} N}(\nu, Q^2)\cos^2\frac{\theta}{2} - W_3^{\bar{\nu} N}(\nu, Q^2)\frac{\epsilon + \epsilon'}{M}\sin^2\frac{\theta}{2}\right).$$

Structure functions

Dimensionless structure functions are introduced as

$$F_1^{\nu,\bar{\nu} N}(x, Q^2) = M W_1^{\nu,\bar{\nu} N}(\nu, Q^2), \tag{B.69}$$
$$F_2^{\nu,\bar{\nu} N}(x, Q^2) = \nu W_2^{\nu,\bar{\nu} N}(\nu, Q^2), \tag{B.70}$$
$$F_3^{\nu,\bar{\nu} N}(x, Q^2) = \nu W_3^{\nu,\bar{\nu} N}(\nu, Q^2). \tag{B.71}$$

In terms of these structure functions the neutrino and antineutrino cross sections are:

$$\frac{d^2\sigma(\nu N)}{dx\, dQ^2} = \frac{G_F^2}{2\,\pi}\left(\frac{M_W^2}{Q^2 + M_W^2}\right)^2 \cdot$$
$$\left[xy^2 F_1^{\nu N} + (1 - y)F_2^{\nu N} + xy\left(1 - \frac{y}{2}\right)F_3^{\nu N}\right], \tag{B.72}$$

$$\frac{d^2\sigma(\bar\nu N)}{dx\, dQ^2} = \frac{G_F^2}{2\,\pi}\left(\frac{M_W^2}{Q^2+M_W^2}\right)^2 \cdot \left[xy^2 F_1^{\bar\nu N} + (1-y)F_2^{\bar\nu N} - xy\left(1-\frac{y}{2}\right)F_3^{\bar\nu N}\right]. \quad \text{(B.73)}$$

with $y = \nu/\epsilon$.

B.3 Compton Scattering

Kinematic variables

Consider the elastic photon scattering process

$$\gamma(k) + N(P) \to \gamma(k') + N(P') \quad \text{(B.74)}$$

on a nucleon with initial and final four-momenta P and P'. The in- and outgoing real photon has four-momentum k and k', respectively, with $k^2 = k'^2 = 0$.

We will generally work in the laboratory frame with $P^\mu = (M, 0)$. We denote the energy of the in- and outgoing photon in this frame by $\omega = |\vec{k}|$ and $\omega' = |\vec{k}'|$, and the photon scattering angle by θ. Energy and momentum conservation gives

$$\omega' = \frac{\omega}{1+\frac{\omega}{M}(1-\cos\theta)}. \quad \text{(B.75)}$$

Polarization vectors

The incident and scattered photons are characterized by polarization vectors ε^μ and ε'^ν. We choose

$$\begin{aligned} \varepsilon^\mu &= (0,\vec\varepsilon\,), \quad \vec{k}\cdot\vec\varepsilon = 0, \\ \varepsilon'^\nu &= (0,\vec\varepsilon\,'), \quad \vec{k}'\cdot\vec\varepsilon\,' = 0, \end{aligned} \quad \text{(B.76)}$$

where the latter conditions reflect the transversality of the photon.

It is convenient to choose a coordinate system such that the z-axis coincides with the direction of the wave vector $\vec{k}$. Photons polarized *linearly* in the plane perpendicular to $\vec{k}$ are described using cartesian unit vectors

$$\vec\varepsilon_x = (1,0,0), \qquad \vec\varepsilon_y = (0,1,0). \quad \text{(B.77)}$$

It is more common to use helicity vectors corresponding to *circularly* polarized photons:

$$\vec\varepsilon_{\lambda=+1} = -\frac{1}{\sqrt{2}}(\vec\varepsilon_x + i\vec\varepsilon_y), \qquad \vec\varepsilon_{\lambda=-1} = \frac{1}{\sqrt{2}}(\vec\varepsilon_x - i\vec\varepsilon_y), \quad \text{(B.78)}$$

with

$$\vec\varepsilon^{\,*}_{\lambda'}\cdot\vec\varepsilon_\lambda = \delta_{\lambda'\lambda}. \quad \text{(B.79)}$$

T-matrix

The Compton scattering T-matrix is given by

$$(2\pi)^4\delta^4(k+P-k'-P')\,T_{fi} = ie^2\int d^4x\int d^4y\; e^{ik'\cdot x-ik\cdot y}\;\varepsilon'^{*\mu}\varepsilon^{\nu}\langle N(P')|\mathcal{T}J_\mu(x)J_\nu(y)|N(P)\rangle, \quad \text{(B.80)}$$

where J_μ is the electromagnetic current and $\mathcal{T}$ denotes the time ordered product.

The differential cross section for Compton scattering is:

$$\frac{d\sigma}{d\Omega_{\text{lab}}} = \left(\frac{\omega'}{8\pi M\omega}\right)^2 |T_{fi}|^2 . \quad \text{(B.81)}$$

For unpolarized nucleons, the average over initial and sum over final spins has to be taken.

The T-matrix can be written:

$$T_{fi} = e^2\;\varepsilon'^{*\mu}\varepsilon^{\nu}T_{\mu\nu}(k',P';k,P), \quad \text{(B.82)}$$

with the Compton tensor

$$T_{\mu\nu} = i\int d^4x\; e^{ik'\cdot x}\langle N(P')|\mathcal{T}J_\mu(x)J_\nu(0)|N(P)\rangle. \quad \text{(B.83)}$$

Current conservation implies $k_\nu T^{\mu\nu} = k'_\mu T^{\mu\nu} = 0$.

Forward scattering amplitude

For forward scattering with $\vec{k}' = \vec{k}$ the T-matrix in the lab frame reduces to:

$$T_{fi}(\omega,\theta=0) = 8\pi M\;\chi_f^+\left[f(\omega)\,\vec{\varepsilon}'^{*}\cdot\vec{\varepsilon} + ig(\omega)\,\sigma\cdot(\vec{\varepsilon}'^{*}\times\vec{\varepsilon})\right]\chi_i, \quad \text{(B.84)}$$

with nucleon Pauli spinors $\chi_{i,f}$. This introduces the spin-flip amplitude g and the non-flip amplitude f.

Optical theorem

The total photon-nucleon cross section is related to the imaginary part of the forward Compton amplitude in the lab frame by

$$\sigma_{\text{tot}}(\omega) = \frac{4\pi}{\omega}\,\text{Im}\,f(\omega). \quad \text{(B.85)}$$

Virtual Compton scattering

The Compton tensor can be generalized to the scattering of virtual photons, replacing the real photon momenta k, k' by q, q' with $q^2, q'^2 \neq 0$.

A special case is the hadronic tensor $W_{\mu\nu}$ for inelastic electron or muon scattering which is related to the forward virtual Compton tensor by

$$2\,\pi M W_{\mu\nu}(\nu, Q^2) = \mathrm{Im}\, T_{\mu\nu}(q, P; q, P). \tag{B.86}$$

Virtual photon cross sections for given polarizations λ and λ' of the in- and outgoing photons are defined as:

$$\sigma_{\lambda'\lambda} = \frac{4\,\pi^2\alpha}{\nu^*}\,\varepsilon^{*\mu}_{\lambda'} W_{\mu\nu}\varepsilon^{\nu}_{\lambda} = \frac{2\,\pi\alpha}{\nu^* M}\,\varepsilon^{*\mu}_{\lambda'}\,\mathrm{Im}\, T_{\mu\nu}\,\varepsilon^{\nu}_{\lambda}, \tag{B.87}$$

with $\nu^* = \nu - \frac{Q^2}{2M}$.

Scattering from a point Dirac particle

Here we give the lab frame differential cross section for unpolarized Compton scattering from a pointlike spin-$\frac{1}{2}$ particle with mass M, charge Ze and anomalous magnetic moment κ:

$$\left(\frac{d\sigma}{d\Omega}\right)_{\text{point}} = \left(\frac{d\sigma}{d\Omega}\right)_{\text{KN}} + \left(\frac{d\sigma}{d\Omega}\right)_{\text{P}}, \tag{B.88}$$

with the Klein-Nishina cross section:

$$\left(\frac{d\sigma}{d\Omega}\right)_{\text{KN}} = \frac{Z^4\alpha^2}{2M^2}\left(\frac{\omega'}{\omega}\right)^2\left(\frac{\omega'}{\omega} + \frac{\omega}{\omega'} - \sin^2\theta\right) \tag{B.89}$$

and the contribution associated with the anomalous magnetic moment (Powell cross section):

$$\begin{aligned}\left(\frac{d\sigma}{d\Omega}\right)_{\text{P}} = \frac{\kappa\alpha^2}{M^2}\left(\frac{\omega'}{\omega}\right)^2\frac{\omega\omega'}{M^2}\Big[&(1-\cos\theta)^2 + \frac{\kappa}{2}\left(\frac{9}{2} - 5\,\cos\theta + \frac{1}{2}\cos^2\theta\right) \\ &+ \kappa^2\,(1 - \cos\theta + \frac{1}{2}\sin^2\theta) + \frac{\kappa^3}{4}\,(1 + \frac{1}{2}\sin^2\theta)\Big],\end{aligned} \tag{B.90}$$

where $Z = 1$ and $\kappa = \kappa_p = 1.793$ for the proton, $Z = 0$ and $\kappa = \kappa_n = -1.913$ for the neutron.

C Grassmann Variables

In order to evaluate Feynman path integrals involving fermions we need a classical analogue of the Dirac field operator (A.19) which preserves Fermi-Dirac statistics. We shall write this classical field as:

$$\psi(x) = \sum_{s=\pm 1/2} \int \frac{d^3p}{2\,\epsilon(2\,\pi)^3} \left[a_{p,s} u_s(p) e^{-ip\cdot x} + b^*_{p,s} v_s(p) e^{ip\cdot x}\right], \tag{C.1}$$

where a_i, b_i^* (with $i = (p, s)$) are elements of a Grassmann algebra. These satisfy

$$\{a_i, a_j\} = \{a_i, a_j^*\} = \{a_i, b_j\} = \{a_i, b_j^*\} = 0, \tag{C.2}$$

where $\{\ ,\ \}$ indicates an anti-commutator.

Equations (C.2) imply that $a_i^n = 0$ for $n > 1$, so that the most general function of two elements, say a_1 and a_2, that one can write down is

$$f(a_1, a_2) = c_0 + c_1 a_1 + c_2 a_2 + c_3 a_1 a_2, \tag{C.3}$$

with c_i normal c-numbers. Note that one could also write the last term in Eq. (C.3) as $-c_3 a_2 a_1$. It is convenient to define the derivative of such a function as

$$\frac{\partial f}{\partial a_1} = c_1 + c_3 a_2. \tag{C.4}$$

while the anti-commuting properties of the a's mean that

$$\frac{\partial f}{\partial a_2} = c_2 - c_3 a_1. \tag{C.5}$$

More formally we find that the ordering of variables in a second derivative matters:

$$\frac{\partial}{\partial a_1 \partial a_2} = -\frac{\partial}{\partial a_2 \partial a_1}. \tag{C.6}$$

The final mathematical operation which we need to define is integration. Its rules are quite unintuitive since for Grassmann variables integration is equivalent to differentiation. Essentially we want the "infinitesimal" quantities da_i to also anti-commute and we define multiple integrals as an integral over the right-most variable followed by the next:

$$\int da_1 da_2 F \equiv \int da_1 \left(\int da_2 F\right), \tag{C.7}$$

and thus, by considering $(\int da)^2$ one easily finds that it must be equal to $-(\int da)^2$ and hence

$$\int da = 0. \tag{C.8}$$

Finally, it is usual to define

$$\int daa = 1. \tag{C.9}$$

With these definitions and derived properties we have all we need in principle to evaluate any expression involving Grassmann variables.

D Dimensional Regularization

The loops encountered in perturbative calculations of Feynman diagrams are commonly ultraviolet divergent. While it was formerly necessary to introduce some high momentum cut-off to regularize such an integral, this tends to destroy the gauge invariance upon which modern quantum field theories, such as QCD, are based. The current technique for dealing with this problem, which *does* preserve gauge invariance, was introduced by 't Hooft and Veltman [Nucl. Phys. B44 (1972) 189]. The idea is to evaluate the integral for D dimensional space-time, with $D < 4$, thus rendering it finite. Having evaluated the integral for $D < 4$, one analytically continues towards $D = 4$. The ultraviolet singularity then enters as a divergence in $\frac{1}{\epsilon}$, where $D = 4 - 2\epsilon$. The renormalization procedure involves the introduction of counter terms which cancel these divergences.

We choose the convention that the D-dimensional volume element is $\dfrac{d^D k}{(2\pi)^D}$, where $d^D k$ is defined by:

$$d^D k = k^{D-1} dk \, d\Omega_D \tag{D.1}$$

with

$$\int d\Omega_D = \frac{2\pi^{\frac{D}{2}}}{\Gamma(\frac{D}{2})}, \tag{D.2}$$

the surface of a D-dimensional unit sphere.

The fundamental integral of interest is

$$I_D(m) = \int \frac{d^D k}{(2\pi)^D} \frac{1}{k^2 - m^2 + i\epsilon}, \tag{D.3}$$

which, using Eqs. (D.1) and (D.2), becomes:

$$I_D(m) = -i m^{D-2} (4\pi)^{-\frac{D}{2}} \Gamma(1 - \frac{D}{2}). \tag{D.4}$$

This can be analytically continued towards $D = 4$ by setting $D = 4 - 2\epsilon$. However, one must first introduce some other dimensionful parameter, μ, so that the ln's appearing in an expansion of Eq. (D.4) have dimensionless arguments. We therefore write

$$m^{2-2\epsilon} = m^2 (\mu^2)^{-\epsilon} \left(\frac{m^2}{\mu^2}\right)^{-\epsilon}, \tag{D.5}$$

and expand $(\frac{m^2}{\mu^2})^{-\epsilon} = 1 - \epsilon \ln \frac{m^2}{\mu^2} + \mathcal{O}(\epsilon^2)$. Then the fundamental integral becomes

$$I_D(m) = \frac{im^2}{16\,\pi^2}\mu^{-2\epsilon}\left(\frac{1}{\epsilon} + 1 - \gamma - \ln\frac{m^2}{\mu^2} + \ln 4\pi + \mathcal{O}(\epsilon)\right), \tag{D.6}$$

where γ is the Euler constant, $0.5772\ldots$.

It is now straightforward to obtain a number of other integrals by differentiating Eq. (D.6) with respect to m^2:

$$\int \frac{d^Dk}{(2\pi)^D}\frac{1}{(k^2 - m^2 + i\epsilon)^n} = i(-)^n\frac{(m^2)^{2-\epsilon-n}}{(4\,\pi)^{2-\epsilon}}\frac{\Gamma(n-2+\epsilon)}{\Gamma(n)}. \tag{D.7}$$

Since the continuation to D-dimensions preserves angular symmetry, we have

$$\int \frac{d^Dk}{(2\pi)^D}\,k^\mu f(k^2) = 0, \tag{D.8}$$

for any function $f(k^2)$ for which $\int d^Dk f(k^2)$ is defined.

By substituting $k \to k + p$ one finds:

$$\int \frac{d^Dk}{(2\pi)^D}\frac{1}{(k^2 + 2k\cdot p - m^2 + i\epsilon)^n} = i(-)^n\frac{(m^2+p^2)^{2-\epsilon-n}}{(4\,\pi)^{2-\epsilon}}\frac{\Gamma(n-2+\epsilon)}{\Gamma(n)}. \tag{D.9}$$

If we differentiate Eq. (D.9) with respect to p^μ we find

$$\int \frac{d^Dk}{(2\pi)^D}\frac{k_\mu}{(k^2 + 2k\cdot p - m^2 + i\epsilon)^n} = \\ i(-)^{n+1}p_\mu\,\frac{(m^2+p^2)^{2-\epsilon-n}}{(4\,\pi)^{2-\epsilon}}\frac{\Gamma(n-2+\epsilon)}{\Gamma(n)}, \tag{D.10}$$

and further differentiating with respect to p^ν:

$$\int \frac{d^Dk}{(2\pi)^D}\frac{k_\mu k_\nu}{(k^2 + 2k\cdot p - m^2 + i\epsilon)^n} = i(-)^n\left[p_\mu p_\nu(n-3+\epsilon)\frac{1}{2}g_{\mu\nu}(m^2+p^2)\right] \\ \times\frac{(m^2+p^2)^{2-\epsilon-n}}{(4\,\pi)^{2-\epsilon}}\frac{\Gamma(n-3+\epsilon)}{\Gamma(n)}. \tag{D.11}$$

Dirac Algebra

The extension of the usual Dirac algebra to D-dimensions is relatively simple, except for problems involving axial currents. One defines

$$g_{\mu\nu}g^{\mu\nu} = D, \tag{D.12}$$

as well as

$$\begin{aligned} \gamma_\mu\gamma^\mu &= D, \\ \{\gamma_\mu, \gamma_\nu\} &= 2g_{\mu\nu}, \\ Tr\gamma_\mu\gamma_\nu &= 4g_{\mu\nu}, \end{aligned} \tag{D.13}$$

and, as usual, the trace of an odd number of gamma matrices is zero.

The major problem with dimensional regularization concerns axial currents. There is no natural continuation of γ_5 to D-dimensions. If one chooses the usual anti-commutation relations for γ_5, namely

$$\{\gamma_\mu, \gamma_5\} = 0, \tag{D.14}$$

the axial anomaly is removed. The solution to this is either to use Eq. (D.14) and then put the axial anomaly back by hand or to modify Eq. (D.14) so that it only applies for $\mu \leq 4$, while for $\mu > 4$ it is the **commutator** of γ_μ and γ_5 which vanishes. This amounts to choosing $\gamma_5 = i\gamma_0\gamma_1\gamma_2\gamma_3$. Even with this choice one needs to take considerable care and we refer to the original paper of 't Hooft and Veltman for details.

E Path Integrals

The formulation of quantum field theory in terms of path integrals is developed in the standard text of Feynman and Hibbs [FH 65]. Our purpose is to summarize the standard formulas used in the path integral approach to derive the Feynman rules of modern gauge field theories such as QCD.

The essential formulae upon which many other results are built are the matrix identity

$$\ln\det M = \operatorname{Tr}\ln M, \tag{E.1}$$

(where Tr denotes the trace) and the standard Gaussian integral

$$\int_{-\infty}^{\infty} \frac{dx}{(2\pi)^{\frac{1}{2}}} \exp\left(-\frac{1}{2}ax^2\right) = \frac{1}{a^{\frac{1}{2}}}. \tag{E.2}$$

The former is usually proven by diagonalizing the matrix M. It can easily be rearranged to give:

$$\det M = \exp\left(\operatorname{Tr}\ln M\right). \tag{E.3}$$

Supposing that M is a real, symmetric, $N \times N$ matrix, and defining $X^T = (x_1, \ldots, x_N)$, it is easy to see (again by diagonalizing M) that Eq. (E.2) generalizes to:

$$\int_{-\infty}^{\infty} \frac{dx_1}{(2\pi)^{\frac{1}{2}}} \cdots \int_{-\infty}^{\infty} \frac{dx_N}{(2\pi)^{\frac{1}{2}}} \exp\left(-\frac{1}{2}X^T M X\right) = \frac{1}{(\det M)^{\frac{1}{2}}}. \tag{E.4}$$

In dealing with real scalar fields we typically encounter path integrals like

$$\int [d\phi] \exp\left(-\frac{1}{2}\int d^4x' d^4x \phi(x') M(x', x) \phi(x)\right). \tag{E.5}$$

As the absolute normalization of the path integral is irrelevant, we drop unimportant overall constants from this point. Approximating the four dimensional integrals over x and x' within a box of dimensions L using an interval ϵ, the integral in Eq. (E.5) can be written in the form

$$\int [d\phi] \exp\left(-\frac{1}{2}\int d^4x' d^4x \phi(x') M(x', x) \phi(x)\right) = \frac{1}{(\det M)^{\frac{1}{2}}}, \tag{E.6}$$

where $N = \left(\frac{L}{\epsilon}\right)^4$ and M is an $N \times N$ matrix. The same expression is used for the formal limit $L \to \infty$ (corresponding to $N \to \infty$) on the right hand side of Eq. (E.6).

In the case of complex scalar fields, Eq. (E.5) becomes

$$\int [d\phi][d\phi^*] \exp\left(-\frac{1}{2}\int d^4x' d^4x \phi^*(x') M(x',x)\phi(x)\right), \tag{E.7}$$

with M Hermitian. Repeating the steps outlined above one finds:

$$\int [d\phi][d\phi^*] \exp\left(-\frac{1}{2}\int d^4x' d^4x \phi^*(x') M(x',x)\phi(x)\right) = \frac{1}{\det M} = \exp\left(-\operatorname{Tr}\ln M\right), \tag{E.8}$$

Fermions

In order to deal with fermions within the path integral formalism we need to use Grassmann variables (Appendix C). That is, we must consider integrals like

$$\int da_1 \ldots da_N \exp\left(-\frac{1}{2}a^T A a\right). \tag{E.9}$$

Here $a^T = (a_1, \ldots, a_N)$ and A must be an anti-symmetric matrix and N even, if the integral is to be non-zero. Expanding the exponential is simple because a_i^n is zero for $n > 1$, so each a_i can appear at most once. On the other hand, the fact that $\int da_i = 0$, means that each a_i *must* appear once. Then using $\int da_i a_i = 1$ and the anti-commuting properties of a_i and a_j one finds:

$$\int da_1 \ldots da_N \exp\left(-\frac{1}{2}a^T A a\right) = (\det A)^{\frac{1}{2}} \tag{E.10}$$

(for N even). For complex Grassman variables this becomes

$$\int da_1 da_1^* \ldots da_N da_N^* \exp\left(a^\dagger A a\right) = \det A. \tag{E.11}$$

In the same way that we related the limit of such integrals for bosons to continuum path integrals, we find for complex fermion (anti-commuting) fields, ψ:

$$\int [d\psi^*][d\psi] \exp\left(-\int d^4x' d^4x \psi^*(x') A(x',x)\psi(x)\right) = \det A = \exp\left(\operatorname{Tr}\ln A\right). \tag{E.12}$$

The generating functions used to derive Feynman rules involve source terms which are linear in the fields. So for complex fermion fields we encounter integrals such as:

$$\int [d\psi^*][d\psi] \exp\left(-\int d^4x' d^4x \psi^*(x') A(x',x)\psi(x) + \left(\int d^4x(\eta^*(x)\psi(x) + \psi^*(x)\eta(x))\right)\right). \tag{E.13}$$

By changing the variables in the functional integral to $\psi + A^{-1}\eta$ and $\psi^* + \eta^* A^{-1}$ (i.e., completing the square within the exponent in Eq. (E.13)), and integrating out the new fields, using Eq. (E.12), we find:

$$\int [d\psi^*][d\psi] \exp\left(-\int d^4x' d^4x \psi^*(x') A(x', x)\psi(x) + \left(\int d^4x(\eta^*(x)\psi(x) + \psi^*(x)\eta(x))\right)\right) \tag{E.14}$$

$$= \exp\left(\mathrm{Tr}\ln A\right) \exp\left(-\int d^4x' d^4x\, \eta^*(x')\, A^{-1}(x', x)\, \eta(x)\right).$$

References

[Ab 96] K. Abe et al., Phys. Rev. Lett. 76 (1996) 587

[ACG 81] R. G. Arnold, C. E. Carlson and F. Gross, Phys. Rev. C 23 (1981) 363

[Ad 64] S. L. Adler, Phys. Rev. 135 (1964) B 963

[Ad 65] S. L. Adler, Phys. Rev. Lett. 14 (1965) 1047

[Ad 66] S. L. Adler, Phys. Rev. 143 (1966) 1144

[Ad 68] S. L. Adler, Ann. of Phys. 50 (1968) 189

[AD 66] S. L. Adler and Y. Dothan, Phys. Rev. 151 (1966) 1267

[AD 68] S. L. Adler and R. Dashen, *Current Algebras and Applications to Particle Physics*, Benjamin, New York (1968)

[Ad 69] S. L. Adler, Phys. Rev. D 177 (1969) 2426

[Adk 84] G. S. Adkins and C. R. Nappi, Phys. Lett. B 137 (1984) 251

[AEL 95] M. Anselmino, A. Efremov and E. Leader, Phys. Reports 261 (1995) 1; ibid. 281 (1997) 399(E)

[AFW 88] J. Ahrens, L. S. Ferreira and W. Weise, Nucl. Phys. A 485 (1988) 621

[AH 82] I. J. R. Aitchison and A. J. G. Hey, *Gauge Theories in Particle Physics*, Adam Hilger, Bristol (1982)

[Ahr 00] J. Ahrens et al., Phys. Rev. Lett. 84 (2000) 5950

[Ahr 88] L. A. Ahrens et al., Phys. Lett. B 202 (1988) 284

[AL 73] E. S. Abers and B. W. Lee, Phys. Reports 9C (1973) 1

[Alf 73] V. de Alfaro, S. Fubini, G. Furlan and C. Rosetti, *Currents in Hadron Physics*, North Holland, Amsterdam (1973)

[All 85] D. Allasia et al., Z. Phys. C 28 (1985) 321

[An 99] K. A. Aniol et al. (HAPPEX Collaboration), Phys. Rev. Lett. 82 (1999) 1096

[Ank 94] H. Anklin et al., Phys. Lett. B 336 (1994) 313

[Ank 98] H. Anklin et al., Phys. Lett. B 428 (1998) 248

[Ant 94] A. Antonelli et al., Phys. Lett. B 334 (1994) 431

[AP 77] G. Altarelli and G. Parisi, Nucl. Phys. B 126 (1977) 298

[AR 88] G. Altarelli and G. G. Ross, Phys. Lett. B 212 (1988) 391

[Ar 94] R. A. Arndt, R. L. Workman and M. M. Pavan, Phys. Rev. C 49 (1994) 272

[Are 98] H. J. Arends et al., in: Proc. Int. Conf. *BARYONS '98*, D. Menze and B. Metsch, eds., World Scientific, 1999, p. 418

[Arm 93] T. A. Armstrong et al., Phys. Rev. Lett. 70 (1993) 1212

[Arn 86] R. Arnold et al., Phys. Rev. Lett. 57 (1986) 174

[ARW 96] R. Alkofer, H. Reinhardt and H. Weigel, Phys. Reports 265 (1996) 139

[Bab 98] D. Babusci, G. Giordano and G. Matone, Phys. Rev. C 57 (1998) 291

[Bak 81] N. J. Baker et al., Phys. Rev. D 23 (1981) 2499
[Bal 60] A. M. Baldin, Nucl. Phys. 18 (1960) 310
[Bar 68] W. Bartel et al., Phys. Lett. B 28 (1968) 148
[Bar 81] G. Bardin et al., Phys. Lett. B 104 (1981) 320
[BAR 92] Proc. Int. Conf. on the Structure of Baryons, *BARYONS '92*, Yale Univ., 1992, M. Gai, ed., World Scientific, Singapore (1993)
[Bar 94] G. Bardin et al., Nucl. Phys. B 411 (1994) 3
[BB 86] W. Broniowski and M. K. Banerjee, Phys. Rev. D 34 (1986) 849
[BCB 87] W. Broniowski, T. D. Cohen and M. K. Banerjee, Phys. Lett. B 187 (1987) 229
[BD 64] J. D. Bjorken and S. D. Drell, *Relativistic Quantum Mechanics* and *Relativistic Quantum Fields*, McGraw-Hill, New York (1964)
[Be 90] R. Beck et al., Phys. Rev. Lett. 65 (1990) 1841
[Be 97] R. Beck et al., Phys. Rev. Lett. 78 (1997) 606
[Bec 99] J. Becker et al., Eur. Phys. J. A 6 (1999) 329
[Bel 75] A. A. Belavin, A. M. Polyakov, A. A. Schwartz and Y. S. Tyupkin, Phys. Lett. B 59 (1975) 85
[Ber 71] C. H. Berger et al., Phys. Lett. B 35 (1971) 87
[Ber 96] J. C. Bergström et al., Phys. Rev. C 55 (1997) 2016
[BF 89] D. S. Beder and H. W. Fearing, Phys. Rev. D 39 (1989) 3493
[Bha 88] R. K. Bhaduri, *Models of the Nucleon: from Quarks to Solitons*, Addison-Wesley Publ. Co., New York (1988)
[BHM 01] V. Bernard, T. R. Hemmert and U.-G. Meissner, Nucl. Phys. A (2001)
[Bij 96] J. Bijnens, Phys. Reports 265 (1996) 369
[Bir 90] M. C. Birse, Prog. Part. Nucl. Phys. 25 (1990) 1
[Bjo 66] J. D. Bjorken, Phys. Rev. 148 (1966) 1476
[Bjo 67] J. D. Bjorken, in Proceedings 3rd International Conference on Electron and Photon Interactions, Stanford, (1967)
J. D. Bjorken and E. A. Paschos, Phys. Rev. 185 (1969) 1975
[BKM 91] V. Bernard, N. Kaiser and U.-G. Meissner, Phys. Rev. Lett. 67 (1991) 1515
[BKM 92a] V. Bernard, N. Kaiser and U.-G. Meissner, Nucl. Phys. B 383 (1992) 442
[BKM 92] V. Bernard, N. Kaiser and U.-G. Meissner, Phys. Rev. Lett. 69 (1992) 1877
[BKM 94] V. Bernard, N. Kaiser, U.-G. Meissner and A. Schmidt, Z. Physik A 348 (1994) 317
[BKM 95] V. Bernard, N. Kaiser and U.-G. Meissner, Phys. Rev. C 52 (1995) 2185
[BL 80] S. J. Brodsky and G. P. Lepage, Phys. Rev. D 22 (1980) 2157
[BL 86] D. Bailin and A. Love, *Introduction to Gauge Field Theory*, Adam Hilger, London (1986)
[Bl 88] J. P. Blaizot and G. Ripka, Phys. Rev. D 38 (1988) 1556
[BLW 95] J. Bartels, H. Lotter and M. Wüsthoff, Z. Phys. C 68 (1995) 121
[Bo 92] P. E. Bosted et al., Phys. Rev. Lett. 68 (1992) 3841
[Bog 67] P. N. Bogoliubov, Ann. Inst. Henri Poincare 8 (1967) 163
[Bor 97] B. Borasoy and U.-G. Meissner, Ann. of Phys. 254 (1997) 192
[Bor 99] C. Boros et al., Phys. Lett. B 468 (1999) 161
[BQ 90] G. T. Bodwin and J.-W. Qiu, Phys. Rev. D 41 (1990) 2755

[BR 79] G. E. Brown and M. Rho, Phys. Lett. B 82 (1979) 177; G. E. Brown, M. Rho and V. Vento, Phys. Lett. B 94 (1979) 383
[Br 95] E. E. W. Bruins et al., Phys. Rev. Lett. 75 (1995) 21
[BRW 86] G. E. Brown, M. Rho and W. Weise, Nucl. Phys. A 454 (1987) 669
[BSW 75] F. Borkowski, G. G. Simon, V. H. Walther and R. D. Wendling, Nucl. Phys. B 93 (1975) 461
[Bu 00] V. Burkert, Nucl. Phys. A 663/664 (2000) 29c
[Bu 89] V. Burkert, in: *Baryon Spectroscopy and Structure of the Nucleon*, eds. G. Adam, N. C. Mukhopadhyay and P. Stoler, World Scientific (1989), p. 17
[Bur 80] A. J. Buras, Rev. Mod. Phys. 52 (1980) 199
[Bur 00] M. Burkardt, hep-ph/0005108
[BWS 91] F. Brandstaeter, U.-J. Wiese and G. Schierholz, Phys. Lett. B 272 (1991) 319
[Ca 63] N. Cabibbo, Phys. Rev. Lett. 10 (1963) 531
[CCFR 93] W. C. Leung et al. (CCFR Collaboration), Phys. Lett. B 317 (1993) 655
[CCFR 99] T. Adams et al. (NuTeV Collaboration), in Proceedings Workshop on *Physics with Electron Polarized Ion Collider - EPIC '99*, Bloomington, (1999) hep-ex/9906038; A. O. Bazarko et al. (CCFR Collaboration), Z. Phys. C 65 (1995) 189
[CGLN 57] G. F. Chew, M. L. Goldberger, F. E. Low and Y. Nambu, Phys. Rev. 106 (1957) 1345
[Ch 74] A. Chodos, R. L. Jaffe, K. Johnson and C. B. Thorne, Phys. Rev. D 10 (1974) 2599
[Cho 93] S. Choi et al., Phys. Rev. Lett. 71 (1993) 3927
[Chu 91] P. L. Chung and F. Coester, Phys. Rev. D 44 (1991) 229; F. Coester (1999), unpublished
[CJ 34] I. Curie and F. Joliot, Compt. Rend. 198 (1934) 254
[Cl 79] F. E. Close, *An Introduction to Quarks and Partons*, Academic Press, London (1979)
[CP-P 99] S. Aoki et al. (CP-PACS Collaboration), Phys. Rev. D 60 (1999) 114508
[Cr 83] M. Creutz, *Quarks, Gluons and Lattices*, Cambridge Univ. Press, Cambridge (1983)
[CR 00] S. Capstick and W. Roberts, Prog. Part. Nucl. Phys. (2000); nucl-th/0008028
[Cre 75] R. J. Crewther, Nucl. Phys. B 87 (1975) 52
[CSP 92] C. Ciofi degli Atti, G. Salme and E. Pace, Phys. Rev. C 46 (1992) 1591
[CT 75] A. Chodos and C. B. Thorn, Phys. Rev. D 12 (1975) 2733
[CT 88] F. E. Close and A. W. Thomas, Phys. Lett. B 212 (1988) 227
[CTEQ 95] CTEQ Collaboration, Rev. Mod. Phys. 67 (1995) 157
[CZ 84] V. L. Chernyak and A. R. Zhitnitsky, Phys. Reports 112 (1984) 173
[Da 70] M. Damashek and F. J. Gilman, Phys. Rev. D 1 (1970) 1319
[Dan 62] G. Danby et al., Phys. Rev. Lett. 9 (1962) 36
[DeG 75] T. DeGrand, R. L. Jaffe, K. Johnson and J. Kiskis, Phys. Rev. D 12 (1975) 2060
[DH 66] S. D. Drell and A. C. Hearn, Phys. Rev. Lett. 16 (1966) 430
[d'Ho 98] N. d'Hose et al., in: *Chiral Dynamics: Theory and Experiment*, Lecture Notes in Physics 513 (1998) 228
[d'Ho 99] N. d'Hose et al., in: *BARYONS '98*, eds. D. Menze and B. Metsch, World Scientific, Singapore (1999), p. 380
[Dia 84] D. I. Diakonov and V. Y. Petrov, Nucl. Phys. B 245 (1984) 259
[DiG 92] A. DiGiacomo and H. Panagopoulos, Phys. Lett. B 285 (1992) 133

[Do 89] J. F. Donoghue, Ann. Rev. Nucl. Part. Sci. 39 (1989) 1
[Do 92] J. F. Donoghue, E. Golowich and B. R. Holstein, *Dynamics of the Standard Model*, Cambridge Univ. Press (1992)
[Dok 77] Yu. L. Dokshitzer, Sov. Phys. JETP 46 (1977) 641
[Dr 00] D. Drechsel, M. Gorchtein, B. Pasquini and M. Vanderhaegen, Phys. Rev. C 61 (2000) 015204
[Dr 92] D. Drechsel, in: *Perspectives in Nuclear Physics at Intermediate Energies*, ed. S. Boffi et al., World Scientific, Singapore (1992)
[Dr 97] D. Drechsel, G. Knöchlein, A. Metz and S. Scherer, Phys. Rev. C 55 (1997) 424
[Dr 99] D. Drechsel, O. Hanstein, S. Kamalov and L. Tiator, Nucl. Phys. A 645 (1999) 145
[DTA 81] L. R. Dodd, A. W. Thomas and R. F. Alvarez-Estrada, Phys. Rev. D 24 (1981) 1961
[Dub 91] D. Dubbers, Prog. Part. Nucl. Phys. 26 (1991) 173
[DWT 87] L. R. Dodd, A. G. Williams and A. W. Thomas, Phys. Rev. D 35 (1987) 1040
[DYW 70] S. D. Drell, T.-M. Yan, Phys. Rev. Lett. 24 (1970) 855;
G. B. West, Phys. Rev. Lett. 24 (1970) 1206
[E143 98] K. Abe et al. (E143 Collaboration), Phys. Rev. D 58 (1998) 112003
[E155 99] P. L. Anthony et al. (E155 Collaboration), Phys. Lett. B 458 (1999) 536
[Ec 95] G. Ecker, Prog. Part. Nucl. Phys. 35 (1995) 1
[Ed 94] T. Eden et al., Phys. Rev. C 50 (1994) R1749
[Ede 00] J. Edelmann, G. Piller, N. Kaiser and W. Weise, Nucl. Phys. A 665 (2000) 125
[Ede 98] J. Edelmann, N. Kaiser, G. Piller and W. Weise, Nucl. Phys. A 641 (1998) 119
[EJ 74] J. Ellis and R. L. Jaffe, Phys. Rev. D 9 (1974) 1444
[ELT 00] T. E. O. Ericson, B. Loiseau and A. W. Thomas, Nucl. Phys. A 663 (2000) 541c
[EMC 88] J. Ashman et al., Phys. Lett. B 206 (1988) 364
[ER 80] A. V. Efremov and A. V. Radyushkin, Theor. Math. Phys. 42 (1980) 97
[ER 00] M. Engelhardt and H. Reinhardt, Nucl. Phys. B 585 (2000) 591
[ESW 96] R. K. Ellis, W. J. Stirling and B. R. Webber, *QCD and Collider Physics*, Cambridge University Press (1996)
[EW 88] T. Ericson and W. Weise, *Pions and Nuclei*, Clarendon Press, Oxford, 1988
[Fed 91] F. J. Federspiel et al., Phys. Rev. Lett. 67 (1991) 1511
[Fei 81] W. Feissler et al., Phys. Rev. D 23 (1981) 33
[Fey 69] R. P. Feynman, Phys. Rev. Lett. 23 (1969) 1415
[Fey 72] R. P. Feynman, *Photon-Hadron Interactions*, Benjamin, Reading (1972)
[FH 65] R. P. Feynman and A. R. Hibbs, *Quantum Mechanics and Path Integrals*, McGraw-Hill (1965)
[FJ 75] G. R. Farrar and D. R. Jackson, Phys. Rev. Lett. 35 (1975) 1416
[FJ 79] G. R. Farrar and D. R. Jackson, Phys. Rev. Lett. 43 (1979) 246
[FL 77] R. Friedberg and T. D. Lee, Phys. Rev. D 15 (1977) 1694, D 16 (1977) 1096
[Fo 58] L. L. Foldy, Rev. Mod. Phys. 30 (1958) 473
[Fog 79] G. L. Fogli and G. Nardulli, Nucl. Phys. B 160 (1979) 116
[FP 67] L. D. Faddeev and V. N. Popov, Phys. Lett. B 25 (1967) 29
[Fu 96] M. Fuchs et al., Phys. Lett. B 368 (1996) 20
[Gal 00] G. Galler et al., LARA collaboration, preprint (2000), to be published
[Gel 54] M. Gell-Mann and M. L. Goldberger, Phys. Rev. 96 (1954) 1433

[Ger 66] S.B Gerasimov, Sov. J. Nucl. Phys. 2 (1966) 908
[GI 85] S. Godfrey and N. Isgur, Phys. Rev. D 32 (1985) 189
[GIM 70] S. I. Glashow, J. Iliopoulis and L. Maiani, Phys. Rev. D 2 (1970) 1285
[GL 60] M. Gell Mann and M. Lévy, Nuovo Cim. 16 (1960) 705
[GL 72] V. N. Gribov and L. N. Lipatov, Sov. J. Nucl. Phys. 15 (1972) 438
[GL 82] J. Gasser and H. Leutwyler, Phys. Reports 87 (1982) 77
[GL 91] J. Gasser, H. Leutwyler and M. Sainio, Phys. Lett. B 253 (1991) 252, 260
[GL 94] J. Gasser and H. Leutwyler, Ann. Phys. (N.Y.) 158 (1994) 165
[GLS 69] D. J. Gross and C. H. Llewellyn Smith, Nucl. Phys. B 14 (1969) 337
[G-M 64] M. Gell-Mann, Phys. Lett. 8 (1964) 214
[Go 61] J. Goldstone, Nuovo Cim. 9 (1961) 154
[Go 62] J. Goldstone, A. Salam and S. Weinberg, Phys. Rev. 127 (1962) 965
[Gö 97] M. Göckeler et al., Nucl. Phys. B (Proc. Suppl.) 53 (1997) 81
[GOR 68] M. Gell-Mann, R. Oakes and B. Renner, Phys. Rev. 122 (1968) 2195
[Gr 76] D. Gromes and I. O. Stamatescu, Nucl. Phys. B 12 (1976) 213
[GR 96] L. Ya. Glozman and D. O. Riska, Phys. Rept. 268 (1996) 263
[Gri 78] V. N. Gribov, Nucl. Phys. B 139 (1978) 1
[GRM 99] T. Gehrmann, R. G. Ross and M. R. Whalley, J. Phys. G 25 (Suppl. 12A) (1999); and for an electronic data base: http://durpdg.dur.ac.uk/HEPDATA or http://www-spires.dur.ac.uk/HEPDATA
[GRV 98] M. Glück, E. Reya and A. Vogt, Eur. Phys. J. C 5 (1998) 461
[GS 96] T. Gehrman and W. J. Sterling, Phys. Rev. D 53 (1996) 6100
[GT 66] M. L. Goldberger and S. Treiman, Phys. Rev. 111 (1966) 354
[Gui 85] P. A. M. Guichon, Phys. Lett. B 164 (1985) 361
[Gui 96] P. A. M. Guichon et al., Nucl. Phys. A 601 (1996) 349
[Gui 98] P. A. M. Guichon and M. Vanderhaeghen, Prog. Part. Nucl. Phys. 41 (1998) 125
[GW 67] M. L. Goldberger and K. M. Watson, *Collision Theory*, Wiley, New York (1967), pp. 518
[GZ 56] S. S. Gerstein and Y. B. Zeldovich, Sov. Phys. JETP 2 (1956) 576;
R. P. Feynman and M. Gell Mann, Phys. Rev. 109 (1958) 193
[Hal 93] E. L. Hallin et al., Phys. Rev. C 48 (1993) 1497
[Has 73] F. J. Hasert et al., Phys. Lett. B 46 (1973) 121
[HDT 96a] O. Hanstein, D. Drechsel and L. Tiator, Phys. Lett. B 385 (1996) 45
[HDT 96b] O. Hanstein, D. Drechsel and L. Tiator, Phys. Lett. B 399 (1996) 13
[He 83] A. J. G. Hey and R. L. Kelly, Phys. Reports 96 (1983) 71
[He 99] C. Herberg et al., Eur. Phys. J. A 5 (1999) 131
[Hei 73] R. L. Heimann, Nucl. Phys. B 64 (1973) 429
[Hem 97] T. Hemmert, B. R. Holstein, G. Knöchlein and S. Scherer, Phys. Rev. D 55 (1997) 2630, Phys. Rev. Lett. 79 (1997) 22,
[Hem 98] T. Hemmert, B. R. Holstein, I. Kambor and G. Knöchlein, Phys. Rev. D 57 (1998) 5746
[Hermes 97] K. Ackerstaff et al. (Hermes Collaboration), Phys. Lett. B 404 (1997) 383
[HJL 99] P. Hoodbhoy, X. Ji and W. Lu, Phys. Rev. D 59 (1999) 014013
[HK 94] T. Hatsuda and T. Kunihiro, Phys. Reports 247 (1994) 221

[Hö 76] G. Höhler et al., Nucl. Phys. B 114 (1976) 505
[Hö 83] G. Höhler, *Pion-Nucleon Scattering*, in: Landolt-Börnstein New Series Vol.I/9b2, ed. H. Schopper, Springer, Berlin (1983)
[Hol 93] G. Holzwarth, ed., *Baryons as Skyrme Solitons*, World Scientific, Singapore (1993)
[Hol 98] B. R. Holstein, in: *Chiral Dynamics: Theory and Experiment*, Lecture Notes in Phys. 513 (1998) 215
[Hoo 76] G. 't Hooft, Phys. Rev. D 14 (1976) 3432
[Hoo 80] G. 't Hooft in: *Recent developments in gauge theories*, ed. G. 't Hooft, Plenum, New York (1980)
[HSV 96] B. W. Harris, J. Smith and R. Vogt, Nucl. Phys. B 461 (1996) 181
[HV 72a] G. 't Hooft and M. J. G. Veltman, Nucl. Phys. B 44 (1972) 189
[HV 72b] G. 't Hooft and M. J. G. Veltman, Nucl. Phys. B 50 (1972) 318
[IK 78] N. Isgur and G. Karl, Phys. Rev. D 18 (1978) 4187
[IK 79] N. Isgur and G. Karl, Phys. Rev. D 19 (1979) 2653
[IL 89] N. Isgur and C. H. Llewellyn-Smith, Nucl. Phys. B 317 (1989) 526
[Io 81] B. Ioffe, Nucl. Phys. B 188 (1981) 317
[IR 88] G. Ingelman and R. Rückl, Phys. Lett. 201 (1988) 369
[Is 99] N. Isgur, Phys. Rev. Lett. 83 (1999) 272
[Is 00] N. Isgur, Phys. Rev. D 62 (2000) 014025
[IZ 80] C. Itzykson and J.-B. Zuber, *Quantum Field Theory*, McGraw-Hill, New York (1980)
[Jaf 81] R. L. Jaffe, Annals Phys. 132 (1981) 32
[Jaf 85] R. L. Jaffe, Lectures presented at the Los Alamos School on Quark Nuclear Physics, Los Alamos, New Mexico, June 10-14, 1985. (Los Alamos Workshop 1985:0537)
[Ji 95] X. Ji, Phys. Rev. Lett. 74 (1995) 1071
[Ji 97] X. Ji, Phys. Rev. Lett. 78 (1997) 610
[Ji 98] X. Ji, J. Phys. G 24 (1998) 1181
[JJ 91] R. L. Jaffe and X. Ji, Phys. Rev. D 43 (1991) 724
[JM 92] E. Jenkins and A. Manohar, in: *Effective Field Theories of the Standard Model*, U.-G. Meissner, Ed., World Scientific, Singapore (1992)
[Jo 00] M. K. Jones et al., Phys. Rev. Lett. 84 (2000) 1398
[Jon 89] G. T. Jones et al., Z. Phys. C 43 (1989) 523
[Jon 96] G. Jonkmans et al., Phys. Rev. Lett. 77 (1996) 4512
[JRT 94] I. Jameson, A. W. Thomas and G. Chanfray, Austral. J. Phys. 47 (1994) 45 and J. Phys. G 18 (1992) L159
[Ken 91] H. W. Kendall, Nobel Lecture, Rev. Mod. Phys. 63 (1991) 597
[Kim 96] H. Kim, P. Pobilitsa, T. Watanabe, Th. Meissner, E. Ruiz Arriola, C. Christov, A. Blotz and K. Goeke, Prog. Part. Nucl. Phys. 37 (1996) 1
[Kit 83] T. Kitagaki et al., Phys. Rev. D 28 (1983) 436
[Kit 90] T. Kitagaki et al., Phys. Rev. D 42 (1990) 1331
[Kl 92] S. P. Klevansky, Rev. Mod. Phys. 64 (1992) 649
[KM 73] M. Kobayashi and M. Maskawa, Progr. Theor. Phys. 49 (1973) 652
[Koc 82] R. Koch, Z. Phys. C 15 (1982) 161
[Koc 86] R. Koch, Nucl. Phys. A 448 (1986) 707
[Kod 80] J. Kodaira, Nucl. Phys. B 165 (1980) 129

[Koe 76] L. Koester, W. Nistler and W. Waschkowski, Phys. Rev. Lett. 36 (1976) 1021
[Koe 88] L. Koester, W. Waschkowski and J. Meier, Z. Phys. A 329 (1988) 229
[Koe 95] L. Koester et al., Phys. Rev. C 51 (1995) 3363
[Kom 00] Y. Koma et al., Nucl. Phys. A 663 (2000) 1027
[Kop 95] S. Kopecky, P. Riehs, J. A. Harvey and N. W. Hill, Phys. Rev. Lett. 74 (1995) 2427
[KR 54] N. M. Kroll and M. A. Ruderman, Phys. Rev. 93 (1954) 233
[Kr 90] A. Krause, Helv. Phys. Acta 63 (1990) 3
[Kre 00] O. Krehl et al., Phys. Rev. C 62 (2000) 025207
[KS 98] S. Kretzer and I. Schienbein, Phys. Rev. D 58 (1998) 0904035
[KSW 95] N. Kaiser, P. B. Siegel and W. Weise, Nucl. Phys. A 594 (1995) 325
[LA 88] E. Leader and M. Anselmino, Z. Phys. C 41 (1988) 239
[La 97] M. Lavelle and D. McMullan, Phys. Reports 279 (1997) 1
[Lai 95] H. L. Lai et al., Phys. Rev. D 51 (1995) 4763
[Lar 94] S. A. Larin, Phys. Lett. B 334 (1994) 192
[Lar 97] S. A. Larin, T. van Ritbergen and J. A. M. Vermaseren, Phys. Lett. B 404 (1997) 153
[LB 79] G. P. Lepage and S. J. Brodsky, Phys. Rev. Lett. 43 (1979) 545, Phys. Rev. D 22 (1980) 2180
[Lee 81] T. D. Lee, *Particle Physics and Introduction to Field Theory*, Harwood Academic Publ., New York (1981)
[Lei 00] D. B. Leinweber et al., Phys. Rev. D 61 (2000) 074502
[Lei 95] D. B. Leinweber, Nucl. Phys. A 585 (1995) 341
[Lei 97] D. B. Leinweber, Annals Phys. 254 (1997) 328
[Lei 99] D. B. Leinweber et al., Nucl. Phys. Proc. Suppl. 73 (1999) 629
[Lev 97] A. Levy, *Low-x Physics at HERA*, in: Lecture Notes in Physics (Springer) 496 (1997) 347
[Lie 99] A. Liesenfeld et al., Phys. Lett. B 468 (1999) 20
[Lip 97] L. N. Lipatov, Phys. Reports 286 (1997) 131
[Liu 96] G. Q. Liu, A. W. Thomas and P. Guichon, Austr. J. Phys. 49 (1996) 905
[Liu 99] K. F. Liu et al., Phys. Rev. D 59 (1999) 112001
[Lle 72] C. H. Llewellyn Smith, Phys. Reports C 3 (1972) 261
[LLT 99] D. B. Leinweber, D. H. Lu, A. W. Thomas, Phys. Rev. D 60 (1999) 034014
[LM 93] G. P. Lepage and P. B. Mackenzie, Phys. Rev. D 48 (1993) 2250
[Lo 54] F. Low, Phys. Rev. 96 (1954) 1428
[Lon 98] J. T. Londergan and A. W. Thomas, Prog. Part. Nucl. Phys. 41 (1998) 49
[LP 85] E. Leader and E. Predazzi, *An Introduction to Gauge Theories and the "New Physics"*, Cambridge University Press, Cambridge (1985)
[LPS 97] A. L'vov, V. A. Petrunkin and M. Schumacher, Phys. Rev. C 55 (1997) 359
[LS 96] J. N. Labrenz and S. R. Sharpe, Phys. Rev., D 54 (1996) 4595
[LT 00] D. B. Leinweber and A. W. Thomas, Phys. Rev. D 62 (2000) 074505
[LTW 00] D. B. Leinweber, A. W. Thomas and S. V. Wright, Phys. Lett. B 482 (2000) 109
[Lu 98] D. H. Lu, A. W. Thomas and A. G. Williams, Phys. Rev. C 57 (1998) 2628
[Lun 93] A. Lung et al., Phys. Rev. Lett. 70 (1993) 718
[LWD 91] D. B. Leinweber, R. M. Woloshyn and T. Draper, Phys. Rev. D 43 (1991) 1659

[LZ 72] B. W. Lee and J. Zinn-Justin, Phys. Rev. D 5 (1972) 3137;
B. W. Lee, Phys. Rev. D 5 (1972) 823
[MacG 95] B. E. MacGibbon et al., Phys. Rev. C 52 (1995) 2097
[Mal 99] K. Maltman, Phys. Lett., B 462 (1999) 195
[Man 74] W. A. Mann et al., Phys. Rev. Lett. 31 (1974) 844
[McK 89] R. D. McKeown, Phys. Lett. B 219 (1989) 140
[Me 86] U.-G. Meissner, N. Kaiser, A. Wirzba and W. Weise, Phys. Rev. Lett. 57 (1986) 1676
[MG 84] A. Manohar and H. Georgi, Nucl. Phys. B 234 (1984) 189
[MH 56] R. W. McAllister and R. Hofstadter, Phys. Rev. 102 (1956) 851
[MH 90] M. J. Musolf and B. R. Holstein, Phys. Lett. B 242 (1990) 461
[Mil 82] K. L. Miller et al., Phys. Rev. D 26 (1982) 537
[MKW 87] U.-G. Meissner, N. Kaiser and W. Weise, Nucl. Phys. A 466 (1987) 685
[MM 94] I. Montvay and G. Münster, *Quantum Fields on a Lattice*, Cambridge Univ. Press, Cambridge (1994)
[MMD 96] P. Mergell, U.-G. Meissner and D. Drechsel, Nucl. Phys. A 596 (1996) 367
[MP 94] A. H. Mueller and B. Patel, Nucl. Phys. B 425 (1994) 471
[MPT 95] W. Melnitchouk, G. Piller and A. W. Thomas, Phys. Lett. B 346 (1995) 165
[MRS 98] A. D. Martin, R. G. Roberts, W. J. Stirling and R. S. Thorne, Eur. Phys. J. C 4 (1998) 463
[MT 93] W. Melnitchouk and A. W. Thomas, Phys. Rev. D 47 (1993) 3783
[MT 96] W. Melnitchouk and A. W. Thomas, Phys. Lett. B 377 (1996) 11
[Mu 97] G. Müller and U.-G. Meissner, Nucl. Phys. B 492 (1997) 379
[Mus 94] M. Musolf et al., Phys. Reports 239 (1994) 1
[Mut 98] T. Muta, *Foundations of Quantum Chromodynamics*, 2nd edition, World Scientific, Singapore (1998)
[Myh 81] F. Myhrer, G. E. Brown and Z. Xu, Nucl. Phys. A 362 (1981) 317
[Na 60] Y. Nambu, Phys. Rev. Lett. 4 (1960) 380
[Na 62] Y. Nambu and D. Lurie, Phys. Rev. 125 (1962) 1429; Y. Nambu and E. Shrauner, Phys. Rev. 128 (1962) 862
[NA51 94] A. Baldit et al., Phys. Lett. B 332 (1994) 244
[NJL 61] Y. Nambu and G. Jona-Lasionio, Phys. Rev. 122 (1961) 345
[NMC 90] D. Allasia et al. (NMC Collaboration), Phys. Lett. B 249 (1990) 366;
P. Amaudruz et al. (NMC Collaboration), Phys. Rev. Lett. 66 (1991) 2712
[NMC 97] M. Arneodo et al., Nucl. Phys. B 487 (1997) 3
[NP 82] H. B. Neilsen and A. Patkos, Nucl. Phys. B 195 (1982) 137
[NuSea 98] E. A. Hawker et al. (NuSea Collaboration), Phys. Rev. Lett. 80 (1998) 3715
[Os 99] M. Ostrick et al., Phys. Rev. Lett. 83 (1999) 276
[Pan 68] W. Panofsky, in Proceedings International Symposium on *High Energy Physics*, Vienna (1968)
[Pas 99] I. Passchier et al., Phys. Rev. Lett. 82 (1999) 4988
[PDG 98/00] Particle Data Group, *Review of Particle Physics*, Eur. Phys. J. C 3 (1998), C 15 (2000)
[Pir 87] H. J. Pirner et al., Nucl. Phys. B 294 (1987) 905
[Pir 92] H. J. Pirner, Prog. Part. Nucl. Phys. 29 (1992) 33

[Pl 90] S. Platchkov et al., Nucl. Phys. A 510 (1990) 740
[Po 76] H. H. Politzer, Nucl. Phys. B 117 (1976) 397
[PP 00] A. Pich and J. Prades, Nucl. Phys. Proc. Suppl. 86 (2000) 236
[PRC 87] J. Praschifka, C. D. Roberts and R. T. Cahill, Phys. Rev. D 36 (1987) 209; R. T. Cahill and C. D. Roberts, Phys. Rev. D 32 (1985) 2419
[PRC 89] J. Praschifka, C. D. Roberts and R. T. Cahill, Int. J. Mod. Phys. A 4 (1989) 4929
[PS 95] M. E. Peskin and D. V. Schroeder, *An Introduction to Quantum Field Theory*, Addison-Wesley, New York (1995)
[PW 00] G. Piller and W. Weise, Phys. Reports 330 (2000) 1
[Rad 96] A. V. Radyushkin, Phys. Lett. B 380 (1996) 417
[RC 53] F. Reines and C. L. Cowan, Phys. Rev. 92 (1953) 830
[RGG 75] A. deRujula, H. Georgi and S. L. Glashow, Phys. Rev. D 12 (1975) 147
[Ri 97] G. Ripka, *Quarks Bound by Chiral Fields*, Clarendon, Oxford (1997)
[Rit 97] K. Rith, *Quark-Gluon Structure of the Nucleon*, in: Lecture Notes in Physics (Springer) 496 (1997) 250
[Ro 00] J. Roche et al., Phys. Rev. Lett. 85 (2000) 708
[Rob 90] R. G. Roberts, *The Structure of the Proton*, Cambridge University Press, Cambridge (1990)
[Roc 82] S. Rock et al., Phys. Rev. Lett. 49 (1982) 1139
[Roh 99] D. Rohe et al., Phys. Rev. Lett. 83 (1999) 4257
[Ros 50] M. N. Rosenbluth, Phys. Rev. 79 (1950) 615
[Ros 90] K. W. Rose et al., Phys, Lett. B 234 (1990) 460
[Rot 97] H. J. Rothe, *Lattice Gauge Theories: An Introduction*, World Scientific Lecture Notes in Physics – Vol. 59, World Scientific, Singapore (1997)
[RS 79] D. A. Ross and C. T. Sachrajda, Nucl. Phys. B 149 (1979) 497
[RW 94] C. D. Roberts and A. G. Williams, Prog. Part. Nucl. Phys. 33 (1994) 477
[Sak 69] J. J. Sakurai, *Currents and Mesons*, Univ. of Chicago Press (1969)
[Sal 69] A. Salam, in: *Elementary Particle Theory*, ed. N. Svartholm, Stockholm 1969, p. 367
[SBS 99] K. Schilling, G. S. Bali and C. Schlichter, Nucl. Phys. Proc. Suppl. 73 (1999) 638
[Sch 91] J. Schmiedmayer et al., Phys. Rev. Lett. 66 (1991) 1015
[Schä 97/98] T. Schäfer and E. V. Shuryak, Ann. Rev. Nucl. Part. Sci. 47 (1997) 359, Rev. Mod. Phys. 70 (1998) 323
[Schr 99] H. Ch. Schröder et al., Phys. Lett. B 469 (1999) 25
[Schw 89] B. Schwesinger, H. Weigel, G. Holzwarth and A. Hayashi, Phys. Reports 173 (1989) 173
[Sil 93] A. F. Sill et al., Phys. Rev. D 48 (1993) 29
[Sim 80] G. G. Simon et al., Z. Naturforsch. 35a (1980) 1
[SMC 97] D. Adams et al. (SMC Collaboration), Phys. Rev. D 56 (1997) 5330
[Spa 00] D. T. Spayde et al. (SAMPLE Collaboration), Phys. Rev. Lett. 84 (2000) 1106
[Squ 64] E. J. Squires, *Complex Angular Momenta and Particle Physics*, W. A. Benjamin Inc., New York (1964)
[ST 97] J. Speth and A. W. Thomas, Adv. Nucl. Phys. 24 (1997) 83
[Ste 33] O. Stern, Nature 132 (1933) 169
[Sto 93] P. Stoler, Phys. Reports 226 (1993) 103

[Sto 99] P. Stoler, in: *BARYONS '98*, eds. D. Menze and B. Metsch, World Scientific, Singapore (1999), p. 408
[SVH 73] P. A. Schreiner and F. Von Hippel, Nucl. Phys. B 58 (1973) 333
[SVZ 79] M. A. Shifman, A. I. Vainshtein and V. I. Zakharov, Nucl. Phys. B 147 (1979) 385, 448
[SWT 98] A. W. Schreiber, A. G. Williams and A. W. Thomas, eds., *Nonperturbative Methods in Quantum Field Theory*, World Scientific, Singapore (1998)
[Tan 97] P. C. Tandy, Prog. Part. Nucl. Phys. 39 (1997) 117
[TFM 00] A. W. Thomas, F. M. Steffens and W. Melnitchouk, Phys. Rev. Lett. 85 (2000) 2892
[Th 80] S. Théberge, A. W. Thomas and G. A. Miller, Phys. Rev. D 22 (1980) 2838
[TTM 81] S. Théberge, A. W. Thomas and G. A. Miller, Phys. Rev. D 24 (1981) 216
[Th 81] A. W. Thomas, J. Phys. G 7 (1981) L283
[Th 84] A. W. Thomas, Adv. in Nucl. Phys. 13 (1984) 1, eds. J. Negele and E. Vogt, Plenum Press, New York (1984)
[Tho 83] A. W. Thomas, Phys. Lett. B 126 (1983) 97
[TK 00] A. W. Thomas and G. Krein, Phys. Lett. B 456 (1999) 5
[TMT 82] S. Théberge, G. A. Miller and A. W. Thomas, Can. J. Phys. 60 (1982) 59
[To 95] I. S. Towner and J. C. Hardy, in: *Symmetries and Fundamental Interactions in Nuclei*, eds. E. M. Henley and W. C. Haxton, World Scientific, Singapore (1995) pp. 183-249
[Tom 66] Y. Tomozawa, Nuovo Cim. 46 A (1966) 707
[Tr 85] S. B. Treiman, R. Jackiw, B. Zumino and E. Witten, *Current Algebra and Anomalies*, Princeton Univ. Press (1985)
[UKQ 99] C. R. Allton et al. (UKQCD Collaboration), Phys. Rev. D 60 (1999) 034507
[VJT 85] E. A. Veit, B. K. Jennings and A. W. Thomas, Phys. Rev. D 31 (1985) 1033
[VTJ 85] E. A. Veit, A. W. Thomas and B. K. Jennings, Phys. Rev. D 31 (1985) 2242
[VW 91] U. Vogl and W. Weise, Prog. Part. Nucl. Phys. 27 (1991) 195
[VZ 72] A. I. Vainshtein and V. I. Zakharov, Nucl. Phys. B 36 (1972) 589
[Wa 89] R. C. Walker et al., Phys. Lett. B 234 (1989) 353
[WDL 92] W. Wilcox, T. Draper and K.-F. Liu, Phys. Rev. D 46 (1992) 1109
[We 66] S. Weinberg, Phys. Rev. Lett. 17 (1966) 616
[We 67] S. Weinberg, Phys. Rev. Lett. 19 (1967) 1264
[We 79] S. Weinberg, Physica A 96 (1979) 327
[We 96] S. Weinberg, *The Quantum Theory of Fields*, Vol. II, Cambridge Univ. Press, Cambridge (1996)
[Weg 71] F. J. Wegner, J. Math. Phys. 12 (1971) 2259
[Wei 65] W. I. Weisberger, Phys. Rev. Lett. 14 (1965) 1047
[Wi 74] K. G. Wilson, Phys. Rev. D 10 (1974) 2445
[Wil 00] D. H. Wilkinson, Eur. Phys. J. A 7 (2000) 307
[Wo 91] C. E. Woodward et al., Phys. Rev. C 44 (1991) R571
[Wil 69] K. Wilson, Phys. Rev. 179 (1969) 1499
[Wil 93] D. H. Wilkinson, Nucl. Instrum. Meth. A 335 (1993) 201
[Wit 83] E. Witten, Nucl. Phys. B 223 (1983) 422, 433
[WW 77] S. Wandzura and F. Wilczek, Phys. Lett. B 72 (1977) 195
[ZB 86] I. Zahed and G. E. Brown, Phys. Reports 142 (1986) 1
[Zi 92] A. Zieger et al., Phys. Lett. B 278 (1992)
[Zu 71] P. A. Zucker, Phys. Rev. D 4 (1971) 3350
[Zw 64] G. Zweig, CERN Reports 8182/TH.401, 8419/TH.412 (1964)

Index